PROBABILITY AND STATISTICS

Dale M. Johnson

Professor of Research
The University of Tulsa
Tulsa, Oklahoma

Copyright © 1989
by
SOUTH-WESTERN PUBLISHING CO.
Cincinnati, Ohio

ALL RIGHTS RESERVED

The text of this publication, or any part thereof, may not be reproduced or transmitted in any form or by any means, electronic or mechanical, including photocopying, recording, storage in an information retrieval system, or otherwise, without the prior written permission of the publisher.

ISBN: 0-538-60015-2

Library of Congress Catalog Card Number: 87-62697

COVER PHOTO: Cover Photo by Melvin L. Prueitt,
Los Alamos National Laboratory

Reviewers

Robert Gerver
North Shore High School
Glen Head, New York

Arnold Lloyd Cutler
Senior High Math Teacher
Independent School District 621
Mounds View, Minnesota

2 3 4 5 6 7 8 RM 5 4 3 2 1 0 9

Printed in the United States of America

PREFACE

Virtually everyone uses or consumes statistical material every day, and most people do so without training in the proper use or potential abuse of statistical information. This book was written to serve as a general-purpose introduction to the field of statistics for a beginning audience. The text is designed for students with a wide variety of interests and future vocational and educational ambitions. It is therefore different from many texts because it is not tailored for a specific academic subject-matter curriculum. To enhance its appeal to a wide audience, applications and examples from the field of education are liberally used throughout the text.

Most students are already familiar with educational settings. This text takes advantage of such experiences, which enables class participants to concentrate on the essentials of probability and statistics instead of investing time and energy in trying to understand a discipline-specific setting or scenario from an unfamiliar field before addressing the statistical issues involved. This textbook is designed for one semester of study, and the ultimate aim is to enable students to be wiser users and more critical consumers of statistical material.

The text has four major divisions: (1) Descriptive Statistics: Concepts and Methods [Chapters 1 – 6]; (2) Probability [Chapters 7 – 8]; (3) Inferential Statistical Concepts [Chapters 9 – 11]; and (4) Inferential Statistical Methodology [Chapters 12 – 16]. Because of the nature of the material, the chapters in the text should be mastered in numerical sequence starting with Chapter 1. Sections designated as "optional" provide important but supplemental material. Therefore, these sections may be omitted without loss of continuity in the material and without missing important prerequisite material for subsequent topics.

Learning objectives provide a preview of each chapter. These can be used as advanced organizers and cues for studying the material in the chapter and should be viewed as learning "targets" for the chapter. At the end of each chapter, key terms and notation remind students of the concepts and terminology discussed in the chapter. Application exercises also are provided that enable learners to check their practical understanding of the material covered and their statistical problem-solving ability.

The use of hand-held calculators for statistical computation is encouraged in this course. Calculators vary in the type and number of functions they perform, ranging from basic arithmetic to complex technical applications. Although statistical calculators are available for relatively modest costs, computations required in this text can be completed with simple and less-expensive calculators. In addition to the four arithmetic operations of addition, subtraction, multiplication, and division, it is recommended that calculators used in this course have a square root ($\sqrt{}$) function and at least one auxiliary memory for storing intermediate numerical calculations. Additional features such as factorial computational keys, exponent keys, statistical functions, and logarithm/anti-logarithm calculation capabilities, while useful in a limited number of situations, are not necessary in this course. Keystroke sequences for performing various calculations are presented throughout the text. These are based on accessibility to a simple calculator with the following functions:

+	addition
−	subtraction
÷	division
×	multiplication
$\sqrt{}$	square root
M+	add to memory
=	equals
MRC	memory recall; displays contents of memory
CLM	clear memory; deletes contents of memory

Each keystroke sequence is supplemented with the procedure and the arithmetic operation involved.

While completion of the material in this text does not depend on computer use, two *optional* features are available for those with access to a microcomputer and an interest in computer applications. First, a computer focus section appears at the end of each chapter. This section provides problem-solving activities using computing concepts (for example, flowcharting) and/or computer capabilities. Second, a computer program disk (called *ACCUSTAT*) is available to solve numerous end-of-chapter problems. These problems are identified with a small microcomputer icon ( ). Students should work the problems by hand, and then check their answers with ACCUSTAT. Students can study for tests by creating new problems (and solutions). Complete operating instructions for ACCUSTAT are included in Appendix I. Help screens appear in all programs. Students can get started with the separate tutorial disk.

Preparation of this textbook has been a lengthy undertaking, and the author is indepted to a number of individuals and organizations who provided assistance, critical reviews, and permission to reprint tables. Gratitude is expressed to my teaching colleagues and the hundreds of students who have reacted to this material in classes during the author's tenure as a statistics teacher. Many no doubt will recognize their ideas that have enhanced this textbook presentation. Thanks are extended to my daughter Pam for her special assistance.

I am grateful to the Literary Executor of the late Sir Ronald A. Fisher, F.R.S., to Dr. Frank Yates, F.R.S., and to the Longman Group Ltd, London for permission to reprint Appendices C, D, E, G, and H from their book *Statistical Tables for Biological, Agricultural and Medical Research* (6th edition 1974). Also, appreciation is extended to William G. Snedecor and George W. Snedecor, authors of *Statistical Methods* (5th edition), and to the Iowa State University Press for permission to reprint Appendix F.

Finally, but most important, I wish to thank my wife for her encouragement, patience, and support. And, as a small token of my appreciation of a pair of great teachers, this text is dedicated to my parents.

Dale M. Johnson

CONTENTS

PART ONE – DESCRIPTIVE STATISTICS: CONCEPTS AND METHODS

PART TWO - PROBABILITY

PART THREE - INFERENTIAL STATISTICAL CONCEPTS

PART FOUR - INFERENTIAL STATISTICAL METHODOLOGY

PART ONE
Descriptive Statistics: Concepts and Methods

1

Introduction to Statistics

CHAPTER OBJECTIVES

Upon completion of the chapter, students will be able to:

1. Provide real or hypothetical examples of routine daily experiences with statistics.
2. Distinguish between two common meanings of the word "statistics."
3. Define three characteristics for four levels of numerical scaling of numbers.
4. Distinguish between discrete and continuous variables.
5. Discuss three popular criticisms of statistical methodology.
6. Identify two broad areas of focus in the science of statistics.

NOTE TO THE STUDENT

Much of the Monday morning between-class discussion focused on Friday night's narrow football victory over Cooper High School's traditional rival.

"I hear the team played almost a perfect game," exclaimed Cindy as she rummaged through her locker in an attempt to locate her chemistry book for the next class.

Austin, star linebacker on the Cooper team, stretched over Cindy to reach his locker on the top row and responded, "It was a well-played game—(pause)—by both teams. I suppose our offense should get most of the credit; they averaged 5.2 yards per play—but we did our part on defense by holding those guys to only 14 points. They've been averaging 27.5 points per game this year." He continued, "Our fans are great too; the crowd was larger than normal, don't you think?"

"Oh, I didn't get to go to the game," Cindy confessed. "I had to attend an organizational meeting of the student mathematics council to help plan a regional mathematics competition."

"I'm impressed; how'd you get involved in that?" queried Austin.

"Well," responded Cindy, "it was just one of those things—however, it probably had something to do with my scoring at the 98th percentile on the mathematics achievement test."

"Hey! It's almost time for class," Austin observed. "Is there a chance that maybe, er . . , uh, maybe . . . how about telling me about it next Saturday night?"

Meanwhile, across town from Cooper High at Central University, Professor Stacey was discussing a recent experiment with several of his graduate students. The well-known experimental psychologist was reporting that "the mean score for the experimental group was 16.43, the mean for the placebo group was 12.96, and the mean for the control group was 8.72—both the experimental and placebo groups scored significantly higher than the control group at the 0.05 level of significance."

"What does that mean?" asked a jittery graduate student named Doug.

Smiling, Dr. Stacey replied, "That's a good question. In fact, it's so good that I'll pose the same question to you—and I'll give you two days to respond to our group. Good luck."

Downtown in City Hall, Mayor Janice Allen and her staff were busily preparing a summary of the achievements of her administration for an afternoon press conference. Among other items, Mayor

Allen wanted some visual aids that depicted the drop in unemployment and the increase in new industry during her term as mayor. However, she was somewhat puzzled as to why in one city-wide poll 56% of the respondents gave her administration high marks, while in another only 52% rated her administration favorably. "Oh, well," she concluded, "at least a majority of the city residents who responded to the polls like what I'm doing." (See Figure 1.1.)

A morning TV news commentator was quoting a certain Eastern European nation's news story about a dual track competition between its athletes and an American track and field team. Although the U.S. had won the meet, the Eastern Euorpean government-controlled news agency reported, "The results of an international track and field meet show that our national team placed second in the competition while the Americans finished next to last."

On first impression, these abbreviated discussions have very little in common and seem totally unrelated. However, the accounts do have a common thread that would be recognized immediately by someone trained in the field of statistics. As a matter of fact, it would be difficult for you to participate in a discussion, read a newspaper or magazine article, listen to a TV news report, or even read a leisure-time novel without coming into contact with a statistical concept. Words, phrases, and concepts such as average, larger than normal, percentile, mean, percentage, rank, and significant difference all have a common origin founded in statistics. Indeed, each day of your life you will be exposed to and will use many principles from the field of statistics. The widespread use of statistics in everyday life is undisputed.

The science of statistics can be intentionally misused to achieve certain goals. Remember the news report about the dual track meet? Statistics can be unintentionally misused by those who are simply ignorant of its appropriate applications. In either case, conclusions and impressions generated from erroneous statistical results are virtually never valid or truthful. A major goal in this statistics course will be to develop an awareness of the potential for error created by the misuse of statistics and to become familiar with the tools required to use and consume statistical information intelligently.

You can take a large step toward "statistical literacy" by mastering each of the topics presented during the course. Many of the concepts will not be new, but the terminology and symbols may require some study. You will be introduced to new ways of processing and analyzing information. The concepts are presented in a "common-sense" approach that will aid your learning efforts.

Figure 1.1—*Daily News*, The School Newspaper

SCHOOL EDITION **Volume 23 Number 2**

Regional Schools Plan Math Contest

Over 20 area secondary schools are making preparations for the first annual "Green County Mathematics Talent Contest." The event is jointly sponsored by area schools and is expected to attract over 100 contestants. A planning committee with representatives from the 20 participating schools met Friday night to make preliminary arrangements for the competition. A tentative date for the event was scheduled for May 16th. Two more meetings of the planning committee have been set to establish guidelines for the contest. Faculty in the mathematics department at Central University have volunteered to serve as advisors for

(continued on page 7)

Your combined school and local community news

Weekly Source of Information

Cooper High Rolls To Victory

Cooper High School posted a 21-14 victory over East Side Academy Friday night before an overflowing homecoming crowd. The 8,500 excited fans witnessed a strong Cooper defense hold the potent Academy offense to their lowest scoring output of the season. The Cooper offense, bolstered by the return of halfback Kenny Jones, ripped Academy's defense for over 350 total yards while Cooper posted its 5th straight win of the season. The Cooper defensive team anchored by all-state linebacker Austin

(continued on page 2)

Professor A. A. Stacey

Local Professor Honored for Research

A Central University professor of psychology has received an award from a national organization for his work on the psychological aspects of weight control. In a news release from the National Psychological Commission headquarters, Dr. A. A. Stacey was named as this year's recipient of the commission's annual award for research excellence. Professor Stacey was selected from a field of 50 finalists at the 32nd annual meeting of the Commission held in San Francisco last month. When contacted at his office at the university, Stacey revealed that he had been involved with the project

(continued on page 4)

Survey Shows Support for Mayor Allen

A recently completed survey by an independent polling agency revealed that most adult citizens approve of Mayor Allen's administration. A random sample of 978 city residents were questioned by telephone interviewers during the month of September. According to a spokesman for the agency, 56% of the respondents

(continued on page 3)

The common myth that extensive mathematics skills are required to learn statistics is unfounded. In this course mathematics has already been used to produce theorems, tables, and formulas. All you will be required to do is some simple arithmetic (addition, subtraction, multiplication, division) and develop some skill in the use of tables. The rest is simple logical reasoning. If you have a fair grasp of arithmetic fundamentals, you have the mathematical prerequisites for the course. In fact, you don't even have to like mathematics to be successful in this course.

Here are several hints to guide you through a successful encounter with the field of statistics:

1. *Study daily.* This is the way to keep current with the material. Most of the topics covered in this text are presented on the assumption that the previous material has been mastered. So, to avoid a "helpless" feeling of being "lost," keep up daily, as the new topics will usually be a simple extension of the previous material.
2. *Review frequently.* Periodic review is the best way to retain what you've learned. If you fail to review previous material, the concepts will fade from your memory. This could make grasping new material more difficult for you.
3. *Practice.* Do all of the exercises at the end of each chapter even when not formally assigned by your teacher. Skill is improved through practice, and becoming proficient in the use and consumption of statistics is no exception. When reading newspapers or magazines, see how many statistical concepts and presentations you recognize. Critique them using the skills you have learned.
4. *Use your common sense.* Most of what is found in a statistics course is merely formalized common sense and reasoning. Use these traits to determine if material and statistical findings are reasonable.
5. *Master terminology.* Each field has its own set of definitions, procedures, symbols, and terminology to facilitate communication; statistics is no exception. Without an understanding of the statistical tools of communication, you'll find the course to be relatively difficult.

Hand-held calculators are acceptable for completing the arithmetic involved in statistics in this text. Although such aids are not necessary for solving most of the problems, a calculator will facilitate the computational tasks. If you do use a calculator, an inexpensive one with square root and logarithm functions will be adequate.

WHAT IS (ARE) STATISTICS?

Ask a dozen people to define the term "statistics," and you could easily get a dozen different responses. Indeed, the word "statistics" can have many different meanings, each of which is usually determined by the context in which it is used.

The plural form of the word is older and more widely used than the singular form. Used as a plural, the word **statistics** is synonymous with the phrase *numerical data.* **Numerical data** are simply a set of numbers such as the shoe sizes of your classmates. Therefore, if reference is made to statistics displayed in the *World Almanac* or in the stock market report, this means that they contain numerical data.

For centuries government leaders have depended on data or information to assist in managing the affairs of state. Information expressed as numbers put together in an orderly form dates back prior to Aristotle and his treatises on the "matters of state." It is believed that the words "statistics" and "state" are derivatives of the same root word. Through the ages, civilized nations have compiled statistics to determine the size of the work force, economic status, and material resource strengths. Today our lives are increasingly influenced by judgments and decisions based on this kind of quantitative information.

When used as a singular noun, statistics refers to a scientific discipline made up of methods and procedures for studying numerical data. A common dictionary definition states: "Statistics is the science dealing with organizing, summarizing, analyzing, and interpreting numerical data." Such a definition assumes that the process starts with numerical data, that is, with one or more sets of numbers. In general, the term "data" refers to a set of facts. More specifically, in the science of statistics, data refers to facts expressed as numbers.

Creating numerical data is synonymous with measurement and is sometimes considered to be a function of statistics. **Measurement** is simply the process of assigning numbers to objects, events, people, or other things according to a specified rule. Hat sizes, social studies achievement, basketball field goals, internal memory of computers, and dimensions of lumber are all assigned numbers (measures) that assist us in knowing and communicating information about the particular object, event, or performance. For example, when a grade-point average is to be computed, the first step is to create numbers. This is typically done by assigning an A four points, a B three points, a C two points, and so on until the letter grades have been coded with numbers.

The creating of numbers supplies numerical data as a starting point for statistics. The quality of statistical results and conclusions depends on the quality of the measurement process. A well-known principle in computer science is stated as: "Garbage In, Garbage Out" (GIGO). The principle is equally applicable to the field of statistics in that the quality of statistical results can be no better than the data that serve as input to the statistical process. In the interest of conciseness, this text will not dwell on the collection (or creation) of original data, but will use real or hypothetical data that are already available.

Organizing Numerical Data

Organizing numerical data is systematically arranging data into formats that will facilitate an examination of a set of numbers. A set of numbers is sometimes called a numerical distribution. When data are collected, they are generally in a very haphazard state. One of the statistician's first tasks is to arrange the data in an orderly fashion suitable for examining how the numbers are distributed. Tables, charts, and graphic or pictorial representations are common means for accomplishing this task.

Summarizing Numerical Data

Summarizing numerical data is describing a numerical distribution in a shorthand form. The earlier example of computing a grade-point average illustrates one way that data can be summarized. Knowing that a student's grade-point average is 2.35 does not mean knowing each letter grade on that person's transcript. However, the quantity 2.35 does efficiently provide useful, if abbreviated, information about the achievement marks earned.

Analyzing Numerical Data

Analyzing numerical data implies that the numbers are going to be arithmetically manipulated. Various techniques and formulas are used in statistics to analyze data for the purpose of extracting certain information from one or more numerical distributions. For example, you may want to know how gas mileage changes as a function of engine power. Statistics provides several techniques for addressing or analyzing data in such problems.

Interpreting Numerical Data

Interpreting numerical data is usually the final task in statistics. Meaningful descriptions are formulated and probable conclusions are drawn based on results of data summaries and data analyses. The validity of the interpretation depends on the accuracy of the data and on the use of appropriate statistical techniques in analyzing the data.

Numerical Data

Numerical data, the last key phrase in the definition of statistics, refers to the numbers that have been collected. Some problems may arise because numbers do not always convey the same meaning. For example, consider the number 10. The number 10 may be assigned as a football jersey number. It also may refer to the 10th President of the United States, express a shoe size, measure a temperature on the Celsius thermometer, or identify an interstate highway.

From this illustration it becomes obvious that a number will not always have the same meaning across various contexts. You need to know the rule that was used to assign meaning to the number and the context in which it was used. Numbers can be categorized according to the operations that can be performed and meaningfully conveyed by the numerical data. A commonly used scaling classification categorizes numbers as: (1) nominal, (2) ordinal, (3) interval, or (4) ratio.

Nominal Numbers. **Nominal numbers** are used to identify or name individuals or groups. This is the most primitive and restrictive way to assign numbers. Nominal numbers:

1. can be used to name or label individuals or groups;
2. cannot be used to rank order individual elements according to some attribute or characteristic;
3. cannot be used to determine how much of a difference exists between individuals on some attribute or characteristic; and
4. cannot be used to determine how much of an attribute or characteristic is present in an absolute sense.

Examples of nominal numbers include numbers on basketball jerseys, interstate highway numbers, postal ZIP codes, and license-plate numbers. Many times nominal numbers are used to identify and distinguish between groups, such as males = 1, females = 2; or students = 1, parents = 2, and faculty members = 3. They are used only for naming or identification purposes and cannot even be used for such simple operations as rank ordering individuals.

Ordinal Numbers. **Ordinal numbers**, on the other hand, can be used to rank order individuals according to some attribute or characteristic. Ordinal numbers:

1. can be used to name or label individuals or groups;
2. can be used to rank order individuals according to some attribute, performance, or characteristic;
3. cannot be used to determine how much difference exists between individuals; and
4. cannot be used to determine how much of an attribute is present in an absolute sense.

A common use of ordinal numbers is centiles (or percentiles) shown on standardized test score results. Other examples are the rankings of how individuals finished in a beauty contest, a 100-yard dash, or an election. Someone finishes first, another second, another third, and the ranking continues relative to whatever attribute, performance, or characteristic is being quantified.

Interval Numbers. **Interval numbers** are sometimes referred to as "equal interval" numbers. Measurements at the interval level retain all the properties of numbers scaled at the nominal and ordinal levels, and, in addition, the intervals between consecutive numbers are assumed to be equal. Therefore, numbers scaled at this level, unlike nominal and ordinal scaled measures, may be used in addition and subtraction operations. Interval numbers:

1. can be used to name or label;
2. can be used to rank order individual elements;
3. can be used to determine differences between individuals on some trait or characteristic; and
4. cannot be used to determine how much of the attribute is present in an absolute sense.

One example of numbers scaled at an ordinal level is temperatures on the Fahrenheit scale. If the high temperature on Monday was 76°F and on Tuesday was 73°F, then the difference $76 - 73 = 3$ is a meaningful operation and tells you that the high on Monday was 3°F warmer than the high on Tuesday.

Although zero is a point on many interval scales such as temperature, a "true" or "absolute" zero does not exist on such scales. Thus a score of zero on an ability test does not imply a total absence of ability, just as a temperature of zero does not mean a total lack of kinetic energy or molecular motion generating heat. Getting zero words correct on a spelling test does not mean a total lack of spelling ability. A scale does exist that has all the properties of the interval scale plus an absolute zero point. Some examples are distance, weight, speed, and elapsed time. The numbers so scaled are anchored to an absolute zero, which implies "none" or a total lack of the characteristic being measured. These numbers are on a ratio scale.

Ratio Numbers. **Ratio numbers** have equally scaled units just as do interval numbers; however, the ratio scale also has an absolute zero. Ratio numbers:

1. can be used to name or label;
2. can be used to rank order individual elements;
3. can be used to determine differences between individual elements; and
4. can be used to determine how much of an attribute exists in an absolute sense.

Many physical measurements can be scaled with ratio numbers. The number of feet from point A to point B, the number of square yards of carpet in a room, the capacity of a container in gallons, and the weight in pounds of a fish caught at the lake all make use of measurements on a ratio scale. This is because each method of quantification illustrated has an absolute zero point that implies a total lack of distance, area, capacity, and weight, respectively.

One consequence of measuring on a scale that has an absolute zero is that the arithmetic operation of division may be used and will yield meaningful results. For example, if you consider height in centimeters, you could determine that June is twice as tall as her younger brother by using a "ratio" or division. Contrast this with the ridiculous notion of dividing with nominal numbers: Interstate Highway 40 divided by Interstate 10 yields Interstate 4, a totally meaningless result because of the scaling level of the measurements.

By examining the characteristics of the various numerical scalings, you can observe that "higher"-scaled numbers may be used as numbers on "lower" levels. That is, ratio numbers may be used as interval, as ordinal, or as nominal numbers. This is because as you consider the characteristics of the scales, ratio numbers can satisfy all of the requirements for numbers on the lower-level scales. However, numbers on lower-level scales generally cannot be transformed into numbers on a higher level of scaling. For example, ordinal numbers cannot provide information necessary to qualify as an interval or ratio scale. Generally there is no way to use the numerals on basketball jerseys (nominal) to determine how tall (ratio) the players are.

VARIABLES

A **variable** is a measurable characteristic. Measurement is the process of assigning numbers to various attributes or characteristics according to a specified rule. Therefore a set of measurements, sometimes called *variates*, taken according to specified conditions and rules is called a *variable*. Later in the text in the context of results or outcomes of probability experiments, they will be called *random variables*. Variables are usually represented by letters or names supplied by the person using the data.

For example, if you measured the heights of all students in your statistics class, you would have compiled a set of numbers scaled in inches or centimeters. You would notice that the heights vary. Further, you could call the height variable some letter, say H or X or Y. Although naming is arbitrary, the letters X and Y have traditionally been popular choices for naming variables. If you choose, you could assign a name instead of a letter to the variable, such as "*height.*" Obviously, if a multi-lettered name is used, it should be descriptive of the attribute; hence the name "*age*" used for the height of a number of individuals would undoubtedly add confusion to an otherwise simple operation.

If more than one set of measurements is taken on a single set of people, objects, or events, several variable names will be necessary to identify the respective sets of numbers. If you took measurements of your classmates' heights (in centimeters), weights (in kilograms), and ages (in months), you have created three variables and could label them X, Y, and Z, respectively. Or you might label them H, W, and A (first letters of the respective attributes) to distinguish among the distributions and to assist in identifying each.

It is often convenient to classify variables into one of two categories. Consider the following examples and think about the answers to the questions posed. The Mets could be $8\frac{1}{2}$ games ahead of the Expos. Could it be $8\frac{1}{2}°$ Celsius outside? Could the Rangers be $12\frac{1}{8}$ games ahead of the Angels? Could the temperature be 12.125°? If the weather report states the temperature is 15°C, how confident would you be that it was *exactly* 15°? If you had a sufficiently accurate and valid thermometer, could the temperature be recorded as 14.1396401°C? Could the Indians be 3.685214 games behind the Yankees? After carefully pondering the differences in baseball standings and temperature measurements, you could conclude that something fundamental distinguishes the two. You are correct.

Discrete Variables

Some variables are *discrete variables*. **Discrete variables** often take on a countable number of values. The number of brothers and sisters of your classmates in statistics expressed as a whole number or integer, (0, 1, 2, 3, . . .), for each student is a discrete variable. However, as illustrated previously with the baseball standings, discrete variables may take on noninteger values. Baseball standings can take on increments of $\frac{1}{2}$. They are discrete because they can only assume certain values (0, $\frac{1}{2}$, 1, $1\frac{1}{2}$, . . .) incremented in intervals of $\frac{1}{2}$. The price of stock as quoted in the stock market report can take on only $\frac{1}{8}$ dollar increments and is therefore discrete.

Continuous Variables

A variable is called a **continuous variable** when the attribute or characteristic continues without interruption along a measurement spectrum. Continuous variables can, at least theoretically, be scaled in infinitely small intervals. Consider a weather report that reveals that the low temperature for a certain day was 48° Fahrenheit and the high was 68°F. Consider two aspects of this report. First, the actual number of possible temperature values between the high and the low is infinite. You can understand that for the temperature to pass from, say, 70° to 71°, it must pass through 70.1, 70.11, 70.115, and so on. Second, even though the low was reported to be 48°, it is understood that the low temperature was probably between 47.5° and 48.5°. For the purpose of weather reports, rounding temperatures to the nearest whole number is usually sufficient.

In general, whether a variable is considered discrete (sometimes called discontinuous) or continuous depends on the nature of the characteristic or attribute represented by the variable rather than on the reported measurement. Arithmetic achievement as measured by standardized tests would be considered a continuous variable because it changes continuously through all values within any interval. However, students' scores are typically reported as whole numbers (for example, the number of items correct or percentile rank).

Many of the variables in this course are treated as continuous even if the measurements are discrete. You need to understand why the term "continuous" is used for variables whose values fall along a continuing scale passing through an infinite number of parts regardless of the measuring scheme used. Generally, if the attribute is considered to be continuous, the measured variable will likewise be treated as continuous. Later in the text when the topic of probability distributions is presented, you will be shown how, under certain circumstances, discrete variables can be adequately approximated with continuous distributions.

CRITICISM AND SKEPTICISM

Statistics, because of its universal application, has a high degree of visibility. Well-publicized instances of the abuse of statistics have generated opposition to its use and skepticism about the validity of statistically based decisions. Three common objections to statistical methodology are: (1) statistics are anti-individual; (2) statistical methodology depends exclusively on numbers; and (3) statistics can be used to prove anything. Simply to deny such accusations would be both deceptive and irresponsible. These objections have emerged from decades of use and abuse of statistics and need to be examined in some detail.

The objection that statistics ignore the individual does have some merit. In this sense, the word "individual" refers to both people and things. During this course you will learn about many statistical techniques and concepts. A collection of many individual cases (people, places, events, objects) will provide the context for the statistical orientation. Indeed, very little reference will be made to individuals within the total collection. Predictions affecting the individual will be made based on the available information but always will contain an element of probability and a degree of uncertainty.

For example, based on various atmospheric and climatic conditions, a meteorologist may predict rain in your area. The prediction will be made with a certain degree of confidence and reported in percentages, that is, 20%, 70%, 90%, or some other degree of probability. Therefore the prediction of rain for a particular day at a particular location (an individual case) is made to include the possibility of error. However, most people would agree that a statement such as "Today there is an 80% chance of local precipitation" is better than no information about the likelihood of rain. In the past, when the same weather conditions existed, 80% of the time it rained. On a particular day, it may rain or it may not rain. You can be fairly confident that over a large number of days for which the chance of rain will be predicted to be 80%, you will get wet on most of those days. The exact outcome for the "individual" element (the day in this case) is addressed, but not with total confidence.

An important question then arises whether information about trends based on many individuals is important in making decisions about a single individual within the set. Would your decision on wearing a raincoat or carrying an umbrella be easier if you were told that the probability of rain in your area for today was 10% (or, alternatively, 90%)? Perhaps a more accurate statement would be that statistical results very seldom pertain to individuals except under conditions specified by probability. Statistics do have implications for individual cases (with a degree of uncertainty), and arguments to the contrary are quite superficial.

The second objection is that statistical methodology depends on numbers. The implication is that numbers are "cold" and "impersonal" and therefore fail to incorporate sufficient information about human beings. While it is true that the procedures used in the field of statistics rely on numerical data, numbers are at least as precise, as meaningful, and as understandable as verbal alternatives. To say that the gas mileage for a particular automobile is "only fair" begs the question. To say that the car averages 18.5 miles per gallon on the open highway and 11.6 mpg in the city provides a much better description.

Three points should be understood about the requirements for numerical data. First, characteristics of objects, events, or places (which are frequently the subject of statistical analyses) can be scaled numerically with precision and accuracy (that is, distances, inflation rates, time intervals, volume, etc.). Second, the statistical orientation should not be faulted for inadequate or inaccurate data collection. Although statisticians are concerned about the quality of data and the reliability of the sources of data to be analyzed, statistical methods analyze the data without regard to their validity. Or, to phrase the stance more briefly, statistical methodology and techniques are used to analyze data that exist; they do not create the data. The third point is more tongue in cheek. You undoubtedly know many people who are identified as "Carol" or "Jim" or "David." But how many other people do you know who have a social security number the same as yours? In this case a number is more personal than even a common name because of its uniqueness.

A third criticism that has been leveled at statistics is that it seems one can prove almost anything using statistical methodology. A statement more consistent with reality is that nothing can be proved with statistics. In fact, as you progress through this text, you will never see the word "prove" used to describe the conclusions of any statistical analysis. Terms such as "estimate," "demonstrate," "support," "confirm," and "refute" will be used rather than "prove," because the phrase "to prove" implies a state of total certainty. Such a concept has no legitimate place in statistics.

Statistics is an applied science, and the methods themselves are not valued as "right" or "wrong." However, the application of certain methods may be appropriate or inappropriate, depending on the circumstances. Apparent contradictions in results that trouble the nonstatistician can usually be explained logically by one trained in statistics. Again, as your sophistication in statistics increases, resolving troublesome results will become easier. In the final analysis, this course should prepare you to deal with both the uses and abuses of statistics, and your ability to come to grips with statistics and many of the related controversies will increase as you progress.

PREVIEW OF STATISTICS

The field of statistics can be viewed as several individual components that merge to form the discipline. On a broad level, statistical methods can be classified as descriptive or inferential. Various subtopics may then be assigned to one of these two major branches. Figure 1.2 provides a broad overview of the field. It should be noted that while the various components depicted in the figure appear as unique and separate topics, they are in fact closely related and depend on each other for continuity and completeness. Each block in the figure represents one or more statistical procedures or techniques that will be addressed in this course.

Descriptive Statistics

The names and titles used for the various components, methods, and results of statistics are generally good descriptors of the respective concepts. The phrase "descriptive statistics" is no exception. **Descriptive statistics** refers to the methods and results used to "describe" numerical data. Organizing and displaying data in a way that facilitates understanding of the distribution can be accomplished in a variety of ways, as indicated in Figure 1.2.

Individual Descriptive Statistics. *Individual descriptive statistics* are used to describe an individual's performance compared to a reference group, which is sometimes called a "norm" group. Results can be shown as some form of rank order such as centiles or used to

Figure 1.2—Components of Statistics

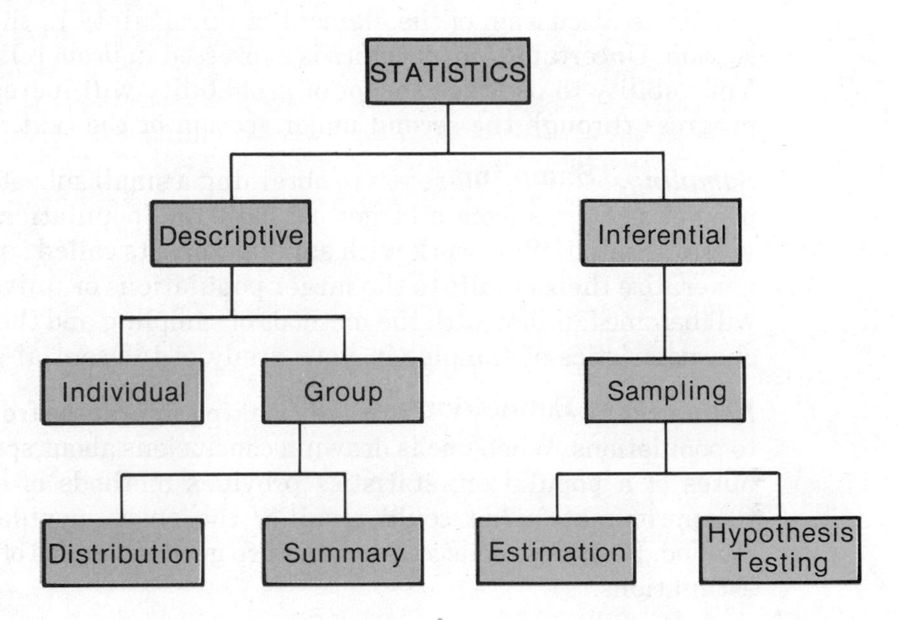

express how well an individual performs in relation to the "typical" group performance. An individual standard score tells whether the performance was superior to or below an "average" performance and by how much.

Group Descriptive Statistics. *Group descriptive statistics* are used to describe an entire set of data. This classification usually takes one or both of two forms: (1) a description of the distribution or (2) a summary of the information. Descriptions of distributions of numbers are often accomplished with tables or graphs representing most or all of the numbers in the set. Summary statistics generally are shorthand versions of the entire set of numbers. They provide an indication of the location of the *central portion* of a distribution of numbers and how much *scatter* or spread exists in the distribution. In some instances, descriptive statistics can be used to summarize how two sets of numbers *relate* to each other. Summary statistics do not customarily show the original data, but use various measures to summarize the data. An example of this is the grade-point average.

Inferential Statistics

Consider the statement: "Descriptive statistics describe the data you have; **inferential statistics** describe data you don't have." While such a statement tends to oversimplify the issue, it does provide a useful conceptualization of inferential statistics. All phases of inferential statistics are based on the science of probability. You may recall the discussion of the element of uncertainty in the previous section. Uncertainty in statistics is expressed in probabilistic terms. Your ability to use the concept of probability will increase as you progress through the second major section of the text.

Sampling. **Sampling** refers to obtaining a small subset of objects, people, or events from a larger set called a "population" or "universe." Statisticians work with smaller subsets called samples and generalize their results to the larger populations or universes. You will become familiar with the methods of sampling and the necessary characteristics of samples in your study of inferential statistics.

Estimation. **Estimation** involves making inferences from samples to populations. When one is drawing conclusions about specific attributes of a population, statistics provides methods of estimation to approximate what could result *if* the entire population were studied. Results of polls and surveys are good examples of statistical estimations.

Hypothesis Testing. **Hypothesis testing** uses various statistical methods to provide information about the relationship between sets of numbers called variables. Used mostly by researchers to analyze two or more sets of data, hypothesis testing is an important aspect of inferential statistics, and a large portion of this text is devoted to hypothesis testing techniques and methods.

The real world consists of people who behave in all kinds of ways, things that change from moment to moment, comparisons that are made, and complex interactions that occur. Such events are reality. The scientific methods and results of statistical analyses are not reality, but an attempt to simplify reality. Reality is never as simple as the simplifier. However, the statistician attempts to reduce the noise, confusion, and mass of reality to comprehensible forms and summaries. Although some distortion accompanies the statistician's efforts, the simplification process is necessary to modern civilizations in today's "information age."

Statistical methods are tools for the thinking person, not a substitute for thinking. Those who are not prepared to use statistics appropriately and to consume statistical information intelligently will be at a distinct disadvantage. You are to be commended for taking this course as a first step toward preparing yourself in this important field. If you treat the course seriously, you will find that statistics have utility, they are sensible and comprehensible, and you may even find them interesting and fun.

COMPUTER FOCUS 1—Planning a Solution: Flowchart Symbols Part I

Introduction

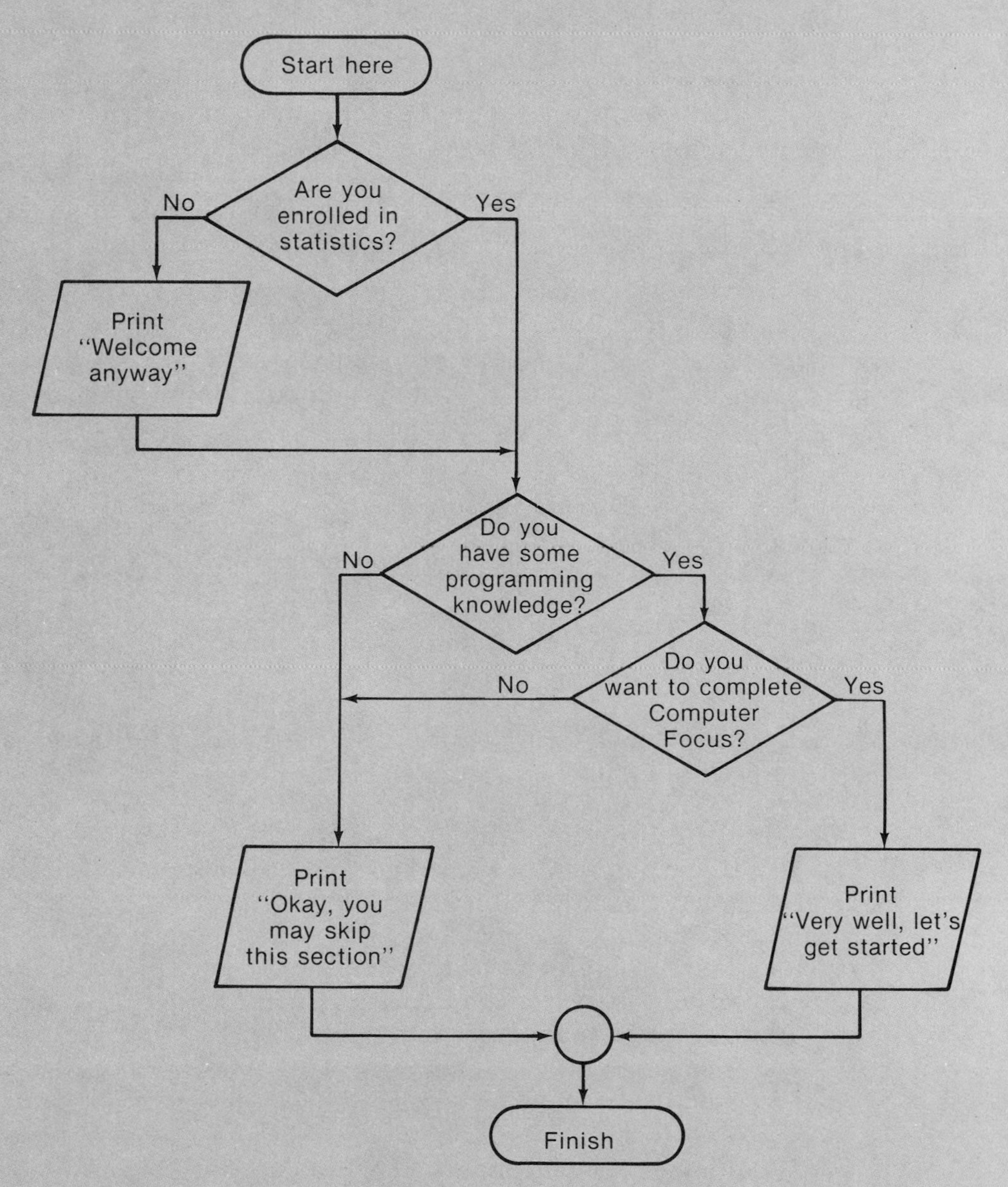

Purpose

The purpose of this Computer Focus is to familiarize you with some symbols that are used to graphically illustrate a plan or strategy or algorithm for solving a problem. The resources of the computer can be used to solve a problem only after its solution has been reduced to a series of discrete, logical steps. The procedural sequence of the steps can be shown by means of a visual device called a flowchart. Were you able to follow the "flow" in the diagram at the beginning of this section?

Symbols

For the present, the steps of a flowchart can be thought of as being one of five types of tasks, as follows:

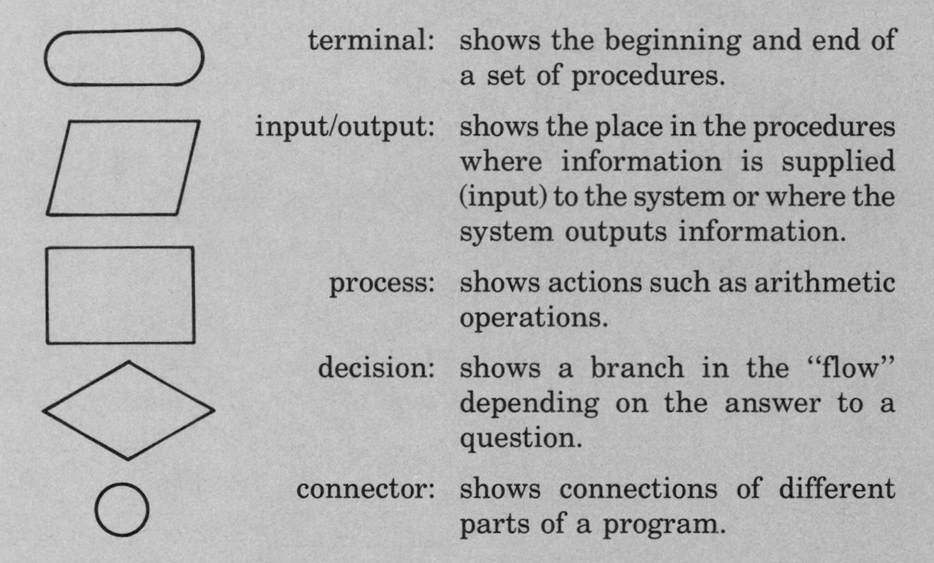

terminal: shows the beginning and end of a set of procedures.

input/output: shows the place in the procedures where information is supplied (input) to the system or where the system outputs information.

process: shows actions such as arithmetic operations.

decision: shows a branch in the "flow" depending on the answer to a question.

connector: shows connections of different parts of a program.

Verify each symbol's role by referring to the introductory flowchart at the beginning of this Computer Focus.

Check Your Understanding

Identify which symbol would most likely be associated with the verbal descriptions that follow.

1. Begin the process at this point.
2. Solve the expression (Length × Width).
3. Are there more cases to analyze?
4. End.
5. Add one number to another.
6. Is $N = 30$?
7. Flow line of A connects flow line of B at point C.
8. Show the solution to the problem.
9. Start.
10. Provide information to the system.

KEY TERMS AND NOTATION

- Analyzing data
- Continuous variables
- Descriptive statistics
- Discrete variables
- Estimation
- Hypothesis testing
- Inferential statistics
- Interpreting data
- Interval numbers
- Measurement
- Nominal numbers
- Numerical data
- Ordinal numbers
- Organizing data
- Ratio numbers
- Sampling
- Statistics
- Summarizing data
- Variable

APPLICATION EXERCISES

1.1 With the help of a dictionary, distinguish between the words "data" and "datum."

1.2 Ask five adults to define "statistics." Classify each response as being a singular noun describing the discipline of statistics or as a plural synonym for "numerical data." Combine your results with those of your classmates and determine which definition is most popular.

1.3 Casey ran the 100-meter dash in 14.2 seconds to win the event for the senior girls. Melinda finished second, and Doris was third. Can you tell if the distance between Casey and Melinda was greater than the distance between Melinda and Doris at the finish line? Why or why not? What kind of scaling is involved?

1.4 A particular brand of automobile tire is rated for a projected life of 60,000 miles. What do you think that means? Will everyone who has a set of these tires get 60,000 miles of life from them? Can you think of some qualifiers that should be assumed if a buyer is expecting 60,000 miles from these tires.

1.5 A recent study on transfer of employees within companies reported that "only 6% of all relocated employees are women." Would that figure be more apt to be a descriptive statistic or an inferential statistical result? Why?

1.6 Another recent report stated that 71% of motorists use self-service pumps rather than full-service islands. Speculate on how the American Automobile Association could have collected data to arrive at that figure.

1.7 Consider these two statistics:
(1) One ounce of Swiss cheese contains 105 calories.
(2) Melanoma (a type of skin cancer caused by excessive exposure to the sun's ultraviolet rays) will strike 1 in 100 persons by the year 2000 unless its accelerating growth rate is slowed.

Write a one-paragraph report addressing these questions. Which of the two do you think is more accurate (precise)? Why? Is the more accurate statistic more "truthful"? Is the more "truthful" statistic more important?

1.8 A group of college students were categorized by major field of study as follows: 1 = technical and scientific; 2 = social and behavioral sciences; 3 = business oriented; 4 = music and fine arts; and 5 = medical/health related. This method of measurement uses which type of numerical scaling?

1.9 A subset of a population is called a(n) __________.

1.10 A __________ variable is one in which there is an infinite number of values that can occur between any two points.

1.11 Using the terms in the boxes in Figure 1.2, how would you classify a numerical figure showing your average daily attendance per month at school during a particular school year?

1.12 A city-limit sign for a midwestern city shows:

POPULATION 63,729

Does that information convey an accurate count of the population? Does it provide a close estimate of the number of people in town? Discuss some assumptions you should make in interpreting the number.

2

Displaying Data

CHAPTER OBJECTIVES

Upon completion of the chapter, students will be able to:

1. Name at least three methods of displaying numerical data in an organized form.
2. Construct a bar graph from given data.
3. Distinguish between a grouped bar graph and a stacked bar graph.
4. Construct a circle graph from given data.
5. Create a pictogram from given data.
6. Identify distortions (illusions) in graphs or picture charts.

Brian and Paige, working as delegates of the student council, had spent most of October conducting a survey in the high school neighborhood. The purpose of the survey was to collect data on how the families in the immediate vicinity of the high school viewed relations between students and the neighborhood families and how the relationships could be improved. Questions on the survey covered many aspects of the school/neighborhood relationship. One example is a school-sponsored clean-up, fix-up day called "Tidy Friday." More than 80 family units responded to the 40-item questionnaire, and the forms were stacked in a box to await processing.

Brian and Paige now faced the task of bringing some order to the unorganized data and eventually making a summary report of the results to the faculty and students in an all-school assembly. Paige had recently completed a statistics class and was eager to apply her newly acquired skills to a practical problem with "real" data. Because Brian had been an outstanding student in the commercial art class, Paige decided to summarize the information in pictorial form for their presentation. In this way she could not only take advantage of her own skill in transforming jumbled data into pictures, but she also could capitalize on Brian's ability to prepare commercial-quality visuals to use in their presentation.

The problem that confronted Brian and Paige is an everyday occurrence for many people. Presentations of organized data in picture form are familiar to every reader of magazines, newspapers, or journals. Such graphic representations can convey a large amount of information very efficiently and effectively. This chapter treats the methodology of graphing nominal variables and could fittingly be titled "Bringing Order to Chaos."

GRAPHIC REPRESENTATION

One of the statistician's first tasks is to present numerical data in an organized fashion. Various methods can be used to accomplish this task. Some popular techniques for presenting data graphically include bar, circle, and picture graphs that can be used with nominal data.

Bar Graphs

A popular method of displaying numerical information is the bar chart or **bar graph**. In a bar graph, rectangles or bars are used to represent the data. Spaces are placed between the bars to emphasize the unordered character of nominal data. A bar graph has two scales or axes; one vertical and one horizontal. One of the scales must be used to indicate units or quantity, while the other scale identifies the type or source of the quantity. When constructing a bar graph, keep these principles in mind:

1. all bars should begin on the same line;
2. all bars should be the same width;
3. spaces between the bars should be equal;
4. the quantitative scale (axis) should begin at the zero point, if possible;
5. a descriptive title should be provided for the graph; and
6. it is necessary to identify what each bar and each scale represents.

Simple Bar Graph. Each year the Gallup Poll of the Public's Attitude Toward Education is conducted on a nationwide basis. In a recent survey, adults were asked: "Generally speaking, do the local public school students in this community have too many rights and privileges, or not enough?" The results (in percentages of responses) were:

Too many	40%
Not enough	12%
Just about right	25%
No opinion	23%

To construct a simple bar graph showing this information, follow these steps:

1. Draw a horizontal line (axis) and label it with the nominal (and discrete) variable name. Place the categories of this variable along the horizontal axis.
2. Draw a vertical line (axis) on the left of and perpendicular to the horizontal axis. Label it "percent of responses." (If actual numbers were used instead of percentages, the vertical axis could be labeled "frequency," "count," "number," or some other appropriate name.)
3. Place a numerical scale along the vertical axis. It should begin at zero. However, when this is not possible or practical, a method does exist for exceptions, which will be illustrated. The scale should consist of equally spaced units and should continue until all percentages are included.
4. Draw a bar over each category on the horizontal line that extends as high on the vertical scale as the percentage (or

frequency) for that category of responses. Space between the bars indicates the discrete and nominal nature of the variable. As will be shown later, graphs for continuous interval and ratio data do not have spaces between bars to explicitly show the continuous nature of such variables.

For the current example, the graph would look like Figure 2.1A.

Figure 2.1A—Simple Bar Graph

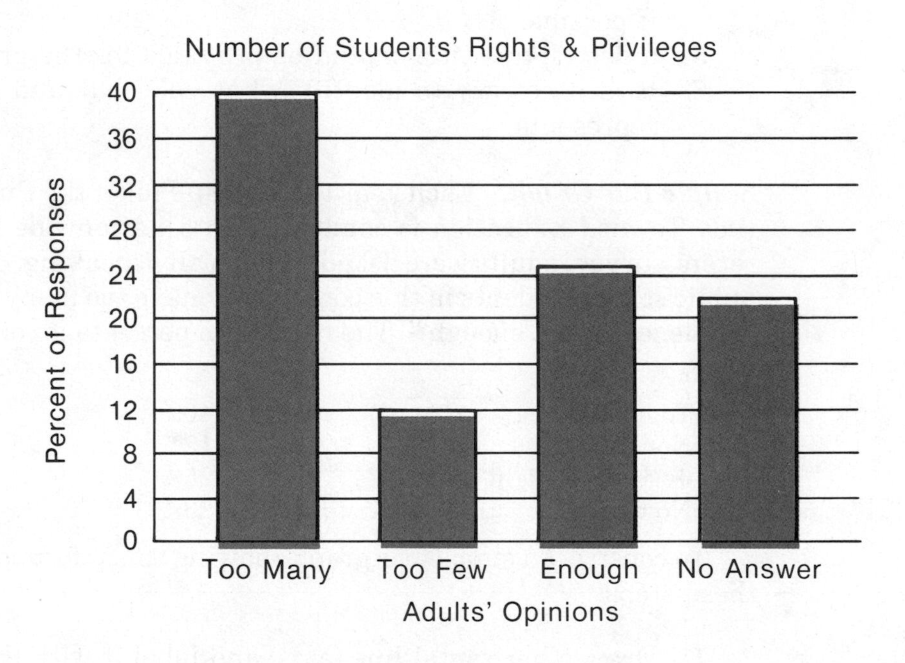

Figure 2.1A illustrates a **vertical bar graph**, because the bars are placed vertically on the graph. By interchanging the vertical and horizontal scales, one would convert the form to a **horizontal bar graph** as shown in Figure 2.1B.

As observed earlier, starting the numbering on the quantitative axis with zero may be impractical or too inconvenient in some situations. This is more likely to be the case when the smallest frequency or percentage is more than half the size of the largest. In such a case it is customary to explicitly show a gap or break in the quantitative axis so that the consumer will be aware of some potential pitfalls (discussed later in this chapter in the section titled "Illusions and Distortions"). This technique is illustrated in Figure 2.1C with the same data set.

Figure 2.1B—Horizontal Bar Graph

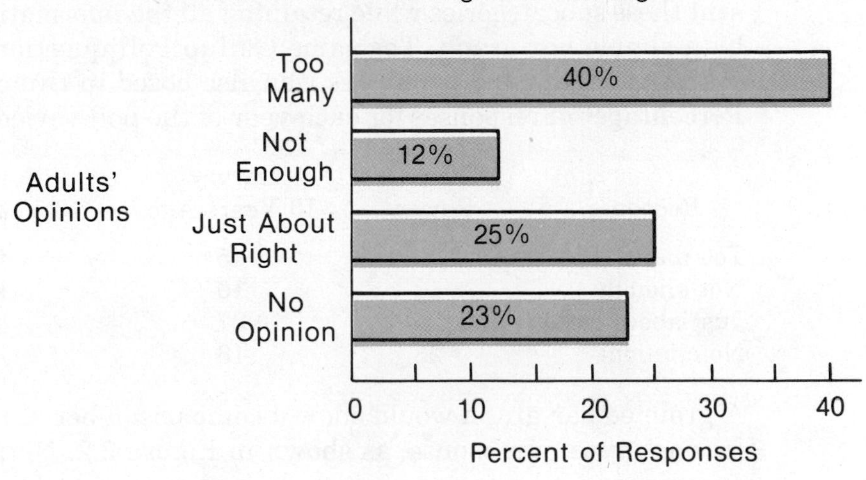

Figure 2.1C—Broken Bar Graph

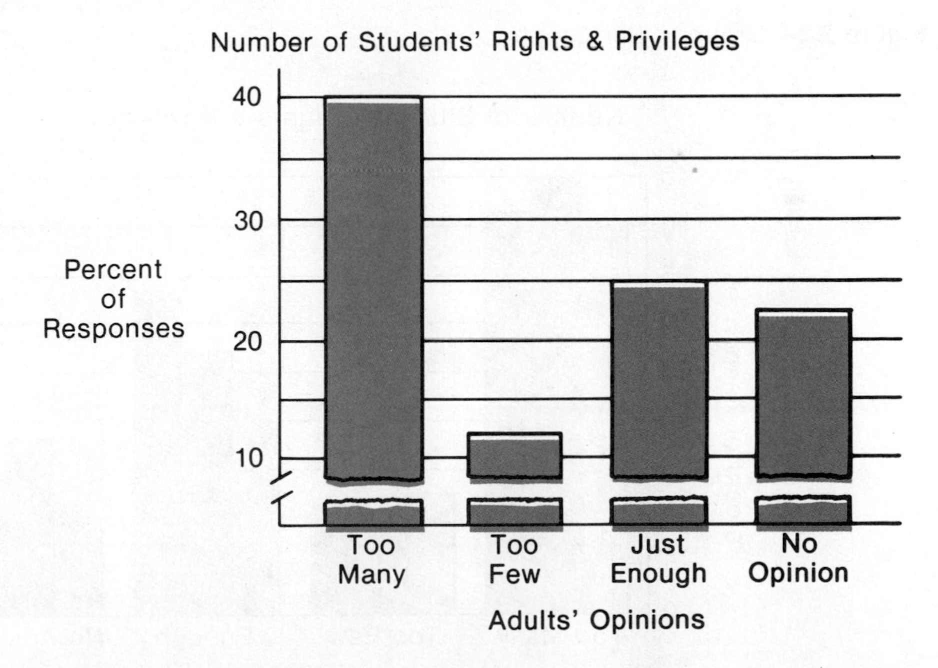

Grouped Bar Graph. One extension of a simple bar graph is a **grouped bar graph**. A grouped bar graph is constructed like the simple bar graph except that two or more subcategories are combined for each category of the discrete variable. That is, when data for each catgegory of the nominal variable can be divided into two or more components, a grouped bar graph can be used to visually represent these subcategories while retaining all the information provided by a simple bar graph. The same Gallup Poll question regarding students' rights and privileges was also posed in two other years. Percentages of responses for each year of the poll varied as follows:

Response	Current Survey	Survey 10 Years Ago	Survey 13 Years Ago
Too many	40	45	41
Not enough	12	10	11
Just about right	25	27	33
No opinion	23	18	15

A grouped bar graph would show a comparison between the years for each type of response, as shown in Figure 2.2. Notice that the groupings or subcategory clusters do not have spaces between them, but the main variable categories are still separated by spaces.

Figure 2.2—Grouped Bar Graph

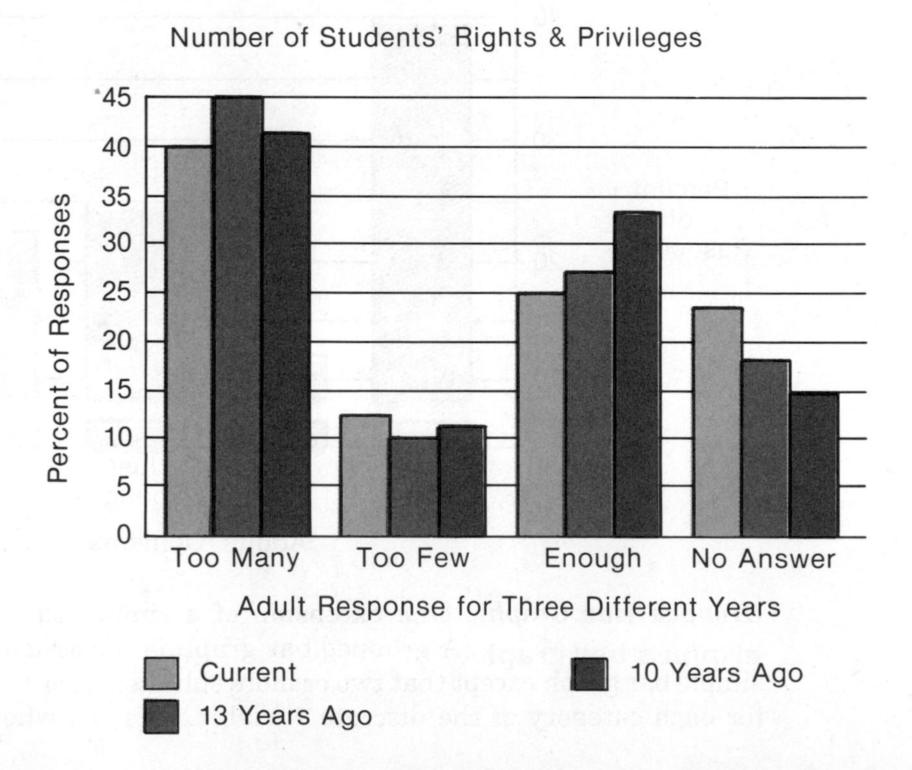

Stacked Bar Graph. Another common variation on the bar graph is referred to as a **stacked bar graph**. Instead of grouping subcategories or components of each variable category for comparison as with a grouped bar graph, one constructs the graph so that the frequencies or percentages representing responses to subcategories are "stacked" vertically. Stacked bars show the cumulative response for two or more subcategories. To envision a stacked bar graph, consider data collected by the Gallup Poll over a five-year period. Respondents were asked to give the public schools a grade—A, B, C, D, or F. The stacked bar graph in Figure 2.3 shows the results of the survey. The percentages of A's and B's, C's, D's and F's, and "no opinions" can be readily compared across the five-year period.

Figure 2.3—Stacked Bar Graph

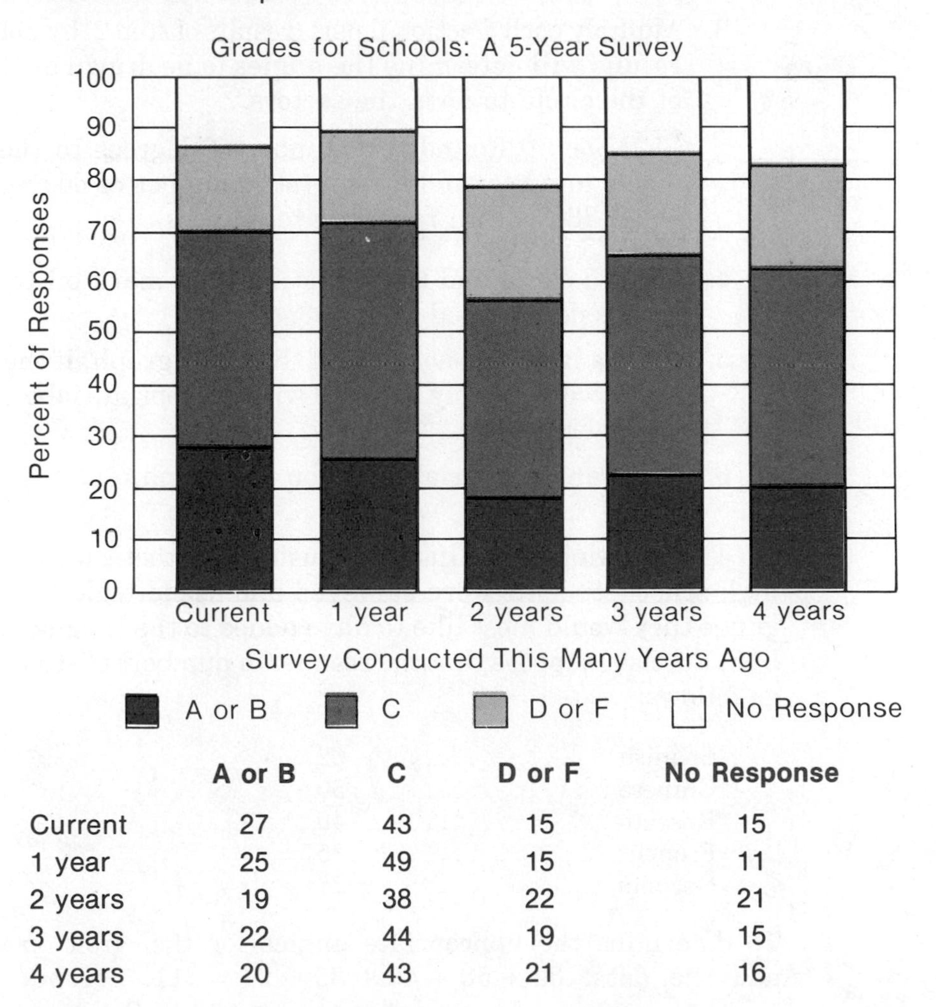

	A or B	C	D or F	No Response
Current	27	43	15	15
1 year	25	49	15	11
2 years	19	38	22	21
3 years	22	44	19	15
4 years	20	43	21	16

Circle Graphs

Another common form of statistical display is the **circle graph** or pie graph. This technique uses a circle to represent the whole of a quantity or a group. The circle is divided into areas or sectors representing the amounts of the respective categories. Circle graphs are useful for visual comparisons of various categories being displayed. Steps for completing a circle graph are:

1. Find the total for the given data. This will be represented by the circle.
2. Compute the fractional part or percentage for each of the given categories. These will be translated into fractional parts or sectors of the circle.
3. Multiply each fractional part (results of step 2) by 360°. These results will determine the angles to be drawn at the center of the circle to form the sectors.

   Note: (a) Round the number of degrees to the nearest whole number; and (b) the total number of degrees should equal 360 within rounding tolerances.

4. Draw a circle and use a protractor to mark off the central angles determined in step 3.
5. Print a label on each part of the circle graph. If the graph is too crowded, supply a legend with appropriate labels adjacent to the circle.
6. Place an appropriate title on the graph.

The following is an illustration using hypothetical data. Suppose high school students were surveyed and asked which foreign language they would most like to have added to the language curriculum. Further, suppose that the results (in numbers of students) were as follows:

Spanish	62
Chinese	50
Russian	40
French	35
German	24

To determine the appropriate angles for the circle graph, first add the data: $62 + 50 + 40 + 35 + 24 = 211$. Second, find the fractional part of the whole for each category (language) and multiply by 360°:

Spanish	$\frac{62}{211} \times 360° = 106°$
Chinese	$\frac{50}{211} \times 360° = 85°$
Russian	$\frac{40}{211} \times 360° = 68°$
French	$\frac{35}{211} \times 360° = 60°$
German	$\frac{24}{211} \times 360° = 41°$

Notice that the final number of degrees in each case has been rounded to the nearest whole degree, because that level of accuracy is sufficient for work with a protractor. Also note that the total number of degrees should add to 360. You can use that as a check of your computations.

Finally, mark off the angles or sectors and place the appropriate labels and title on the graph. Your circle graph should look like Figure 2.4.

Figure 2.4—Circle Graph

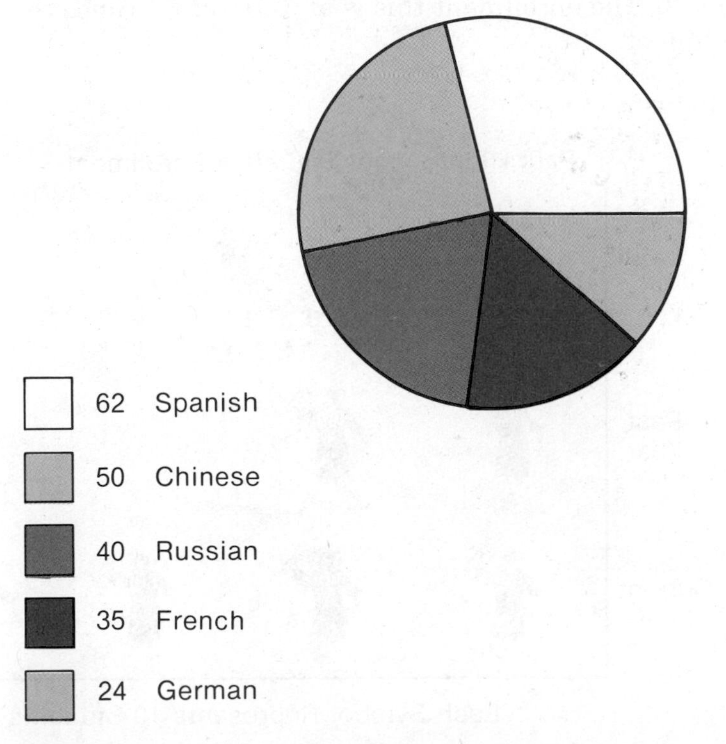

Pictograms

Bar graphs and circle graphs, when properly constructed, provide a reasonably accurate and meaningful visual display of numeric data. However, bar graphs are not very interesting, nor do they tend to attract the attention of the casual reader. Consequently, creators of graphs and charts often use pictures of objects to enhance the entertainment value of their charts. Visual presentations using pictures to portray quantities are called picture graphs or **pictograms**. Because pictograms generally appear more interesting than their more traditional graphic counterparts, popular magazines and newspapers make extensive use of this technique.

Pictures used to represent various quantities usually identify the units being graphed. Money bags, animals, containers, humans, buildings, automobiles, and other common images are used to represent the item being depicted. To show comparisons, the number of images is varied in proportion to the variation in quantities. Pictograms are no more accurate than other types of graphs, but visual esthetics justify their use.

The graph in Figure 2.5 depicts a comparison of the number of students enrolled in a high school statistics class over a three-year period. The enrollment two years ago (10) was doubled in the past year (20). The enrollment this year (35) was $3\frac{1}{2}$ times the first-year enrollment.

Figure 2.5—Pictogram

Notice that the images are all the same size and have equal spacing between them. The quantity represented by one image in this case was 10 students.

Although the amount represented by each symbol or picture is arbitrary, you should select some value that is simple to multiply and add. This will allow readers who are seriously examining the graph to determine the totals easily. For example, if each image or symbol represents 10 or 100 or 10,000, the totals can be readily determined. On the other hand, if each symbol represented 8.34, 96, or 9,920, interpretation would become difficult and the utility of the pictogram would be diminished.

ILLUSIONS AND DISTORTIONS

As visual aids, graphs are efficient and effective means of organizing data. Because they function well, even to the point of being too "user friendly," many consumers of statistics scan the graphs rather than analyzing them in detail. First impressions are often treated as factual even though illusions may cause the user to misinterpret the data being presented. An illusion is simply a false perception, conception, or interpretation of what one sees. Consequently, it is important for users to be aware that illusions may distort the facts whether created on purpose or unintentionally.

The data previously used to illustrate pictograms provide some examples of how figures may be accurately presented while potentially creating an inaccurate first impression.

Figure 2.6A shows a bar graph of statistics class enrollment for a three-year period at Park High School. The graph provides an accurate impression of the facts. The bar lengths are in a ratio of 10:20:35, which accurately reflects the enrollments. The bar graph in Figure 2.6B shows the same enrollment data. However, the impression may be that the enrollment for the past year was three times the enrollment of two years ago because the middle bar is three times longer. Similarly, the current enrollment at a glance appears to be six times the enrollment two years ago because of the respective lengths of the bars. The difference is that the vertical axis (the one with the number scaling) for the graph in Figure 2.6B does not start at the zero point; it originates at 8. Finally, Figure 2.6C gives the impression that the increase in enrollment over the three-year period was much more modest than it actually was. Even though the origin is at zero, unequal units along the vertical axis create an illusion for the casual observer that is not consistent with the data.

Figure 2.6A—Park High School Statistics Enrollment

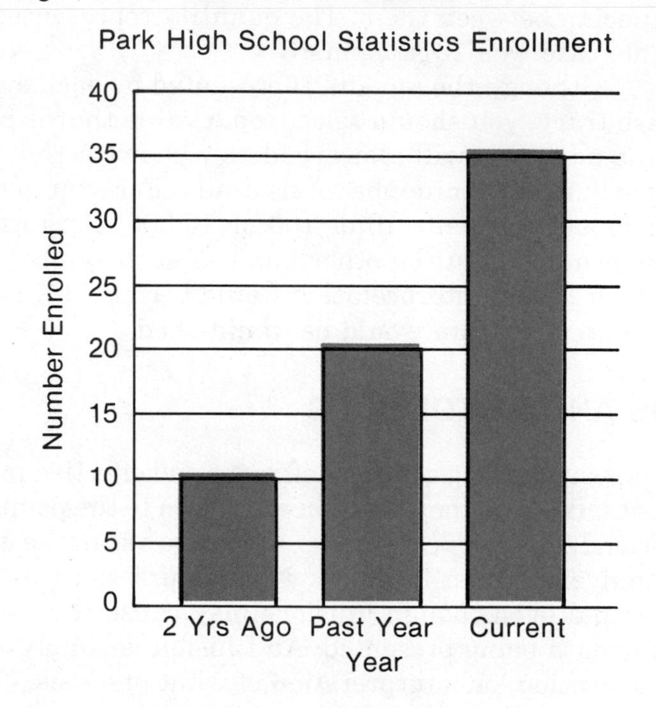

Figure 2.6B—Park High School Statistics Enrollment

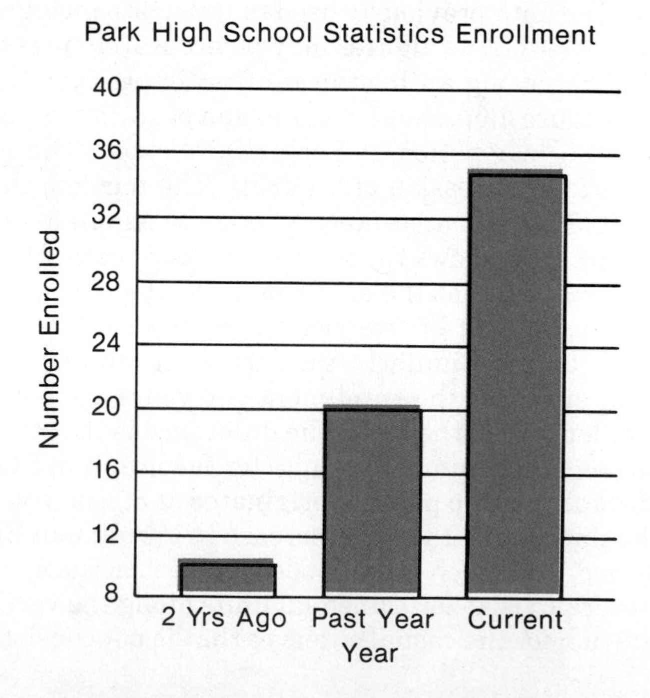

Figure 2.6C— Park High School Statistics Enrollment

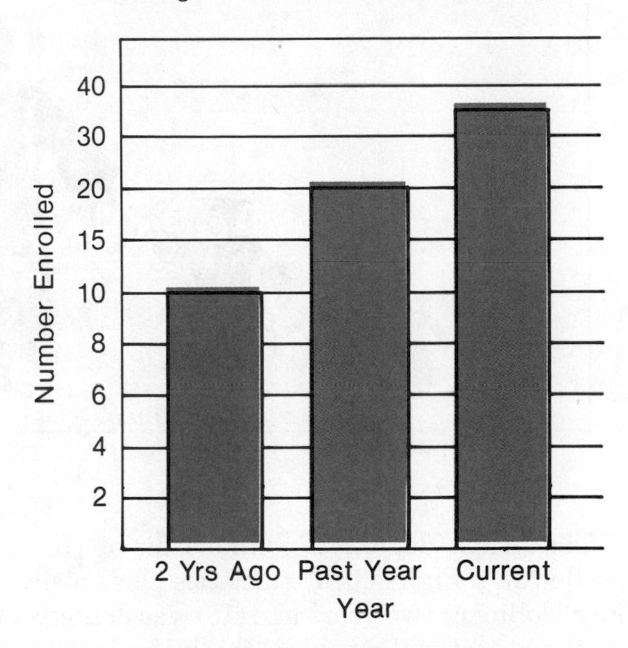

Although all three graphs provide exactly the same information when examined closely, Figure 2.6A is least likely to be misinterpreted by a reader. When creating graphs or when consuming graphic information, be aware of the possibility of misinterpretation by casual observers.

There are even more problems associated with pictograms. First, they are generally difficult to construct without the use of a template. Second, to show comparisons, the sizes of the images often are varied in proportion to the variation in quantities. When the height of an image is varied to show different quantities, the width (and thus the area of the figure) and the presumed third dimension (and thus the volume or capacity) of the object also increase if the images are similar in shape and maintain original proportions.

The third problem is a direct result of the second relating to proportional size variation. The impression formed by a reader about the relative comparisons in a pictogram is almost always exaggerated beyond the factual interpretation. Consider the pictogram in Figure 2.7, which shows the same enrollment data shown in Figure 2.6A.

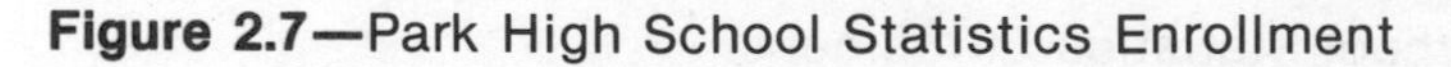

Figure 2.7—Park High School Statistics Enrollment

The graph depicts a comparison of the number of students enrolled in a high school statistics class over a three-year period. The enrollment two years ago (10) was doubled last year (20). Therefore the height of the student image for last year is twice the height of the image for two years ago. The enrollment this year (35) was $3\frac{1}{2}$ times larger than two years ago. Thus the current-year image is $3\frac{1}{2}$ times as high as the first image. At this point, everything seems to be in order.

First impressions of the graph are influenced by appearances and mental images. Because the widths of the images were changed, the area consumed by the respective images changed even more dramatically. For example, the area of the image for last year is no longer twice the area of the image representing enrollment two years ago; it is closer to four times the area! Because the world in which we live is three dimensional, one tends to visualize objects that way even when the images are flat. But the mental tendency to add depth to the images distorts the data even more. The image representing the current enrollment is $3\frac{1}{2}$ times as tall as the figure for the enrollment two years ago. However, in three dimensions, the volume (or capacity if the pictogram used money bags or containers) does not maintain the same proportions. In fact, the image for the current year appears to be about 40 times as large as the image representing the enrollment of two years ago. Obviously, such an interpretation of the enrollment increase would be erroneous.

COMPUTER FOCUS 2—Flowchart Symbols Part II

Purpose

The purpose of this Computer Focus is to provide an opportunity for you to expand the use of flowchart symbols by connecting them in sequence with lines and arrows showing the direction of "flow." This will be accomplished using a nonmathematical problem so that you can concentrate on following the logic of a set of procedures with which you are already familiar instead of spending time on a mathematical problem. Numerical problem solving will come later.

Problem

Your problem is to match the five types of capabilities required to make a local telephone call (using seven digits). The logic has been provided below. Copy the flowchart form on a sheet of paper and insert the appropriate capabilities (tasks) from the list into the respective symbols to illustrate the "Logic of a Local Telephone Call."

Problem Specifications

Given these capabilities, complete the flowchart by inserting one capability in each of the symbols. (Each capability is used only one time.)

Input/Output:

1. Talk.
2. Dial a number.

Process:

1. Lift receiver to hear dial tone.
2. Hang up phone.
3. Wait a few minutes.
4. Wait a moment.

Decisions:

1. Do you hear a dial tone?
2. Have you dialed seven times?
3. Do you get a busy signal?
4. Do you want to try again?
5. Has the party answered?
6. Will the party answer?

Terminal/Interrupt:

1. Begin (already inserted).
2. Finish and hang up.

Flowchart
Logic of a Local Telephone Call

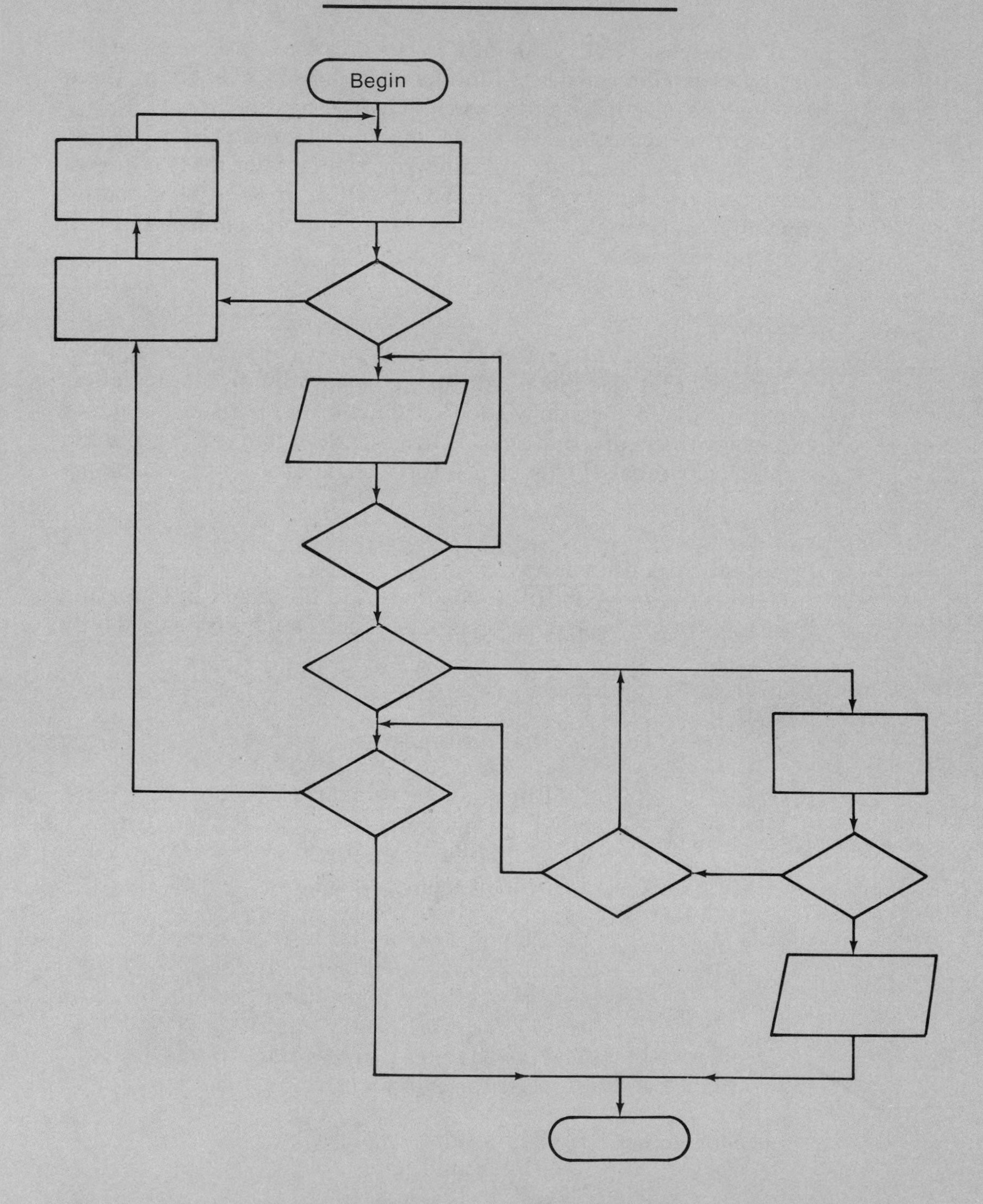

KEY TERMS AND NOTATION

Axis (axes)	Horizontal bar graph
Bar graph	Pictogram
Circle graph	Stacked bar graph
Grouped bar graph	Vertical bar graph

APPLICATION EXERCISES

2.1 In a Gallup Poll survey, adults were asked: "I'd like your opinion about extracurricular activities such as the school band, dramatics, sports, and the school newspaper. How important are these to a young person's education?" Results were as follows:

Very important	39%
Fairly important	41%
Not too important	14%
Not at all important	3%
No opinion	3%

Construct a vertical bar graph showing these results. Scale the percentages on the vertical axis in intervals of 5; scale the five categories of responses along the horizontal axis.

2.2 Men and women graded the public schools in the United States as follows:

Grade	Men (%)	Women (%)
A	8	10
B	33	34
C	32	29
D	11	5
F	3	10
Don't know	13	12

Construct a *grouped* bar graph. Scale the vertical axis in percentages from 1 to 35 in units of 5. The six response categories will be represented along the horizontal axis. The responses of men and women should be compared in the respective groupings.

2.3 American adults were asked: "How important is a college education today?" During three different years, the percentage responses were:

Rating	Year		
	Current	2 Years Ago	7 Years Ago
Very important	64	58	36
Fairly important	27	31	46
Not too important	7	8	16
Don't know	2	3	2

Construct a *stacked* bar graph using the vertical axis to represent percentages 0-100. Place the three years along the horizontal axis. What is the apparent trend in adult thought about the importance of a college education over the past seven years?

2.4 Causes of accidental deaths (per 100,000) for young adults ages 15-24 are:

Motor vehicle accidents	42
Drowning	6
All other types	12

Construct a circle (pie) graph that depicts these data.

2.5 Recently Gallup conducted a nationwide survey of adults who were asked this question: "Do you feel that teachers or school authorities should or should not be allowed to open students' lockers or examine personal property if they suspect drugs, liquor, or stolen goods are hidden there?" Responses by percentage were:

Should	78
Should not	18
Don't know	4

Construct a pie graph showing the results.

2.6 Use the data from Application Exercise 2.5 to construct a horizontal bar graph.

2.7 The average television viewing times in hours per week for various age groups were reported for the month of November and were:

Age	55+		20–24		Teens	
Gender	Male	Female	Male	Female	Male	Female
Hours	37	42	20	27	23	22

Prepare a grouped bar graph showing these data.

2.8 Use the data in Application Exercise 2.7 to construct a stacked bar graph.

2.9 The data that follow show the educational attainment for persons ages 20 to 22 (in percentages). Use this information to prepare a circle graph.

Less than high school	22
High school (4 years)	45
College (1 to 3 years)	26
College (4+ years)	7

2.10 The leading uses of home computers in percentages are as follows:

Video games	51
Business/office homework	46
Child's learning tool	46
Adult's learning tool	42
Checkbook and budget	37
Word processing/mailing lists	34
In-home business	27
Other	36

Prepare a pictogram using a microcomputer symbol to depict the computer use data.

2.11 Why is the constant 360 used in the process of constructing a pie graph?

2.12 For the data that follow, decide which type of graph would best display the data to satisfy the stated purpose in a, b, and c. The data show the average ages of recipients of degrees in various scientific disciplines in a recent year.

Discipline	Average Age in Years Bachelor's	Master's	Doctoral
Physics and astronomy	22.5	26.2	29.4
Biological/earth science	23.1	25.8	28.6
Mathematics/statistics	22.2	26.2	30.4
Computer science	21.9	27.0	30.1
Engineering	23.6	25.5	31.2

a. Purpose: The graph should provide a comparison of the average ages of recipients of doctoral degrees across the scientific disciplines.
b. Purpose: The graph should provide direct comparisons between the ages of master's and bachelor's degree recipients across the science fields.
c. Purpose: The graph should provide direct comparisons among the ages of all three degree levels across the scientific disciplines.

2.13 Given the data that follow, create a graph that would be a deliberate attempt to distort the data to try to influence your readers. You want the readers to conclude that athletics is by far the favorite extracurricular activity. The data were collected from a survey of students who were asked to identify their favorite extracurricular activity.

Activity	Male	Female	Total Responses
Athletics	22	17	39
Drama	15	18	33
Band/music	18	20	38
Student organization	13	12	25
Other	12	13	25

2.14 Use the data in Application Exercise 2.13 to construct a graph that would emphasize (even distort for the casual observer) the difference between male and female responses.

2.15 Collect information from your classmates on a topic of your choice and display the data in an appropriate graph.

3

Frequency Distributions

CHAPTER OBJECTIVES

Upon completion of the chapter, students will be able to:

1. Organize original data into a frequency distribution table.
2. Distinguish between grouped and ungrouped frequency distributions.
3. Transform a frequency distribution into a relative frequency distribution and into a cumulative frequency distribution.
4. Transform a relative frequency distribution into a relative cumulative frequency distribution.
5. Distinguish between "real" and "apparent" class interval limits.
6. Construct a joint frequency contingency table from two categorical variables.
7. Construct a histogram, frequency polygon, and frequency curve from a frequency (or relative frequency) distribution.
8. Construct an ojive given a relative cumulative frequency distribution.
9. Plot points on a scattergram when given a bivariate distribution.
10. (Optional) Distinguish between a histogram and a stem-and-leaf diagram.

TABLES

Consider a list of all the telephone numbers in your town. They are arranged in alphabetical order in your local telephone book for utility. The telephone numbers could be arranged in numerical order, or they could be entered in a random or haphazard way. Although the various lists would contain exactly the same information, the usefulness of the three arrangements of phone numbers can scarcely be compared.

Similarly, statisticians with a collection of numbers must organize the data in such a way that they can be readily understood. The methods presented throughout this chapter have the single purpose of organizing and displaying large masses of numerical data. These sets of data are unorganized in their original form and require systematic arrangement to assist users in making sense of the numbers.

Several statistical methods are available for arranging data in a compact, comprehensible, and accurate form so that essential characteristics of the data can be easily recognized and interpreted. Although the previous data display chapter dealt with discrete variables, most of the data of concern in this text are assumed to be continuous. Therefore methods included in this chapter are appropriate for use with continuous data. One way to organize and display data is in a table referred to as a **frequency distribution**.

Frequency Distribution

Pamela had a unique idea for her high school civics class term project. As the result of letters of inquiry followed by a meeting with the local police commissioner, she had received permission to conduct a study on the speed of vehicles traveling on a recently completed boulevard. The commissioner was so interested in the project that Pamela was allowed to go on patrol with a police officer in an unmarked radar-equipped vehicle for a period of one hour immediately after school. Her topic for the term project was "Conformity to Posted Speed Limits." When the project was finished, the civics teacher gave Pamela permission to add a subtitle to her original project title: "Prior to Rush-Hour Traffic."

Pamela and Officer Pauline Mahaffey stationed themselves close to the central part of the new four-mile boulevard. Officer Mahaffey's job for this period was to "clock" the vehicles and report the speed of every fourth automobile passing the radar site. Pamela recorded the data as Officer Mahaffey called them out. At the end of the period, Pamela had recorded 120 datum points (speeds) rounded to the nearest whole number. Her field record is shown in Table 3.1.

Table 3.1 Unorganized Listing of Vehicle Speeds

58	53	50	56	55	52	53	60
52	47	53	41	51	51	55	52
49	52	46	54	53	54	50	53
54	45	48	50	52	60	51	55
51	55	54	55	56	50	50	53
53	44	59	51	52	54	57	52
61	52	53	48	53	55	53	55
52	56	52	56	47	52	49	52
54	50	54	49	42	58	54	51
53	49	47	57	51	50	62	56
57	51	50	44	46	52	52	54
52	48	43	48	55	46	57	50
51	54	51	49	45	53	51	54
53	47	53	40	51	58	52	53
55	55	59	49	52	52	48	56

Ungrouped Frequency Distribution. Next the data needed to be organized into some systematic form so that the speed of the vehicles could be easily comprehended. Pamela decided to arrange the numbers from highest to lowest and then tally the frequency of occurrence of each recorded speed. Pamela had not taken a statistics class, but common sense directed her procedures. She started the organization process as shown in Table 3.2. She labeled "speed" with a variable named X.

By counting the tally marks for the respective speeds, Pamela recorded the frequency for each value of the variable X. She created a table known to statisticians as a *frequency distribution*, or, more specifically, an *ungrouped frequency distribution*. Notice that the table consists of two columns: a column labeled X (a variable representing speed), and another column labeled f that provides the frequency of occurrence of each value of X. These data are shown as a frequency distribution in Table 3.3.

When the data were organized in a frequency distribution, the interpretation became easier. The picture was much clearer than when the data were simply a collection of unorganized numbers. Notice that from the frequency distribution, the extreme speeds (high and low) can be easily identified. Speeds that occur the most frequently are also obvious. Further, you can observe how the speeds are distributed along the entire scale. It is easy to tell if the

Table 3.2 Tally of Vehicle Speeds

Speed (X)	Tally
62	/
61	/
60	//
59	//
58	///
57	////
56	~~////~~ /
55	~~////~~ ~~////~~
54	~~////~~ ~~////~~ /
53	~~////~~ ~~////~~ ~~////~~
52	~~////~~ ~~////~~ ~~////~~ ///
51	~~////~~ ~~////~~ //
50	~~////~~ ////
49	~~////~~ /
48	~~////~~
47	////
46	///
45	//
44	//
43	/
42	/
41	/
40	/

Table 3.3 Frequency Distribution of 120 Vehicle Speeds

Speed (X)	Frequency
62	1
61	1
60	2
59	2
58	3
57	4
56	6
55	10
54	11
53	15
52	18
51	12
50	9
49	6
48	5
47	4
46	3
45	2
44	2
43	1
42	1
41	1
40	1

frequencies are uniform along the range of speeds or if gaps exist at certain locations. All of this can be determined at a glance when the data are organized.

The frequency distribution in Table 3.3 shows the data in exact detail. That is, every possible speed, rounded to the nearest integer, between the highest and lowest speed is displayed. A distribution as shown in Table 3.3 is more explicitly referred to as an **ungrouped frequency distribution**, because all values are listed in the array of speeds. For display purposes, when the data have a wide range of values, an ungrouped frequency distribution may become a cumbersome means of showing the distribution.

Grouped Frequency Distribution. A **grouped frequency distribution** provides an efficient and economical presentation of the data; but, as a result of the grouping process, some information is lost. This is because exact values, speeds in this case, are lost in the groupings of the class intervals. However, the efficiency gained usually more than offsets the rather small loss of information. Class intervals of 3, 5, 7, 9, etc., depending on the magnitude of the variables, are popular choices. The smaller the width of the class interval, the less information is lost. Before pursuing the procedures further, examine the grouped frequency distribution shown in Table 3.4. Notice that class intervals of 3 units (miles per hour) are used.

Table 3.4 Grouped Frequency Distribution of Auto Speeds (N = 120)

Class Intervals	f
61 - 63	2
58 - 60	7
55 - 57	20
52 - 54	44
49 - 51	27
46 - 48	12
43 - 45	5
40 - 42	3
Total N =	120

Usually, if the difference between the highest value and lowest value in a distribution, called the *range*, is less than 20 units, an ungrouped frequency distribution can easily be used. But if the span is over 20 units, the rule of thumb is to group the values into class intervals and proceed with a grouped frequency distribution. Generally the number of class intervals used should be from 8 to 15. The procedures in Table 3.5 were used for the grouped frequency distribution in Table 3.4.

Notice that the frequencies shown in a grouped frequency distribution cannot be converted back to an exact speed. For example, without examining the ungrouped data, you cannot tell which speed each of the five frequency tallies goes with in the interval 43 – 45. The original data cannot be recreated from the grouped frequency distribution. Consequently, some of the original information has been lost in the transition, but gross interpretations of the data are not affected.

Table 3.5 Grouped Frequency Distribution

Procedure	Application
1. Subtract the smallest number from the largest number in the distribution to determine the range.	1. 62 − 40 = 22
2. The range is divided by the desired number of intervals and the quotient rounded to the nearest *odd* integer. Divide by any number you choose between 8 and 15. Larger ranges generally call for larger numbers of divisions. For the present example, divide by 8.	2. 22 ÷ 8 = 2.75 rounded to 3
3. Arrange the numbers into intervals of the desired width and form a column of class intervals. Using an interval width that is an odd number (3 in this case) will cause the midpoint of an interval to be a whole number.	3. 61 - 63 58 - 60 55 - 57 . . . . . .
4. Form a second column showing the sum of the frequencies falling in the respective intervals. Label the column f.	4. Class Intervals f 61 - 63 2 58 - 60 7 55 - 57 20 52 - 54 44 . . . .
5. Provide an appropriate title for the table.	5. Grouped Frequency Distribution of Automobile Speeds

An important point about class intervals for continuous data is a distinction between *apparent limits* and *real limits* of each class interval. The **apparent limits** for the interval 43 – 45 are the lowest number (43) and the highest number (45) in the class interval. However, these data are continuous even though the measurements were made to the nearest whole number. Conceptually, speed can take on an infinite number of values within the possible range. Therefore the **real limits** for the interval 43 – 45 are actually 42.5 and 45.5. This is true because any speed that has been precisely measured that is larger than 42.5 and smaller than 45.5 will be included in the 43 – 45 class interval.

Even when fractional measures are not used, it is useful to think of each possible value as being represented in the interval. Consider the speed of 45 as being included in the interval from 42.5 to 45.5. Notice that the width of a class interval of 43 – 45 in the grouped frequency distribution includes 43, 44, and 45, or three numbers. The class interval width also can be computed by subtracting the real lower limit from the real upper limit: $45.5 - 42.5 = 3$. However, this method doesn't work with apparent limits; it can only be used with real limits.

One more important point about grouped frequency distributions is that the midpoint of an interval will be a whole number if the interval width is an odd number. Although such a condition is not a requirement for a grouped frequency distribution, having integers as midpoints is convenient and customary in statistical work. The middle number often will be used to stand for all the values in the interval. As you can see, the respective midpoints of the class intervals shown in Table 3.4 are 41, 44, 47, 50, and so on until the highest midpoint (62) is reached.

Continuing with the same example, vehicle speeds that were obtained by Pamela for her civics project can be further treated in table form. Customarily statisticians construct three additional columns for displaying their results: (1) cumulative frequency distribution (cf), (2) relative frequency distribution (rf), and (3) relative cumulative frequency distribution (rcf). These distributions are shown in Table 3.6 as extensions of the frequency distribution (f) developed in Table 3.4. Examine Table 3.6 and then follow the subsequent discussion of the procedures used to construct such a table.

Table 3.6 Frequency, Cumulative Frequency, Relative Frequency, and Relative Cumulative Frequency of Auto Speeds

Class Intervals	f	cf	rf	rcf
61 - 63	2	120	0.02	1.00
58 - 60	7	118	0.06	0.99
55 - 57	20	111	0.17	0.93
52 - 54	44	91	0.37	0.76
49 - 51	27	47	0.22	0.39
46 - 48	12	20	0.10	0.17
43 - 45	5	8	0.04	0.07
40 - 42	3	3	0.03	0.03

Cumulative Frequency Distribution

The frequency (f) column has already been addressed and forms the basis for the remainder of the columns. The **cumulative frequency (cf)** distribution column is constructed by adding the respective frequencies for all class intervals representing speeds less than or equal to a particular interval. By definition, the cf of a particular score in a frequency distribution is the number of scores that are less than or equal to that score. The cf of a class interval in a grouped frequency distribution is the sum of the frequency of scores in that interval and in all lower intervals. Note that the final cf will always be equal to the total number of scores in the distribution. The name "cumulative frequency" actually describes the procedure. You can tell very quickly how many autos were traveling at a *certain speed or less* by looking at the cf distribution.

Relative Frequency Distribution

The **relative frequency (rf)** distribution shows the proportion of frequencies in the respective class intervals. For example, 0.10 or 10% of the total number of observations fall in the class interval 46 – 48. That is, the frequency (12) is 0.10 of the total (120).

The formula for calculating the relative frequency is:

$$\mathbf{rf} = \frac{\mathbf{f}}{N}$$

where f is the frequency for an interval and N stands for the total number of observations.

Thus the rf for the class interval 52 – 54 is computed this way:

$$\text{rf} = \frac{44}{120} = 0.366\ldots \text{ or } 0.37 \text{ rounded to two decimal places}$$

The remainder of the values of rf for the class intervals are computed similarly: $\frac{f}{N}$ rounded to two decimal places.

The rf distribution may be converted to percentages by multiplying each proportion by 100. The importance of the rf distribution is that it should be relatively constant or unchanging no matter how many total observations are made. Think about it: If Pamela had made 500 observations instead of 120, the f and cf columns would have changed drastically. On the other hand, the rf, because it represents a proportion, would look very much the same with regard to the numerical values for the class intervals. As you will see, the same rationale applies to the relative cumulative frequency (rcf) distribution.

Relative Cumulative Frequency Distribution

The **relative cumulative frequency (rcf)** distribution is the last column constructed in Table 3.5.

The formula for determining the rcf is:

$$\mathbf{rcf} = \frac{\mathbf{cf}}{N}$$

where cf is the cumulative frequency and N is the total number of observations.

The rcf can be determined by dividing the cf for a class interval by the total number of observations (N). Or perhaps rcf can be determined more easily by adding the rf distribution from the lowest class interval to the highest. Note that the rcf for the highest class interval will always be 1.00 or 100%. The rcf shows immediately what proportion or percentage of the auto speeds was equal to or below a particular class interval. Those of you who are familiar with percentile scores on standardized test results may recognize rcf as being similar to a percentile (or centile) score. You can easily see with the aid of the rcf distribution in Table 3.5 that 93% of the vehicles were traveling at 57 mph or less. Or, more precisely, using real limits, 93% were traveling 57.5 mph or slower.

To help keep the f, cf, rf, and rcf distributions in perspective, think about an analogy: f is to cf as rf is to rcf; or, alternatively, f is to rf as cf is to rcf. Better yet, focus on the name of each distribution and notice how the names actually describe the corresponding procedures for obtaining the respective distributions.

Joint Frequency Distribution

The concept of a frequency distribution may be easily extended to a situation that involves two variables, which is referred to as a **bivariate** distribution. In such a circumstance, there exist two measurements on each person or case and all of the information may be reported in a single simple table for comparison purposes. Several names are used to identify such a table, including **joint frequency distribution**, bivariate frequency table, contingency table, and frequency crosstabulation table.

Suppose for the sake of illustration that Pamela had obtained a second measure on the vehicles in her study. Assume that she classified each car as a "sports model" or "not a sports model" according to some accepted criterion. For this illustration this measure is a discrete variable, although it could have been continuous if the cars could have been rated in degrees from "nonsport" to "full sport" models along a continuum. Table 3.7 makes use of both variables by showing a single *joint frequency distribution*.

Table 3.7 Joint Frequency Distribution—Type of Vehicle by Velocity (Class Intervals)

Class Intervals	Type: Sports Cars	Type: Nonsports Cars	Row Marginals (Totals)
40 - 42	1	2	3
43 - 45	2	3	5
46 - 48	4	8	12
49 - 51	12	15	27
52 - 54	21	23	44
55 - 57	12	8	20
58 - 60	4	3	7
61 - 63	1	1	2
Column Marginals (Totals)	57	63	120

The frequency shown in each *cell* or intersection of a row and column tells how many vehicles of a particular type (one variable) were traveling at what speed (another variable). For example, 12 sports cars in the survey were traveling in the interval 49 – 51 mph. The row and column totals are called *marginal frequencies*. More specifically, the sums of the frequencies in the rows of a joint frequency table are called row marginals. The sums of the frequencies in the respective columns are called column marginals. Although joint frequency distributions or contingency tables will be used for more advanced statistical analysis in Chapter 16, for the present this example has simply illustrated how a bivariate frequency distribution can be used to describe data.

VISUAL PRESENTATIONS

The usual step following the construction of a frequency distribution table is to present data in pictorial form. This enables readers and statistical consumers to easily and quickly apprehend the essential characteristics of the data. These pictures, called graphs, should not be considered as a routine substitute for statistical analyses of data, but rather as *visual aids* that complement more sophisticated treatment of the numerical information.

Histogram

Frequency distributions are often shown in graphic form. The information conveyed is the same whether a table or a graph is used. A common method of graphing a frequency distribution is the *histogram*. A **histogram** is a vertical bar graph presentation of a frequency distribution. The vertical distance, or height of the bars, represents the frequency or relative frequency. The class intervals, or their respective midpoints, are distributed along the horizontal axis. To denote the continuous nature of the speed variable, the bars do not have spaces between them. This is the physical feature of a histogram that immediately distinguishes it as a special kind of vertical bar graph.

The data in Table 3.6 are graphically shown in the histogram in Figure 3.1. Notice that the frequency distribution was used, although the relative frequency could have been graphed. Also note that the midpoints of the class intervals have been used to represent the entire range of the interval. For example, the interval 40 – 42 is represented by 41, its midpoint.

Figure 3.1—Histogram of Vehicle Speeds

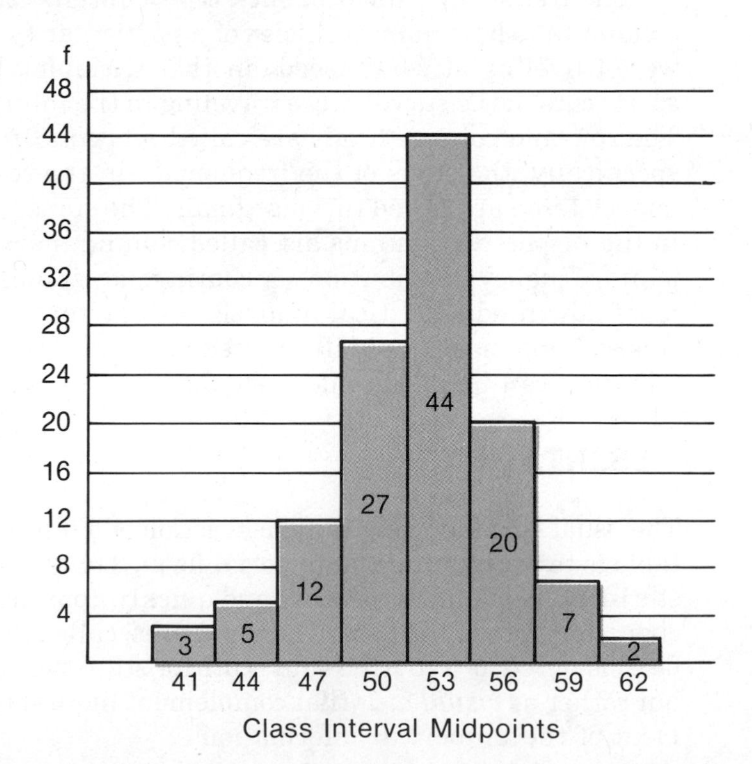

Generally the rules for constructing a histogram are the same as for any vertical bar graph except for the absence of spaces between the bars. Pamela could use a histogram in her civics report to show the speeds of the vehicles in an attractive and comprehensible form. She also has other choices to show the same information.

Frequency Polygon

One of the other means Pamela could use is called a line graph or, more technically, a *frequency polygon.* To construct a frequency polygon, start by setting up the vertical and horizontal axes of the graph in the same way as for setting up a histogram. However, instead of using the length of bars to show the frequency for a class, a **frequency polygon** uses points that are connected by straight lines. The next step, after you have set up the two axes, is to mark points vertically above the class midpoints at a height that represents the respective frequencies on the vertical axis. It is as though you placed dots in the middle of the top of the bars on an imaginary histogram. Then connect these dots from left to right. Figure 3.2 shows the resulting frequency polygon for the same data used in Figure 3.1.

Figure 3.2—Frequency Polygon of Vehicle Speeds

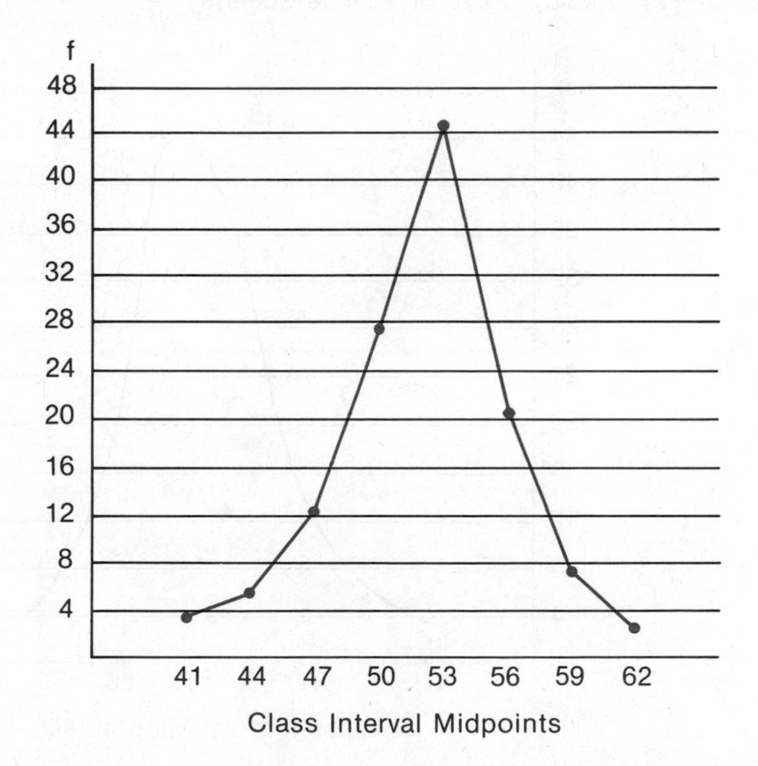

When you look at a frequency polygon, the origin of the name is no longer a mystery. After all, the graph shows the frequency of the various class intervals. It does so with the visual aid of a polygon or a many-sided figure; thus the name "frequency polygon."

Curve

To further emphasize the continuous nature of the variable, "smooth-line" graphs rather than "broken-line" polygons are common. Such figures are generally called "smooth-line curves" or, more simply, **curves**. Curves are created in much the same way as frequency polygons and provide the reader essentially the same information. When one constructs a curve, the points are "fitted" to a smooth continuous line rather than to short, straight-line segments. A curve is somewhat more appealing in appearance than a polygon. Figure 3.3 displays the vehicle speeds with a curve.

Histograms, frequency polygons, and curves are used to present frequency distributions in pictorial form. They are also used to display relative frequency distributions. This is accomplished by simply changing the vertical axis scaling from f (frequency) to rf (relative frequency) proportions.

Figure 3.3—Frequency Curve of Vehicle Speeds

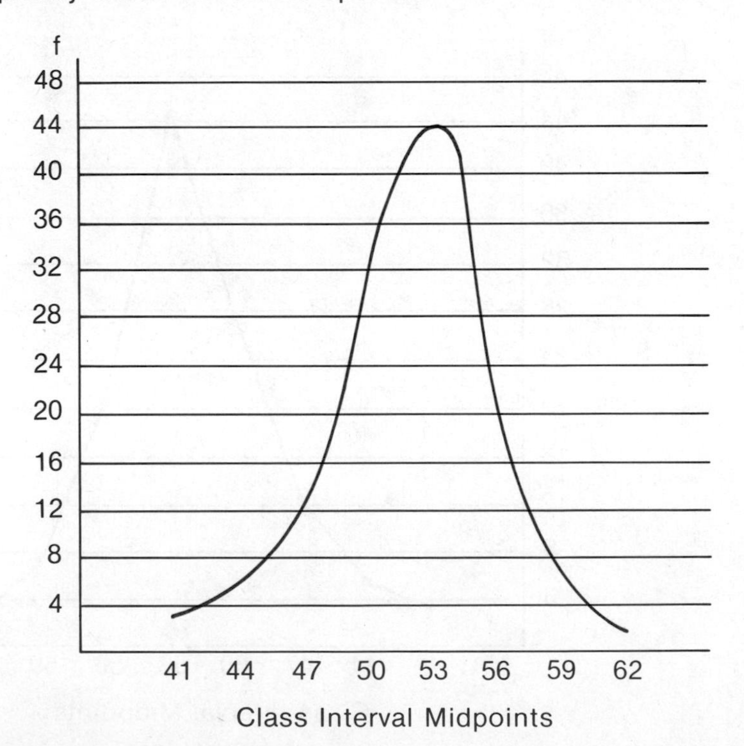

Ojives

Graphic representations of the cumulative frequency (cf) and relative cumulative frequency (rcf) distributions are identical to each other in shape, but are very different from graphs of f and rf. Generally cf or rcf distributions are graphed with a curve. These curves are called **ojives** (sometimes spelled "ogives"). Usually ojives take on the shape of a "lazy S" as shown in Figure 3.4. Figure 3.4 shows an ojive of the same data used throughout this section.

Figure 3.4—Ojive Showing Vehicle Speeds

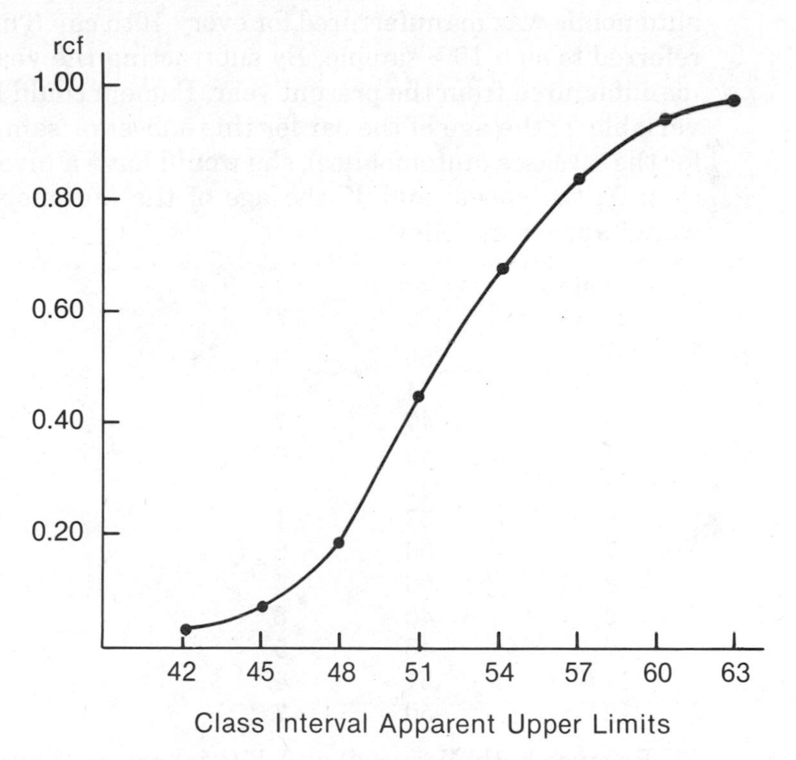

One difference in the scaling of the horizontal axis for an ojive should be noted. Because cf (or rcf) distributions represent frequencies equal to or below a particular class interval, either the apparent upper limits or the real upper limits are more accurate scalings on the horizontal axis for an ojive. All of the frequencies in a class interval would not generally fall at or below the midpoint, but they would be equal to or below the upper limit of the class interval.

Notice that the ojive in Figure 3.4 shows the relative cumulative frequency. If the vertical axis scaling was changed to cumulative frequency, the shape would remain similar–a small slope at the lower class intervals, an accelerating slope toward the central portion, and finally a slowing of the vertical rise in the upper ranges. You can very quickly examine an ojive and tell approximately how many automobiles were traveling at a certain speed or less. For example, by moving up the graph from 54, you can see that about 0.76 (or 76%) of the velocities were 54 mph or less.

Bivariate Plot

For illustration purposes, suppose that Pamela and Officer Mahaffey not only had recorded the speed of each automobile that passed their unmarked radar site, but also had noted the year the automobile was manufactured for every 10th car. This set would be referred to as a 10% sample. By subtracting the year an auto was manufactured from the present year, Pamela could have created a variable Y: the age of the car for this subset or sample. Therefore, for the 12 cases (automobiles), she would have a bivariate distribution: X, the speed; and Y, the age of the automobiles. The data would appear as follows:

Automobile # (an ID #)	Velocity (X)	Age (Y)
1	50	3
2	54	3
3	47	7
4	49	6
5	52	3
6	57	1
7	56	2
8	53	5
9	45	8
10	48	5
11	51	4
12	50	7

Because both X (speed) and Y (age) are continuous, a **scattergram** or bivariate plot would be an appropriate means for showing both variables simultaneously on a graph. Steps for completing a scattergram are as follows:

1. Set up two perpendicular axes—one to represent each variable. In the present example, the vertical axis represents age (Y) and the horizontal axis represents velocity (X) in miles per hour.
2. Scale both axes with equal interval distances so that the full ranges of values of both variables are included. For the

example, Y-values ranged from 1 to 8 years old; X-values ranged from 45 to 57 mph. (Scattergrams are generally an exception to the rule that the scales must start at zero. The primary consideration is to include the full range, that is, accommodate the minimum and maximum values.)

3. Plot the points on the graph. Each point represents two values—one an X-value and the other a Y-value. For automobile #1, find 50 on the X-axis; then move up vertically until you reach 3 years old. The intersection of the two values defines the position of the point (or plot) as shown in Figure 3.5. Each of the 12 cases in the sample is plotted similarly.

Figure 3.5—The Bivariate Point $X = 50$, $Y = 3$

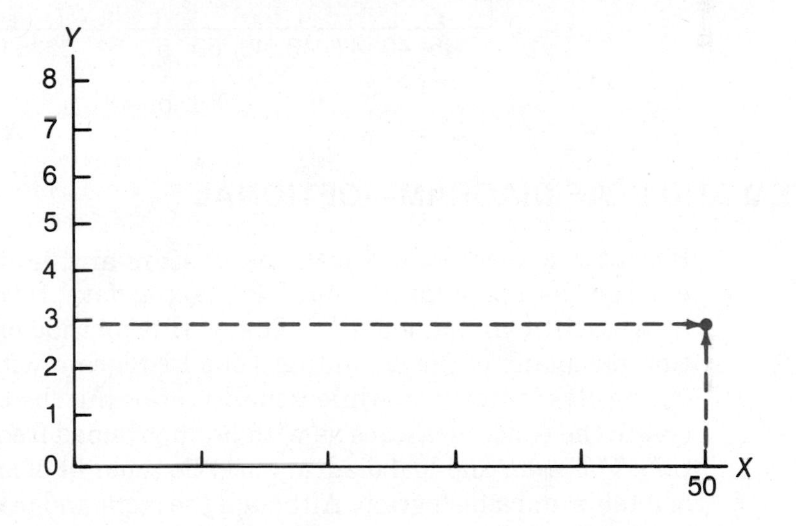

4. Label the axes and provide a title for the scattergram.

Figure 3.6 shows the completed bivariate plot or scattergram. Notice that each point represents two values for a particular car. There will be as many points (plots) on a scattergram as there are cases. Can you detect a trend in the scattergram? How does the speed of the autos tend to vary with the age of the car? Scattergrams play a major role in the topic of correlation, which is presented in Chapter 12. For the present, consider the bivariate plot as a graphic method of displaying a bivariate distribution when both variables are assumed to be continuous.

Figure 3.6—Bivariate Plot (Scattergram) Showing Age and Speed for a Sample of Automobiles

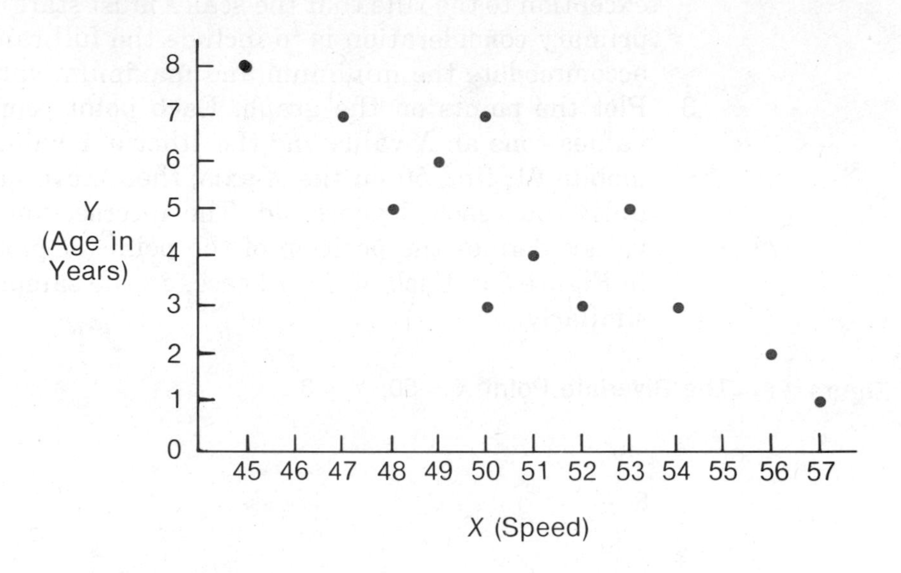

STEM-AND-LEAF DIAGRAM—OPTIONAL

Recently a technique known as a **stem-and-leaf diagram** has emerged as a popular means of displaying data. Introduced in 1977 by a statistician named J. W. Tukey, this method of data presentation has many of the advantages of a histogram with grouped data (using class intervals) while usually retaining the capability to recreate the exact measures as with an ungrouped frequency distribution. The stem-and-leaf diagram may be thought of as a combination of a table and a histogram. Although the stem-and-leaf display is generally not as attractive or as visually pleasing as a well-done graph, it performs all of the important functions of a grouped frequency histogram and of an ungrouped frequency distribution, such as:

1. revealing the range of the data
2. showing where the data are concentrated
3. providing a visual sense of the symmetry or uniformity of the distribution
4. showing gaps in the distribution, and
5. maintaining the exact values of the original data.

The stem-and-leaf diagram reliably performs the functions of organizing and displaying univariate distributions. For a particular numerical value, the higher-ordered place holder (usually the tens or hundreds place) is called the *stem*. The lower-ordered place holder (usually the units) is known as the *leaf*. So, for a quantity such as 68, 6 would represent the stem; 8 the leaf. While methods of constructing stem-and-leaf diagrams for large numbers and for fractions are beyond the scope of this text, an examination of Figure 3.7 will show the basics of the process. The data used for the stem-and-leaf diagram in Figure 3.7 represent horsepower for a sample of compact cars.

Figure 3.7—Stem-and-Leaf Diagram of Horsepower Rating of Domestic and Imported Compact Cars ($N = 92$)

Stems Units = 10	Leaves
16	0
15	0000002556
14	
13	
12	0000555555
11	00000222244
10	288
9	0222222235566777
8	2224444555555566666668
7	444888
6	44
5	2222226666

The stems in Figure 3.7 represent the 10's; therefore, they represent 50, 60, 70, and so on until the largest of 160 is reached. The units (1's or units place holder) are not shown as part of the stems; rather they form the leaves of the diagram. Consequently, the stem 10 represents a 100-horsepower rating. Notice that there were three compact automobiles with horsepower between 100 and 110.

The leaves corresponding to the stem of 10 are 2, 8, and 8. The three compacts have horsepower ratings of 102, 108, and 108, respectively. One compact had 160 horsepower, six were rated at 150, one at 152, two at 155, and one at 156. None of the 92 compact cars had horsepower ratings in the 130's or 140's. The remainder of the diagram is interpreted in the same manner.

In addition to permitting re-creation of the original individual values (horsepower), notice that the leaves create the appearance of a horizontal bar graph or histogram. So, whether the purpose of a presentation is to provide exact data or to show the data pictorially for casual perusal, the stem-and-leaf diagram is appropriate.

COMPUTER FOCUS 3—Planning a Solution: Flowcharting

Purpose

Computing can be thought of as automating a set of well-defined procedures in a computational sequence. Such a position has been described by A. M. Turing in his classic article "Can a Machine Think?" which was included in *The World of Mathematics* compiled by James R. Newman. This Computer Focus provides you with a "machine." The aim of the Computer Focus is to give you an opportunity to "teach" the machine to "think" by specifying a computational sequence (a set of well-defined procedures) by using a flowchart.

Problem: Program a Vending Machine

Before actually using numerical data for calculation purposes, you should complete a flowchart for solving a problem on your own. You have been introduced to the symbols of flowcharting (Computer Focus 1) and have followed the "flow" through a structured flowchart exercise (Computer Focus 2). Now see if you can illustrate an algorithm or strategy for solving a problem. Your task is to develop a logical sequence for the operation of a vending machine. There is *no single correct solution*; in fact, there are hundreds of correct solutions. Therefore your flowchart may illustrate a different logic for solving the problem than do the flowcharts of your classmates. Compare notes with others in your class. Your problem specifications are as follows:

Vending Machine Capabilities

Given: A candy machine that delivers a 45-cent product plus change for any combination input of nickels, dimes, and/or quarters. The "Nickel Counter" records input of "equivalent nickels" until the count is 9 or more. At that point, the machine delivers a product and the proper change.

Input/Output:

1. Collect a coin.
2. Reject a coin.
3. Deliver product.
4. Deliver change.

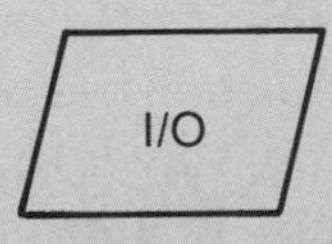

Process:

Processes

1. Clear Nickel Count to zero.
2. Add 1 to Nickel Count.
3. Add 2 to Nickel Count.
4. Add 5 to Nickel Count.
5. Compute change = 5 cents(Count – 9)

Decision:

Decisions

1. Is coin a penny or slug?
2. Is coin a nickel?
3. Is coin a dime?
4. Is Nickel Count less than 9?

Problem: Construct a flowchart that describes the logic of the machine. That is, display how the machine would work for any combination of coins entered in any order. (*Note:* There are many "correct" flowchart possibilities.)

Testing: To check your solution to the candy machine problem, trace the action taken in your flowchart for the various inputs shown below (this is called a walk-through). You should be able to account for the contents of the "Nickel Count" when each coin is deposited in sequence. Further, you should be able to describe how much change will be delivered (if any) at the conclusion of the total deposit.

Test Data—Coins

nickel, nickel, nickel, quarter, dime
nickel, dime, dime, dime, dime
dime, nickel, quarter, quarter
nickel, nickel, dime, dime, dime, dime
dime, dime, dime, dime, dime
quarter, quarter
nickel, quarter, dime, nickel
nickel, nickel, quarter, quarter
dime, quarter, dime

KEY TERMS AND NOTATION

Apparent limits
Bivariate
Class interval
Contingency table
Cumulative frequency (cf)
Curve
Frequency distribution (f)
Frequency polygon
Grouped frequency distribution
Histogram
Joint frequency distribution
Ojive (ogive)
Real limits
Relative cumulative frequency (rcf)
Relative frequency (rf)
Scattergram
Stem-and-leaf diagram (optional)
Ungrouped frequency distribution

APPLICATION EXERCISES

3.1 Following are the results of a final exam taken by 50 students at North Shore High School. The scores are the number of correct items out of 100 multiple-choice questions.

88	89	87	86	89	88	90	88
89	90	89	91	88	84	88	89
92	87	90	88	85	88	86	90
89	85	89	89	88	92	91	90
87	89	91	94	90	87	89	89
89	93	89	90	88	90	88	89
87	90						

Using these test scores, construct an ungrouped:

a. frequency distribution
b. cumulative frequency distribution
c. relative frequency distribution
d. relative cumulative frequency distribution

3.2 The following data represent scores of a sample of students on the College Board tests:

520	610	570	600	570	610	670
520	630	490	580	625	565	680
640	540	585	580	585	580	545
510	640	580	600	690	545	720
555	630	620	585	540	560	450
620	610	610	565	500	480	570
665	585	630	670	570	470	580
620	600	545	585	510	610	560
600	665	600	530			

Create class intervals of width 25 starting at 450, that is, 450 – 474, 475 – 499, 500 – 524, 525 – 549, and so on until you reach the interval 700 – 724. Construct a grouped:

a. frequency distribution
b. cumulative frequency distribution
c. relative frequency distribution
d. relative cumulative frequency distribution

3.3 Using the results from Application Exercise 3.2, construct the following graphs with the frequencies of the class intervals. Show each class interval midpoint such as 462, 487, 512, 537, and so on through 712.

a. histogram
b. frequency polygon
c. ojive

3.4 During a marketing survey, 50 men were asked to choose their favorite soft drink from among three brands. Twenty chose brand A, 13 favored brand B, and 17 chose brand C. Similarly, 50 women made their choices: 15 preferred brand A, 17 favored brand B, and 18 chose brand C. Construct a joint frequency distribution using gender and brand as the two variables. Show the marginal frequencies for the rows and for the columns.

3.5 The estimated population by age and sex in the year 2000 is shown (in hundred thousands) for the United States.

Age	Under 5		5 – 17		18 – 24		25 – 44		45 – 64		65+	
Sex	M	F	M	F	M	F	M	F	M	F	M	F
No.	9.0	8.6	25.5	24.3	12.5	12.1	40.2	39.9	29.4	31.4	13.7	21.3

Construct a joint frequency distribution table.

3.6 At the beginning of the year, the second-period physical education class members at South High School were assessed for weight (pounds) and height (inches). Measurements of the 20 boys in the class were:

Height	Weight	Height	Weight
63	130	71	190
65	135	73	195
64	150	65	155
72	185	63	135
62	120	70	145
71	150	70	150
63	145	68	165
64	155	67	145
69	155	70	165
69	160	68	155

Construct a bivariate plot (scattergram) displaying the data.

3.7 What are the real limits of the following apparent limits for a class interval?

a. 33 – 35

b. 20 – 29

c. 1 – 2

3.8 Suppose apparent class limits for an interval were 10.5 – 11.5. What do you think would be the real class limits? (*Hint:* Make use of the second decimal place—that is, hundredths.)

3.9 *Optional* For the data that follow, construct a stem-and-leaf diagram.

20	68	32	54	46	52	78	50
36	45	33	12	55	51	54	60
82	63	49	29	82	79	35	67
56	39	66	50	59	70	24	75
80	58	60	31	69	44	20	65

4

Summarizing Data: Measures of Central Tendency

CHAPTER OBJECTIVES

Upon completion of the chapter, students will be able to:

1. Compute the mean and determine the median and mode for a given set of numbers.
2. Describe common characteristics of the mean, median, and mode.
3. Locate the relative positions of the mean, median, and mode on a skewed frequency distribution.
4. Determine an appropriate measure of central tendency for data scaled on nominal, on ordinal, and on interval and ratio levels.
5. Explain why the mean is influenced by extreme values in a distribution while the median is relatively unaffected by extreme values.
6. (Optional) Compute harmonic and geometric means for a set of data.

Mr. Richardson taught 10 students to program in the Pascal computer language. To prepare for a statewide programming contest, Mr. Richardson gave the students a standardized 50-item multiple-choice test covering the Pascal programming content. The scores on the test were the number of items correctly answered out of a possible 50. Scores for the 10 students were as follows: 48, 43, 50, 45, 46, 45, 44, 50, 45, and 44.

Mr. Richardson was interested in the "typical" performance of his group of students. On the average, how did the students perform? To determine the "typical" or "average" performance, Mr. Richardson computed a quantity called the arithmetic mean. How and why was the arithmetic mean computed? Before examining Mr. Richardson's problem in more detail, consider the concept of central tendency or average more generally.

No statistical concept is more popular in day-to-day affairs than "the average." The evening news reports the "average" New York and American Stock Exchange prices, teachers summarize the results of exams using averages, the number of students in your school is reported as "average daily attendance," baseball players measure their hitting performance with the "batting average," and so on.

Most of the time the term "average" means "typical." The phrase "central tendency" usually refers to a single value that is typical of a collection of numeric values. The "average man" is a typical male whose height, weight, shoe size, education level, intelligence, and personality traits are commonly found in men. Such physical characteristics and mental traits of the "average man" would fall somewhere close to the *middle* of the distribution of measures for all men. "Typical," "common," and **average** are concepts that are statistically and numerically represented by measures of **central tendency**.

At one time or another, everyone has computed the average for a distribution of numbers. That average is a shorthand method to represent all numbers in that distribution. In this chapter you will find that there are several kinds of averages and that each has certain characteristics and uses. The three most widely used measures of central tendency (average) are the mean, median, and mode. It is probable that you are already familiar with them, although you may not have called them by these particular names. You should be aware, however, that the mean, median, and mode are only three of many measures of central tendency. Because they are the most popular, they will be emphasized in this chapter.

ARITHMETIC MEAN

The most widely used measure of central tendency for most data is the **arithmetic mean**, or **mean** for short. To find the mean of a set of numbers, simply add the numbers and divide the sum by the number of values in the distribution. You've undoubtedly performed that operation many times. Using Mr. Richardson's data, his procedures for finding the mean were as follows. First, he added (summed) the numbers:

$$48 + 43 + 50 + 45 + 46 + 45 + 44 + 50 + 45 + 44 = 460.$$

Then he divided the sum (460) by the number of scores (10):

$$\text{Arithmetic Mean} = 460 \div 10 = 46.$$

(Although this illustration resulted in a whole number for the mean, that will not always occur, and two-decimal-place accuracy is generally sufficient when expressing a mean.)

As you can tell by visual inspection of the set of 10 scores, 46 is located in the central portion of the distribution relative to the size or magnitude of the scores. Therefore the term "central tendency" is numerically and explicitly illustrated by Mr. Richardson's example.

Notation

Now let us formalize Mr. Richardson's problem in standard statistical notation. First, assign a variable name, X, to the scores. Then X (note that the uppercase X is used) stands for the 10 scores on the Pascal test. Statistical analyses often require the sum of the numbers in the distribution of interest. It is desirable to have a simplified system of notation to indicate summation as well as to identify the values that are being added. Consequently, if X is the name of the variable, then successive observations of that variable are written with a subscript as X_1 (read "X sub-one"), X_2, X_3, . . . and so on to the last observation X_N. The general ith observation is written as X_i. (Think of the "general ith observation" as being any arbitrary value in a set of values that represents the generalized case, and i can represent any integer value from 1 through N.) The number of values or observations in a set or collection of numbers is universally denoted as N. The Greek letter **sigma** (Σ) indicates the sum of N observations; that is:

$$\sum_{i=1}^{N} X_i = X_1 + X_2 + X_3 + \ldots + X_N.$$

The symbol $\sum_{i=1}^{N} X_i$ is read: "sum X sub i, for i from 1 to N." The letter below the summation sign Σ is called the summation index and is a symbol representing integers from 1 through N. The 10 values (scores) in Mr. Richardson's numerical distribution could be designated as follows:

$X_1 = 48$	$X_4 = 45$	$X_7 = 44$	$X_{10} = 44$
$X_2 = 43$	$X_5 = 46$	$X_8 = 50$	
$X_3 = 50$	$X_6 = 45$	$X_9 = 45$	

In this case,

$$\sum_{i=1}^{10} X_i = 48 + 43 + 50 + 45 + 46 + 45 + 44 + 50 + 45 + 44$$
$$= 460.$$

Because all statistical techniques used in this book require each value in a defined distribution to be included in the sum, simplified notation will be used; namely, ΣX will be used to mean $\sum_{i=1}^{N} X_i$. So the summation index and variable subscripts will be dropped from the summation notation symbols, and ΣX, ΣY, etc., will be used to designate summation of the values of the variable being analyzed. Next, relate this notation to the computation of the mean for Mr. Richardson's data.

Computation of the Mean

To indicate the total, use a summation sign Σ (the Greek letter sigma). The symbolic notation ΣX means the sum or total of all the values in the set called X. In the present example you have seen that $\Sigma X = 460$. Second, the number of values in a set (the number of Pascal test scores) is $N = 10$. Finally, a common symbol for the *arithmetic mean* of a set of numbers, in this instance for the variable X, is $\overline{X}$ (pronounced "X-bar"). Whatever variable name is used to designate the set of numbers, a bar over the top of the name indicates the mean of the distribution. Another commonly used symbol for the arithmetic mean is the Greek letter mu (μ).

Combining the notation concepts, a concise formula for computing the mean is shown in the following equation.

The formula for determining the mean is:

$$\overline{X} = \frac{\Sigma X}{N}$$

where $\overline{X}$ stands for the mean, ΣX is the sum, and N is the number in the set.

So a more formal and concise setup for Mr. Richardson's problem would take this form:

$$\Sigma X = 460$$
$$N = 10$$

Therefore

$$\overline{X} = \frac{\Sigma X}{N}$$

$$= \frac{460}{10}$$

$$= 46.$$

Notice that the arithmetic mean is algebraically determined with this equation. Further note that every value in the set contributes to the size of the mean—if even one value was altered, the value of the mean would be changed. A discussion of the characteristics of the arithmetic mean is presented later in this chapter. The procedure illustrated in Figure 4.1 shows the computation of the mean for Mr. Richardson's scores using a calculator.

Figure 4.1—Calculator Keystrokes

PROBLEM:

Sum a set of scores and divide by *N* to find the arithmetic mean

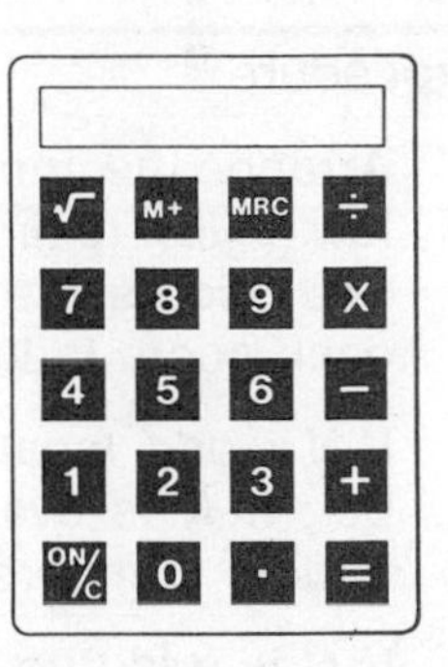

Procedure	Arithmetic Operation	Calculator Keystrokes	Display
1. Add the values: ΣX	1. 48 + 43 + 50 + 45 + 46 + 45 + 44 + 50 + 45 + 44 = 460	1. 48 [+]	48
		43 [+]	91
		50 [+]	141
		45 [+]	186
		46 [+]	232
		45 [+]	277
		44 [+]	321
		50 [+]	371
		45 [+]	416
		44 [+]	460
2. Divide the sum (ΣX) by *N* to get $\overline{X}$	2. $\frac{460}{10} = 46$	2. [÷] 10	10
		[=]	46

MEDIAN

A second important measure of central tendency is the median. The **median** divides the distribution such that half the values are above the median and half below. The median is not algebraically determined, so several different definitions may be found in various textbooks. For the purposes of this course, the following definition will be used: **The median of a distribution of *N* numbers arranged in ascending (or descending) order of size is defined as the middle number if *N* is odd; the sum of the two middle numbers divided by 2 if *N* is even.** The procedures for determining the median for Mr. Richardson's data as implied in the definition are shown in Table 4.1.

Table 4.1 Determining the Median

Procedure	Application
1. Arrange the numbers from high to low (or low to high). Notice that the tied scores are repeated so that each score is listed.	1. 50 50 48 46 45 45 45 44 44 43
2. If *N* is odd, locate the middle number; if *N* is even, locate the two middle numbers.	2. 50 50 48 46 **45 45** 45 44 44 43
3. If *N* is odd, the middle number is the median; if *N* is even, add the two middle numbers and divide by 2 to obtain the median.	3. $\frac{45 + 45}{2} = 45$

The median, because it falls exactly in the middle of the distribution, again illustrates the concept of "central tendency" or "typical" performance.

The median is determined entirely by the rank order of the numbers, not the actual size or magnitude of the numbers. Also, the actual value of each number in the distribution is not particularly important for determining the median. If a number close to either the high or low extreme was changed, unlike the mean, the median would remain the same. For example, if the score 43 in the Pascal test score distribution of Mr. Richardson's class was recorded in error and should have been 38, the mean would change but the median would stay at 45. Characteristics of the median are discussed further later in this chapter.

MODE

The **mode** is the number in a distribution that appears most frequently. The mode may be a unique number, or it may occur at two or more locations in the distribution. In some instances it may not even exist, as in the case where all the values in a distribution occur only one time. Thus when someone says the "average person" is a high school graduate, he or she is probably using the mode to indicate "average" or "typical."

Again using the test scores from Mr. Richardson's class, the number that appears most frequently is 45, as shown: 50, 50, 48, 46, **45**, **45**, **45**, 44, 44, 43. Because the score of 45 occurs most frequently, 45 is referred to as the mode.

Before examining the characteristics of these measures of central tendency, consider one more example. Find the mean, median, and mode of a set of numbers called Y (already arranged in order of size):

Y
8
10
12
12
14
15
15
18
20

$$\Sigma Y = 124$$

$$N = 9$$

$$\overline{Y} = \frac{124}{9} = 13.78$$

Median = 14

Note: This distribution does not have a single mode as both values, 12 and 15, appear an equal number of times. There are two modes; therefore such a distribution is called **bimodal**.

CHARACTERISTICS OF THE MEAN, MEDIAN, AND MODE

You may have noticed that the mean, median, and mode are not always the same value even though they all are supposed to measure the same entity—average. What are their similarities and differences? Before examining specific attributes of the mean, median, and mode, let us explore how they relate to each other in three types of distributions.

Recall from your study of Chapter 3 how frequency curves are developed. Data are initially collected and arranged in a frequency distribution. Then the frequency distribution (f) or the relative frequency distribution (rf) is converted to a graphic image in the form of a histogram, a frequency polygon, or a frequency curve.

First, consider a frequency curve that is sometimes referred to as a *bell curve* because of its physical shape. Such a curve might display the relative frequency of a large number of test scores and would assume a form as shown in Figure 4.2. The baseline in the figure represents scores, and the height of the graph represents relative frequency as described in Chapter 3. Notice that the mean, median, and mode all coincide.

Figure 4.2—Symmetric Distribution Showing Mean, Median, and Mode at Same Location

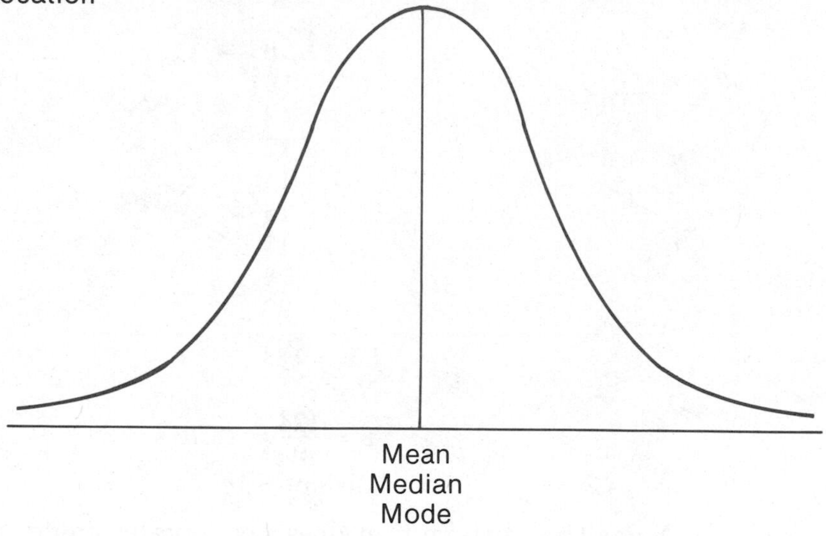

The kind of distribution shown in Figure 4.2 is called **symmetric** This means that the curve to the left of the mean (median or mode) is a mirror image of the right side. More simply, if you folded the distribution curve vertically down the middle, the two sides of the curve would coincide.

Now consider a frequency curve of test scores on an "easy" exam on which many of the students made high scores and tended to "top out" toward the ceiling of the exam. This implies that more of the exam scores would tend to be located toward the right or high side of the distribution. The data that follow represent scores on a review exam for 50 bright students. The numbers represent the number correct of a total of 10 questions.

7	9	10	7	9	8	10	7	9	8
5	7	6	5	8	8	9	8	8	5
6	7	9	6	7	7	8	6	8	4
3	9	8	4	6	9	7	6	7	7
7	8	9	7	9	8	8	8	9	6

Examine these data in three ways. First, determine the mean, median, and mode. Second, construct a frequency distribution and use it to graph a smooth-line frequency curve following the procedures described in Chapter 3. Finally, plot the mean, median, and mode along the baseline and notice their relative positions.

The general form of this type of curve is shown in Figure 4.3 and is referred to as a negatively skewed curve. The **negative skew** is shown by the slope tailing off to the left. Customarily the left side of the horizontal axis where the lower scores are located is referred to as the negative direction; therefore the curve has a *negative skew*.

As shown in Figure 4.3, the three measures of central tendency are no longer located at the same point, as was the case with the symmetric curve. As you discovered with the test score data, the mean (7.32) is lower (further to the left) than the median (7.5), which in turn is lower than the mode (8.0). Although they do not exactly coincide, they all still tend to be in the central portion of the distribution.

Figure 4.3—Asymmetric (Negatively Skewed) Distribution Showing Separation of Mean, Median, and Mode

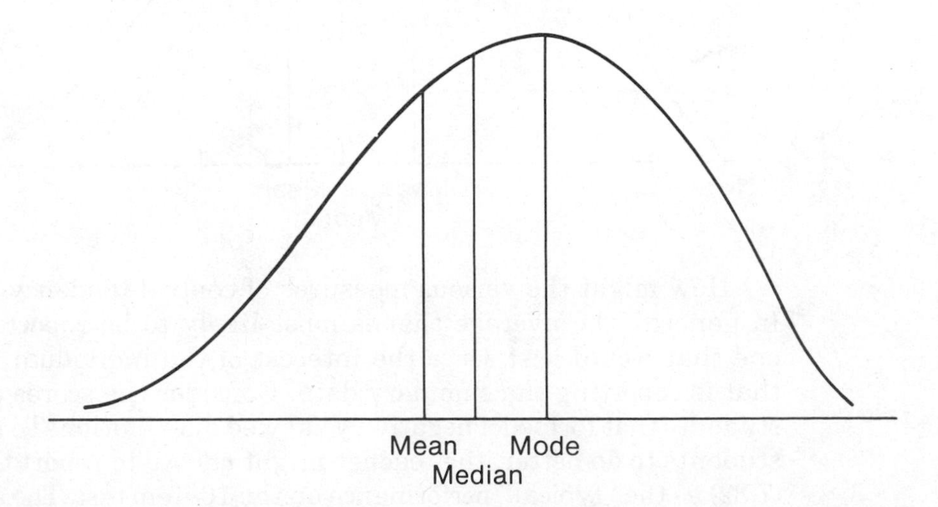

To explain these results, you would expect the mode to be in the portion of the distribution where larger numbers of scores were located; in the negatively skewed example, this is toward the higher (right) side. The mean is determined by all of the values in the distribution, including the "extreme" scores. Low scores are "extreme" in a negatively skewed distribution because they are not part of the larger mass of scores. Therefore the mean is located to the left of the median, which is not influenced by extremes. So the relationship among the three measures of central tendency as illustrated with the negatively skewed test scores can be logically explained.

On the other hand, if students took a difficult exam, the general form of the resulting frequency curve would have a **positive skew**, as shown in Figure 4.4. In a positively skewed distribution, the mode is lowest, the median is in the middle, and the mean is the highest. Can you use the same logic as that used for explaining the relationship among the three measures of central tendency in a negatively skewed distribution to explain the reversal in the sizes of the three averages in a positively skewed distribution?

Figure 4.4— Asymmetric (Positively Skewed) Distribution Showing Separation of Mean, Median, and Mode

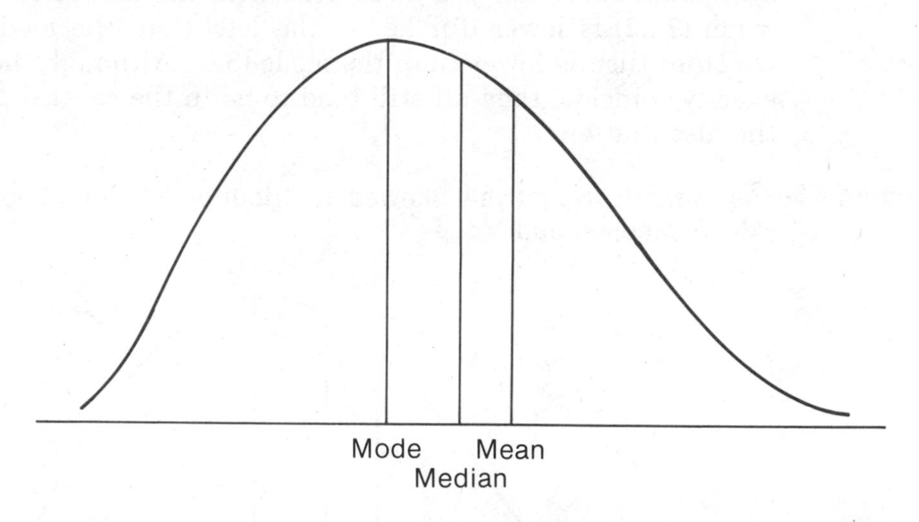

How might the various measures of central tendency be used? In general, the average that is most likely to be reported is the one that would best serve the interest of the individual or group that is reporting the summary data. Consider the scores of the 50 students that formed a negatively skewed distribution. To motivate students to do better, the teacher might choose to report the mean (7.32) as the "typical" performance on the 10-item test. The students, to document how well they are doing, would probably choose the mode (8.0) to report as the "typical" performance.

Consequently, confusion can arise as a result of reporting different averages from the same set of data unless the reader is skilled in statistical use and abuse of averages. Actually, the median is usually preferred for expressing the "average" of skewed data because it is not influenced by extreme data. Some summary statements about the symbols, calculation procedures, and characteristics associated with each measure of central tendency are provided in the following section.

Summary Statements for Measures of Central Tendency

Mean:

1. Commonly used symbols: $\overline{X}$, μ, (Greek letter mu)
2. Calculation: $\overline{X} = \frac{\Sigma X}{N}$
3. Important characteristics:
   a. The mean is the most widely used measure of central tendency.
   b. Computation of the mean uses all of the data in the set or distribution.
   c. The mean is most meaningful when used with data scaled on an interval or ratio level and should not be used with data scaled on lower levels.
   d. The mean is the most reliable or stable measure of central tendency, a characteristic dealt with at length in later material on inferential statistics.
   e. The mean is sometimes known as the "center of gravity" of a set of numbers, which implies that the mean is the point on a baseline at which a frequency curve would balance.
   f. The mean is influenced by extreme values in the distribution.
   g. Because of the algebraic process for computing the mean, it serves well as a defined measure of central tendency for input into subsequent analyses.

Median:

1. Commonly used symbols: Mdn, Q_2
2. Calculation: The median of a set of N numbers arranged in ascending or descending order of size is the middle number if N is odd; if N is even, the median is the sum of the two middle numbers divided by 2.

3. Important characteristics:
    a. The median is the second most commonly used measure of central tendency.
    b. The median divides a numerical distribution equally with half of the values above and half below.
    c. It is not unduly influenced by extreme values and therefore is generally the preferred measure of central tendency for skewed distributions.
    d. It is simple to determine.
    e. The median may be effectively used with ordinal, interval, or ratio scaled data.

Mode:

1. Commonly used symbols: M_o
2. Calculation: The mode (if it exists uniquely) is the number in a distribution that occurs most frequently.
3. Important characteristics:
    a. Of the three measures of central tendency, the mode is the least reliable or stable.
    b. It may not exist, or there may be more than one mode.
    c. The mode is easily determined by observing a frequency distribution or a frequency graph.
    d. The mode may be used with any type of numerical scaling but is most commonly used with nominal data.
    e. There is no guarantee that the mode, if it exists, will represent the central portion of a distribution.

Now consider this description of a typical teacher: "The average teacher is female, 40 years old, works 7 hours and 13 minutes a day, has been a teacher for 15 years, and has worked in the same school district all of that time." This statement uses the term "average" in several different ways. Using the central tendency concepts learned so far, the "average" teacher can be described more precisely as follows. The *mode* for the gender variable is female, meaning there are more female than male teachers. Finding the mean for gender would serve no meaningful purpose. The *mean* age is 40, and the *median* length of a working day is 7 hours, 13 minutes. Finally, the *mean* number of years worked is 15, and most teachers (*mode*) have remained in the same school.

From this illustration you can see that the "average" or "typical" is determined by various techniques. The important concept here is that the particular measure of central tendency used should be typical of the collection of numbers it represents. In general, if the data are scaled only on a nominal level, the mode is the appropriate average. If the data are scaled at an ordinal (rank order) level, the mode or median may be used. With interval or ratio data, the mean, median, or mode may be used for reporting central tendency. Generally, if the data distribution is skewed, the median will be the most accurate and representative of the central portion of the distribution. Occasionally it may be more meaningful to report two or all three measures of central tendency. The main consideration is that the measure of central tendency should do the job for which it was intended, that is, be a good representative of the numbers in the collection.

OTHER MEASURES OF CENTRAL TENDENCY—OPTIONAL

Suppose you drove from Dallas to Kansas City at an average speed of 50 miles per hour. Then you drove back to Dallas and averaged 40 miles per hour on the return trip. A logical question might be: "What was your average speed for the entire round trip?" An answer that would be obvious to most people is 45 miles per hour as determined by the arithmetic mean of 50 and 40. It happens that what appears to be obvious at first glance is *incorrect*. The correct method of solving this problem is to use another measure of central tendency called the *harmonic mean*.

Harmonic Mean

The **harmonic mean** is one of several averages that are not as popular as the mean, median, and mode. The purpose here is to introduce less frequently used measures of central tendency so that you will be aware that other ways of averaging do exist and that the obvious is not always accurate.

The harmonic mean is a more accurate technique for averaging speeds when the distances for which each speed is applicable are the same. Because speed is a function of time, for example, miles per hour, and because the times were not equal in the example of the Dallas to Kansas City round trip, the arithmetic mean is inaccurate. (Incidentally, if an average speed is to be calculated where the *times, but not the distances*, are the same, then the arithmetic mean is the appropriate average.)

The formula for the harmonic mean is given by the following equation.

The formula for harmonic mean is:

$$\overline{X}_h = \frac{N}{\sum \frac{1}{x}}$$

where $\overline{X}_h$ = harmonic mean (subscript h stands for harmonic)
N = the number of values to be averaged, and
$\sum \frac{1}{x}$ = the sum of the reciprocals of the values or the sum of 1 divided by the respective values.

For the example cited earlier, the average speed (harmonic mean) for the round trip would be calculated as follows:

$$\begin{aligned}\overline{X}_h &= \frac{2}{\frac{1}{50} + \frac{1}{40}} \\ &= \frac{2}{0.020 + 0.025} \\ &= \frac{2}{0.045} \\ &= 44.44 \text{ miles per hour.}\end{aligned}$$

Because a longer amount of time driving at 40 miles per hour was required than at 50 miles per hour, the "average" speed would reasonably be expected to drop below 45 miles per hour. This is confirmed by the harmonic mean of 44.44 mph.

Geometric Mean

Another type of mean for computing a measure of central tendency is the *geometric mean.* The **geometric mean** is applicable to data for which the ratio of any two consecutive numbers is constant or nearly constant. For example, the geometric mean would be appropriate for finding the average value of a sum of money that is increasing periodically because of the addition of compound interest to the account. It also would be suitable to use for data representing the size of a population at consecutive time intervals.

For two numbers, say 2 and 8, the geometric mean ($\overline{X}_G$) is the square root of the product of the two numbers: $\sqrt{(2)(8)} = \sqrt{16} = 4$. The mathematical techniques for efficiently computing the geometric mean when more than two values are involved are more complicated than finding the square root of a product. Consequently, unless you are familiar with the use of logarithms, you should go ahead and omit the remainder of the discussion of the geometric mean with several values.

The geometric mean of 2, 5, and 8 is the cube root of the product of the three numbers: $\sqrt[3]{(2)(5)(8)} = \sqrt[3]{80} = 4.31$. If four values are involved, the geometric mean is the fourth root of the product of the four numbers. In general, for a set of N positive numbers X_1, X_2, X_3, . . . X_N, the geometric mean is defined as the Nth root of the product of the numbers:

$$\overline{X}_G = \sqrt[N]{(X_1)(X_2)(X_3)\cdots(X_N)}$$

For computational purposes, it is often more convenient to use the logarithmic form:

$$\log \overline{X}_G = \frac{\log(X_1) + \log(X_2) + \ldots + \log(X_N)}{N}$$

$$= \frac{\Sigma \log(X_i)}{N}$$

Suppose the student government invested \$500 in an interest-bearing savings account on January 1. If 2% interest is added each month, calculate the geometric mean for the balance of the account for the first six months. The values to be averaged are as follows:

January 1 (starting date)	\$500.00
February 1: \$500 + (0.02)(\$500)	\$510.00
March 1: \$510 + (0.02)(\$510)	\$520.20
April 1: \$520.20 + (0.02)(\$520.20)	\$530.60
May 1: \$530.60 + (0.02)(\$530.60).	\$541.21
June 1: \$541.21 + (0.02)(\$541.21)	\$552.03

Then

$$\log X = \frac{\log(500.00) + \log(510.00) + \log(520.20) + \ldots + \log(552.03)}{6}$$

$$= \frac{2.6990 + 2.7076 + 2.7162 + 2.7248 + 2.7333 + 2.7420}{6}$$

$$= \frac{16.3219}{6}$$

$$= 2.7203$$

Therefore

$$\overline{X}_G = 10^{2.7203} = \$525.17$$

The average (geometric mean) for the six dollar amounts is \$525.17 compared to a value for the arithmetic mean of \$525.67. Although the difference in the two is not dramatic, the constant (2%) increase in the values yields data more suitable for the geometric mean to serve as the measure of central tendency.

Geometric Relationship Among the Three Means

The arithmetic mean, geometric mean, and harmonic mean of two values, say X and Y, may be illustrated geometrically. Study Figure 4.5 and locate the segments on the diagram as you follow the discussion.

Figure 4.5—Geometric Illustration of the Arithmetic Mean, Geometric Mean, and Harmonic Mean

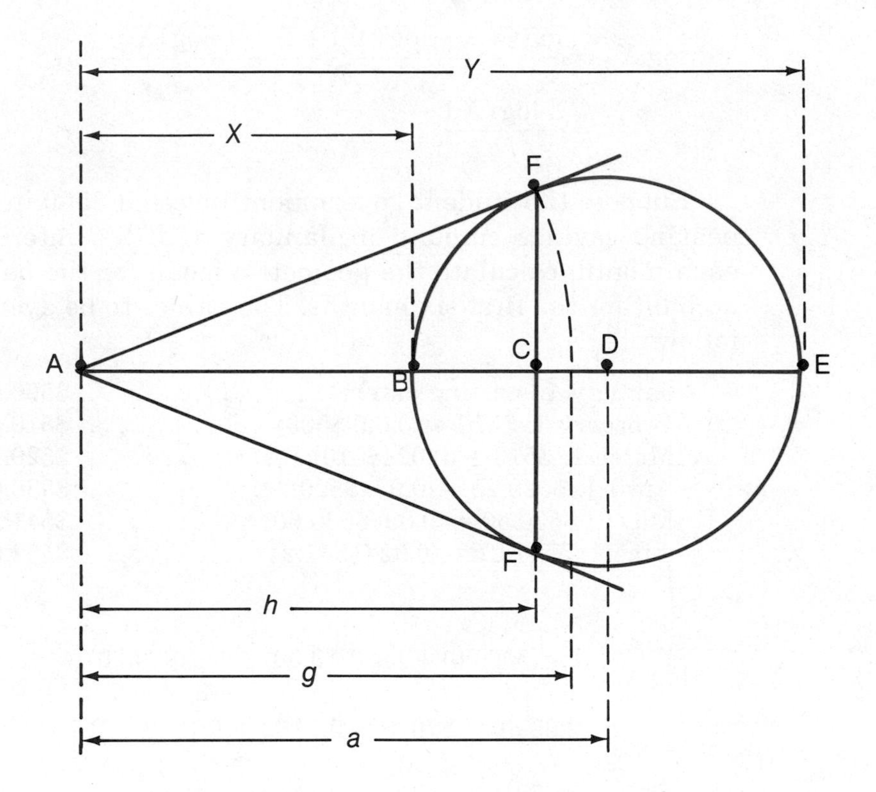

Let segments AB and AE be measured along one line in the same direction. The respective lengths are X and Y units. Using BE as a diameter, construct a circle having D at the center. Draw tangents AF and AF′ and chord FF′, which crosses AE at C. Notice that AC is of length h, AF is length g, and AD is length a. The lengths (in units) of h, g, and a, respectively, are the harmonic, geometric, and arithmetic means of the two values X and Y.

COMPUTER FOCUS 4—Implementing an Algorithm for Computing the Mean

Purpose

Your goal in this Computer Focus is to verify a BASIC computer program for calculating the arithmetic mean of a set of numbers. You will accomplish this goal by studying the flowchart and matching the corresponding BASIC statements with the flowchart symbols. Follow the algorithm through the sequence illustrated by the flowchart. Then key the BASIC program into your computer and test it with several sets of numerical data.

Program Specifications

Input:
1. The value of N (number of values)
2. N values of the variable X; that is, X_i for $i = 1$ to N

Process:
1. *Input* the value of N at step 10.
2. *Initialize* the values of I and SX to zero at steps 15 and 20.
3. *Input* a value of X_i at step 25.
4. *Increment* the "counter" i by 1 at step 30.
5. *Accumulate* the "sum" (SX) by X_i at step 35.
6. *Test* to see if N values have been input at step 40.
7. *Compute* the arithmetic mean (M) at step 45.

Output:
1. N, the number of cases
2. SX, the sum of the values of X_i
3. M, the mean of the distribution

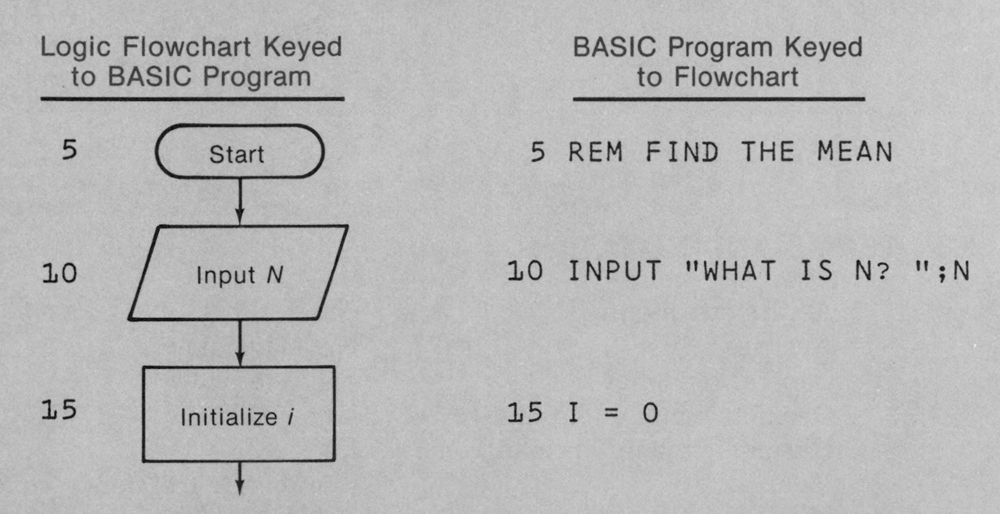

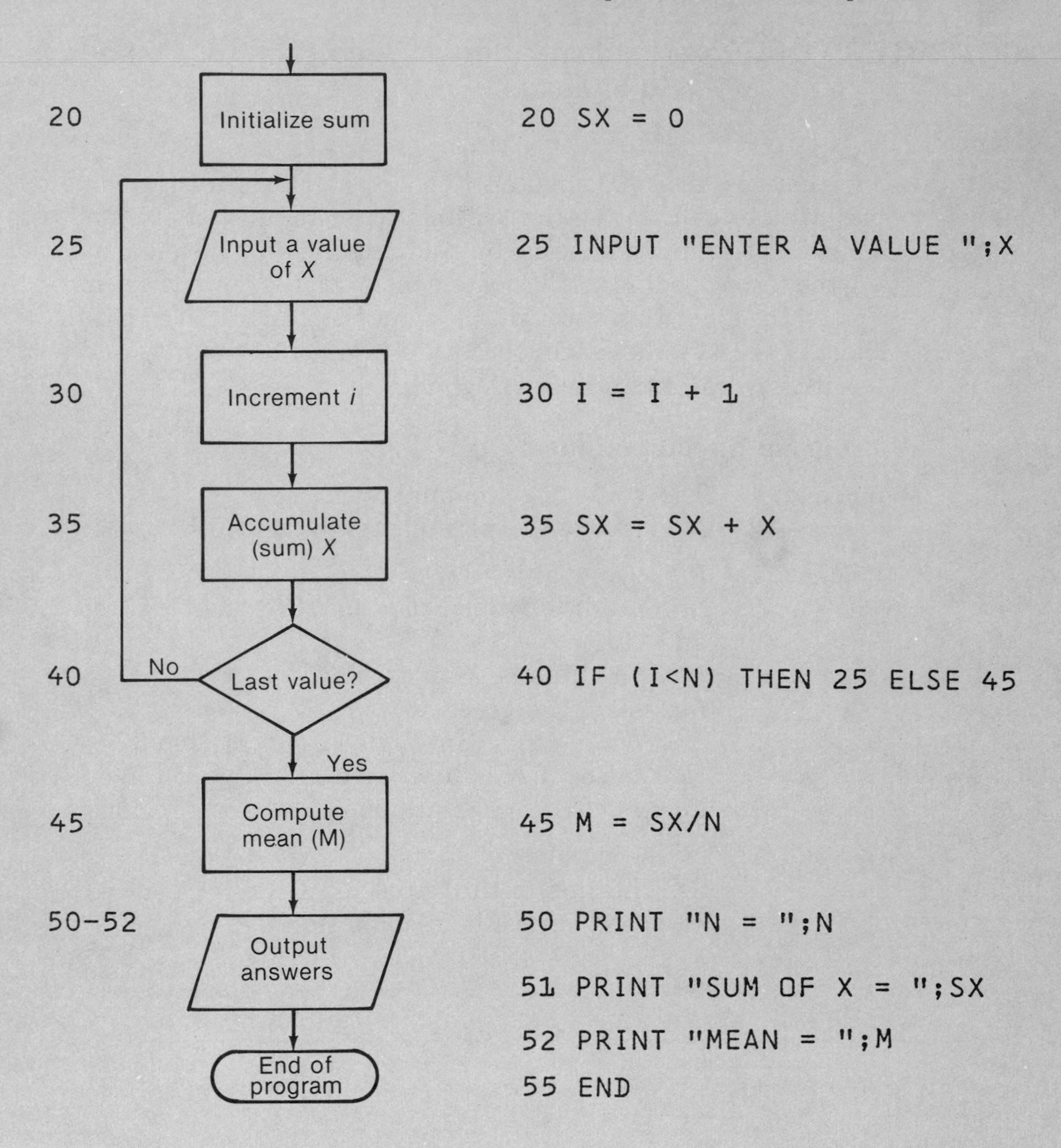

KEY TERMS AND NOTATION

Arithmetic mean
Average
Central tendency
Geometric mean (optional)
Harmonic mean (optional)
Mean
Median
Mode
N
Negative skew
Positive skew
sigma (Σ)
ΣX
Symmetric frequency curve
$\overline{X}$ ("X-bar")

APPLICATION EXERCISES

4.1 For the following sets of numbers, find the ΣX and ΣY.
Variable X: 6, 5, 3, 5, 4, 7, 5, 6, 4
Variable Y: 12.4, 8.6, 5.2, 4.8, 7.1, 3.9

4.2 For the two sets of numbers (X and Y) in Application Exercise 4.1, what is the value of N for each set?

4.3 For the variables X and Y in Application Exercise 4.1, find the mean ($\overline{X}$ and $\overline{Y}$) and the median for each set.

4.4 For variable X in Application Exercise 4.1, find the mode.

4.5 The heights of basketball players on the Glenn View High School varsity (measured in inches) were as follows: 74, 68, 75, 72, 71, 76, 70, 74, 69, 67, 75, 74, 71, and 72.
a. Find the mean height in inches.
b. Determine the average height in feet.

4.6 Find the median and mode height (in inches) of the basketball players in Application Exercise 4.5.

4.7 Of the mean, median, and mode, which one always divides the set of numbers such that half are above and half are below that point?

4.8 The Chamber of Commerce in a resort city, in an attempt to attract tourists during early summer months, reported in a pamphlet an average high temperature for June of 84°F. The local school district, in an attempt to get funding for air conditioning systems for buildings used during the summer, told news reporters that the average high temperature for June was 89°. Meanwhile, the TV weather reporter told viewers that the average high for June was 86°. Inasmuch as they all used the same National Weather Bureau readings, explain how they could arrive at different averages. Can you tell which "average" was used by each group?

4.9 If your teacher used an average as the cutoff between a B grade and a C grade on your first statistics exam, which was quite easy and resulted in a negative skew for the frequency curve, which average (mean, median, or mode) would give you the best opportunity for earning above a C grade?

4.10 A nine-member faculty committee serving as judges consisted of five men and four women. The committee rated the gymnastic team's performance on a 10-point scale from 1 (poor) to 10 (excellent). In a particular event, the men's ratings were 4, 5, 5, 5, 6. For the same event, the women's ratings were 5, 7, 6, 8.

a. What was the mean rating awarded by the men?
b. What was the mean rating awarded by the women?
c. What was the mean rating given by the total committee?
d. Is the overall mean of the committee the same as the mean of the men's and women's mean ratings? Explain your answer.

4.11 Use the following set of data, which represent scores of 65 students on their driver's education performance test, to respond to the questions.

18	17	18	17	18	17	23	17	16	17
22	20	21	19	21	19	18	18	18	24
18	19	18	20	18	20	18	16	18	19
25	19	22	19	21	19	18	19	17	19
18	18	17	18	20	21	20	22	18	18
20	24	20	20	19	23	18	23	19	21
18	19	17	18	19					

a. Compute the mean, median, and mode of the distribution.
b. What would you estimate the shape of the frequency curve of this distribution to be from your knowledge of the three measures of central tendency?
c. Construct a frequency curve to either confirm or refute your response to (b).

4.12 The responses to a statement on an opinionnaire were coded as follows: 1 = agree with the statement; 2 = disagree with the statement; 3 = no opinion. Results of a survey were:

Response	Frequency
1	193
2	26
3	14

a. Which of the three common measures of central tendency would give the most appropriate "typical" response?
b. Why would the arithmetic mean be an inappropriate average to use?

4.13 Given the class size for the following locations in Wilson High School, which average would be the best indication of the central tendency of class size?

Location	Class Size	Location	Class Size
Room 101	28	Room 142	30
Room 115	32	Room 145	29
Room 122	27	Room 150	31
Room 135	28	Room 152	26
Room 140	25	Gym	93

4.14 The notation $\Sigma(X - 5)$ means to subtract 5 from each value in the set and then add the resulting numbers (remember: subtract first, then add the newly created values). For a set of numbers, say X: 3, 3, 7, 6, 6, and 5, find the value of $\Sigma(X - 5)$. Explain why you got this result. (*Hint:* Remember the mean is the "center of gravity.")

4.15 *Optional* Find the harmonic means (to two-decimal-place accuracy) for each of the following sets of data:

a. 2, 8, 4, 4, 6
b. 30, 24, 48
c. 7, 9, 5, 8
d. 70, 60, 40
e. 20, 25, 30, 50

4.16 *Optional* Bobby Max drove to school in heavy traffic and averaged 20 miles per hour. After school he returned home at an average speed of 25 mph. What was his average speed for the round trip?

4.17 *Optional* Find the geometric mean for each of the following sets of numbers:

a. 4, 9
b. 6, 24
c. 4, 16, 8
d. 12, 3, 6
e. 2, 4, 2, 16

4.18 *Optional* For each of the sets of data in Application Exercise 4.17, compute the arithmetic mean. Compare the relative magnitudes of the geometric and arithmetic means. Then speculate on the relative sizes of the two measures of central tendency for any set of numbers.

4.19 *Optional* As recorded by the biology class, the bacteria count in a certain culture increased by 60% per day. If the count was 1,000 on the first day, what were the counts for the second and third days? What was the average (geometric mean) count per day for the three days?

5

Summarizing Data: Measures of Dispersion

CHAPTER OBJECTIVES

Upon completion of the chapter, students will be able to:

1. Distinguish among three definitions of the range of a set of data.
2. Compute the sum of squares of the deviation scores using two methods.
3. Compute the mean deviation of a set of numbers.
4. Calculate the variance of a set of numbers.
5. Determine the standard deviation from a given value of the variance for a variable.
6. Describe how measures of dispersion differ from measures of central tendency.
7. Describe the role of measures of dispersion in summarizing data.

In the previous chapter you learned that the mean, median, and mode are summary statistics representing a typical value and provide an indication of the location of the central portion of a numerical distribution. However, these measures of central tendency provide limited information about the entire distribution of numbers in the collection. Central tendency tells only part of the descriptive story. If you sit on a block of ice with your head in an oven, *on the average* you should be comfortable.

More realistically, consider this example. The East Central High School basketball team has four members who are each 183 centimeters (cm) tall, and the team's center is 203 cm tall, yielding a mean height of 187 cm for the team. The West Side Academy players are 184, 185, 188, 190, and 188 centimeters tall, which averages 187 cm. Are the teams equivalent with respect to height? Notice that although both teams average 187 cm in height, East Central has only one player taller than the shortest player for Academy. Common sense indicates that something about the distributions is different. This illustrates that when an average is the only summary description provided, important information is missing.

At least one additional statistical measure is required to summarize data. Such measures are referred to as *measures of dispersion* and indicate how the data are "scattered" along the spectrum of possible values. As is true of central tendency, there are several measures of dispersion. Each is determined by a different procedure, but all are used for the same basic purpose: to provide an indication of how much spread or variability exists in a distribution. When a set of data is summarized with both a measure of central tendency and a measure of dispersion, a typical value in the distribution is identified as well as how much variation from the average exists.

Using two sets of data with variables X and Y, suppose X represents the number of sit-ups performed by the senior class officers and suppose Y represents the number of sit-ups performed by the varsity cheerleaders. The two sets of hypothetical data and summary calculations are shown in Table 5.1.

The sums of the scores, the means, and the N's for both groups are identical. If only central tendency was reported, the distributions would be summarized as being the same. But by visual inspection it becomes obvious that there is something very different about the distributions of sit-ups. The class officers (X-variable) varied in the number of sit-ups they performed. On the other hand, the cheerleaders (Y-variable) each performed the same number of sit-ups. The distribution referred to as the Y-variable is actually not a variable; Y is a constant in this case.

Table 5.1 Number of Sit-ups for Two Groups of Senior Students

Class Officers	X	Cheerleaders	Y
Representative	15	Head cheerleader	27
Vice-president	22	Yell leader	27
Treasurer	28	Routine captain	27
President	25	Squad member	27
Secretary	34	Squad member	27
Parliamentarian	38	Squad member	27

$\Sigma X = 162$
$N = 6$
$\overline{X} = 27$

$\Sigma Y = 162$
$N = 6$
$\overline{Y} = 27$

This trivial case involving a constant showing no variability is included for the purpose of contrasting a distribution with no dispersion with a variable (X) that does display variability or scatter. Consequently, consider Y, a rather unrealistic distribution for the number of sit-ups, as an extreme case of a lack of dispersion to be used for comparison purposes. Thus the example points out the need for an indication of variability or dispersion when summarizing distributions of numbers. Four techniques that help to more adequately and accurately describe and summarize a set of data are the range, mean (or average) deviation, variance, and standard deviation. The range and mean deviation are used strictly for descriptive and summary purposes, but the variance and standard deviation have continued application throughout the study of statistics.

RANGE

The **range** of a set of numbers is simply a statement of the difference between the smallest and largest numbers in a distribution. There are three methods of reporting the range of a set of data. Consider the values of the X-variable shown in Table 5.1 (sit-ups performed by class officers).

One way of expressing the range is to report the smallest number and the largest number. Thus the range of X is from 15 to 38 sit-ups.

Another way to express the range is to subtract the smallest number from the largest number and report the difference. Using this method, the range for the class officer sit-up data is $38 - 15 = 23$, and the range for the X-variable in Table 5.1 would be reported as 23. This method explains how far apart the minimum and maximum values are located.

A third method for determining and reporting the range of a set of data is to add 1 to the difference between the highest and lowest values in the set. The difference in the minimum and maximum values of the X-variable was 23. If 1 is added to 23, the range would be 24. This method indicates the number of whole number "slots" in the range, *including* the minimum and maximum values.

Verify for yourself the range of the Y-variable in Table 5.1 as shown by the respective methods that follow.

Method 1–The range of Y is from 27 to 27.

Method 2–The range of Y is 0 $(27 - 27 = 0)$.

Method 3–The range of Y is 1 $[(27 - 27) + 1]$.

Any of the three methods can be used because the range disregards N and is entirely determined by only the two extreme values. The first technique has the advantage of indicating the actual minimum and maximum values in the set of numbers and is generally recommended, although the other two techniques will be commonly found in statistical literature and reports.

MEAN DEVIATION

A second measure of dispersion is the **mean deviation** or *average deviation*, as it is sometimes called. Because the aim of this technique is to express an amount of variability, the emphasis is on the word *deviation*. Be careful not to let the term "mean" (or average) cause confusion regarding this technique. Actually, the mean deviation provides an answer to this question about a set of numbers: "On the average, how far do the numbers deviate from the mean of the distribution?" (Reread that question several times until it makes sense in terms of a measure of dispersion or scatter.)

To compute the mean deviation, refer to the X-variable in Table 5.1. The data have been duplicated in Table 5.2, along with a new column showing the absolute values of the *difference between each value and the mean* of the distribution. This difference (between each value and the mean) is called a **deviation score** and is symbolized with the lowercase letter of the original variable name; that is, $x = X - \overline{X}$. Note that x is signed with a plus or minus sign; that is, it is not only an expression of how far each number is from the distribution mean, but also indicates which direction. Because the mean is also known as the "center of gravity" of a set of numbers, the distribution of deviation scores will always add to zero.

Table 5.2 Sit-up Performance and Deviations for Class Officers

Values X	Value − Mean $(X - \overline{X})$	Deviations $(X - \overline{X})$ x	Absolute Values of Deviation Scores $\lvert x \rvert$
15	(15 − 27)	−12	12
22	(22 − 27)	−5	5
28	(28 − 27)	+1	1
25	(25 − 27)	−2	2
34	(34 − 27)	+7	7
38	(38 − 27)	+11	11
$\Sigma X = 162$		$\Sigma x = 0$	$\Sigma \lvert x \rvert = 38$
$N = 6$			$N = 6$
$\overline{X} = 27$			$\lvert \overline{x} \rvert = \text{MD} = 6.33$

In the case of the mean deviation, the only concern is how much difference exists between each number and the mean—not the direction. Hence, of interest is the **absolute value**, which ignores the algebraic sign of x. Notice that the sum of the absolute values of x from Table 5.2 is 38. When 38 is divided by N (6 in this case), the result is the mean deviation (6.33).

The formula for the mean or average deviation is:

$$\mathbf{MD} = \frac{\Sigma \lvert \boldsymbol{x} \rvert}{\boldsymbol{N}}$$

where $\Sigma \lvert x \rvert$ is the sum of the absolute values of x, and N is the number of cases in the collection of numbers.

The computation of the mean deviation for the X-variable in Table 5.1 is shown in Figure 5.1.

In this case the values of the X-variable are, on the average, 6.33 units (sit-ups) away from the mean of X. As with the range, the greater the magnitude of the measure of dispersion, the more scatter or spread exists in the number set.

You can easily verify that the Y-variable in Table 5.1 has a mean deviation of zero in that trivial case. The obvious reason the mean deviation for Y is zero is that, on the average, the values of Y do not deviate from the mean, or they deviate zero units from the mean on the average.

Because the mean deviation uses all of the values, it is generally a preferred measure of dispersion relative to the range. Even though it has some advantages over the range and appeals to common sense, it is not as popular nor as useful as the *variance* or the *standard deviation.*

Figure 5.1—Calculator Keystrokes

PROBLEM:
Compute the mean deviation of a set of data

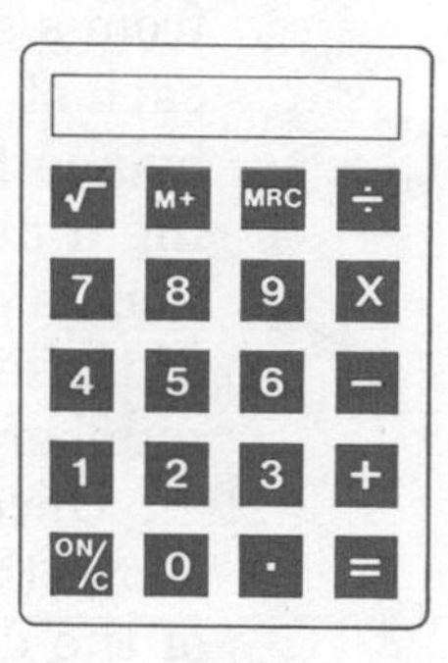

Procedure	Arithmetic Operation	Calculator Keystrokes	Display
1. Compute the mean of the data: $\frac{\Sigma X}{N}$	1. (15 + 22 + 23 + 25 + 34 + 38) ÷ 6 = 27	1. 15 [+]	15
		22 [+]	37
		28 [+]	65
		25 [+]	90
		34 [+]	124
		38 [÷]	162
		6 [=]	27
2. Compute sum of the absolute values of deviation scores: $\Sigma\lvert x\rvert = \Sigma\lvert X - \overline{X}\rvert$	2. 12 + 5 + 1 + 2 + 7 + 11 = 38	2. [−] 15 [=] [M+]	12
		27 [−] 22 [=] [M+]	5
		28 [−] 27 [=] [M+]	1
		27 [−] 25 [=] [M+]	2
		34 [−] 27 [=] [M+]	7
		38 [−] 27 [=] [M+]	11
		[MRC]	38
3. Calculate mean deviation by $\frac{\Sigma\lvert x\rvert}{N}$	3. $\frac{38}{6} = 6.33$	3. [÷] 6 [=]	6.333333
		(round to 2 decimals)	6.33

VARIANCE

Before considering the formula for computing the measure of dispersion known as the variance, it is necessary to understand a quantity known as the "sum of squares." This phrase is short for the "sum of the squares of the deviations from the mean." Recall that the magnitude of the sum of the positive values of x is equal to the sum of the negative values within a set of deviation scores, and therefore $\Sigma x = 0$. One way of eliminating the negative values from a set of deviation scores is to square each x; that is, multiply each of the deviation scores by itself. A negative value multiplied by another negative value results in a positive number, thus creating a distribution with all positive (at least nonnegative) values. Such a distribution is made up of x^2 values, or $(X - \overline{X})^2$. The sum of these values is known as the **sum of squares** for that set of numbers.

Once again, consider the X-variable from Table 5.1. Table 5.3 shows the original sit-up scores, the deviation scores, and the squares of the deviation scores.

Table 5.3 Sit-up Scores, Deviation Scores, and Squares of Deviation Scores

# of Sit-ups X	$(X - \overline{X})$	$(X - \overline{X})$ x	$(X - \overline{X})^2$ x^2
15	15 − 27 =	−12	144
22	22 − 27 =	−5	25
28	28 − 27 =	1	1
25	25 − 27 =	−2	4
34	34 − 27 =	7	49
38	38 − 27 =	11	121
$\Sigma X = 162$		$\Sigma x = 0$	$\Sigma x^2 = 344$
$N = 6$			
$\overline{X} = 27$			

As shown in Table 5.3, the sum of squares, Σx^2, is 344. In the example, the formula for determining the sum of squares is $\Sigma(X - \overline{X})^2$. This procedure involves subtracting the mean from each score (or value), squaring each of those differences, and then summing the squared deviation scores. Because the mean of a distribution of numbers is very seldom an integer (whole number), the

$\Sigma(X - \overline{X})^2$ method becomes somewhat inefficient. A more efficient formula for the calculation of the sum of squares is as follows:

The formula for finding the sum of squares is:

$$\Sigma x^2 = \Sigma X^2 - \frac{(\Sigma X)^2}{N}$$

where ΣX^2 is the sum of the squared values of the raw scores and ΣX is the sum of the scores.

The most efficient setup for the data is shown in Table 5.4. The first column contains the original X-variable, and the second column contains the square of each number in the X-variable set.

Table 5.4 Variable X and X^2 from Table 5.1

X	X^2
15	225
22	484
28	784
25	625
34	1,156
38	1,444
$\Sigma X = 162$	$\Sigma X^2 = 4,718$

Using the values from Table 5.4, the sum of squares is computed as shown in Figure 5.2. Note that the value of the sum of squares computed is the same value shown in Table 5.3. Be aware that the phrase "sum of squares" *does not* refer to the sum of the squares of the original variable values. The sum of squares concept will be used extensively through this text. This concept provides a foundation on which most other statistical techniques are built. It will be the basic step that is required to use many of the formulas and equations in the application of statistical techniques to numerical problems.

Once the value for the sum of squares for a set of data has been determined, the **variance** is found by dividing the sum of squares quantity by N, the number of cases.

The variance is symbolized by the square of the Greek letter sigma (σ^2).

Figure 5.2—Calculator Keystrokes

PROBLEM:

Compute the sum of squares for a set of data

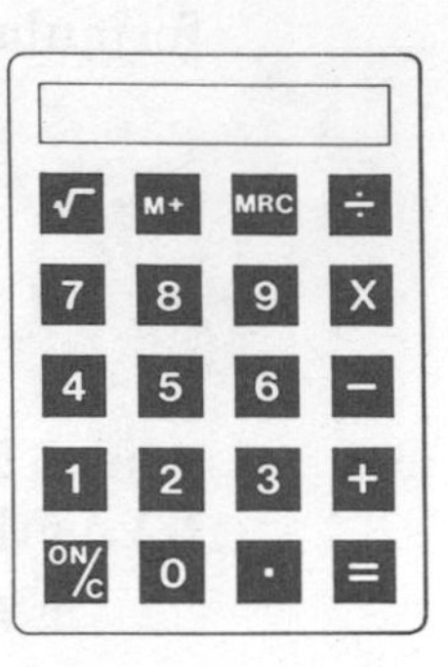

Procedure	Arithmetic Operation	Calculator Keystrokes	Display
1. Sum the original values of the variable: ΣX	**1.** $15 + 22 + 28 + 25 + 34 + 38 = 162$	**1.** 15 [+]	15
		22 [+]	37
		28 [+]	65
		25 [+]	90
		34 [+]	124
		38 [=]	162
2. Sum the squares of the original values of the variable: ΣX^2	**2.** $15^2 + 22^2 + 28^2 + 25^2 + 34^2 + 38^2 = 4{,}718$	**2.** 15 [×] 15 [=] [M+]	225
		22 [×] 22 [=] [M+]	484
		28 [×] 28 [=] [M+]	784
		25 [×] 25 [=] [M+]	625
		34 [×] 34 [=] [M+]	1,156
		38 [×] 38 [=] [M+]	1,444
		[MRC]	4,718
3. Compute the "sum of squares": $\Sigma x^2 = \Sigma X^2 - \frac{(\Sigma X)^2}{N}$	**3.** $\Sigma x^2 = 4{,}718 - \frac{(162)^2}{6}$ $= 4{,}718 - 4{,}374$ $= 344$	**3.** [CLM]	0
		162 [×] 162 [÷] 6 [=]	4,374
		[M+]	4,374
		4,718 [−] [MRC] [=]	344

The formula for variance is:

$$\sigma^2 = \frac{\Sigma x^2}{N}$$

where σ^2 (lowercase Greek letter sigma) represents the variance, Σx^2 is the sum of squares in the set, and N is the number of cases (values).

This equation is illustrated using the X-variable data from Table 5.4.

$$\sigma^2 = \frac{\Sigma x^2}{N} = \frac{344}{6} = 57.33$$

The variance of a distribution of numbers is not as easily interpreted as the range and mean deviation. At this point the variance may be considered a comparative value. When two or more distributions are being compared, the larger the value of the variance, the more spread that exists. Figure 5.3 shows a comparison of two hypothetical frequency curves. Both have the same mean, but one curve shows more dispersion (variability) than the other. As the variance of a distribution increases in size, the distribution is referred to as being more heterogeneous. Conversely, smaller values of variance indicate a more homogeneous (less dispersion) distribution.

Figure 5.3—Two Frequency Curves Showing Differences in Variation

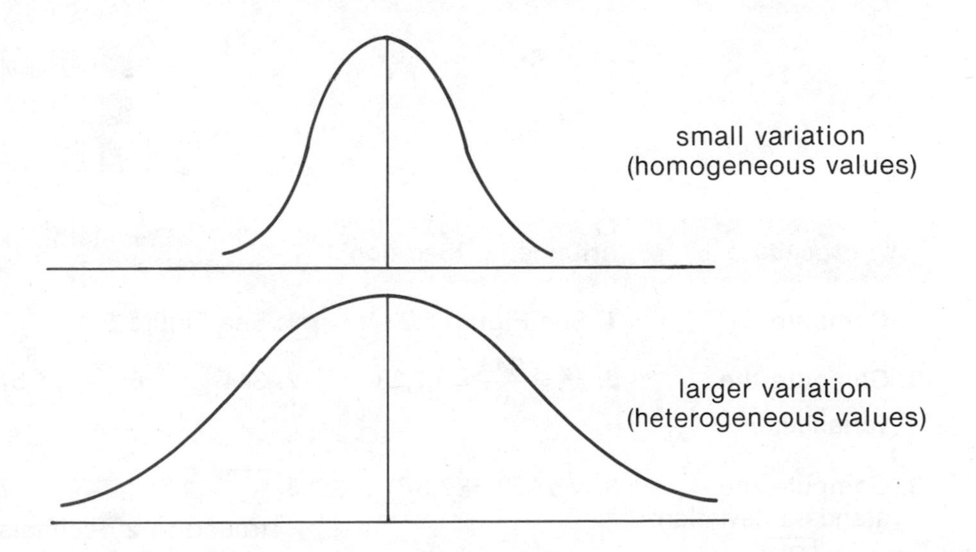

STANDARD DEVIATION

The variance will be used most frequently as an intermediate step for finding the standard deviation, which is the measure of dispersion most widely used by statisticians. The **standard deviation** is defined as the square root of the variance. The symbol for the standard deviation is the lowercase Greek letter sigma (σ). The steps for computing a standard deviation for a set of values are:

1. Sum (add) the original values of the variable to get ΣX.
2. Square each of the original values, creating a new column, and label it X^2.
3. Sum the column of values to get ΣX^2.
4. Using the results of step 1 and step 3, compute the sum of squares—Σx^2.
5. Divide the results of step 4 by N to determine the variance of the distribution.
6. Find the square root of the variance found in step 5.

Figure 5.4 uses the X-variable data from Table 5.1 again to illustrate how to find the variance and standard deviation.

Figure 5.4—Calculator Keystrokes

PROBLEM:
Calculate variance (σ^2) and standard deviation (σ)

Procedure	Arithmetic Operation	Calculator Keystrokes	Display
1. Compute Σx^2	1. See Figure 5.2	1. See Figure 5.2	344
2. Compute the variance: $\sigma^2 = \frac{\Sigma x^2}{N}$	2. $\sigma^2 = \frac{344}{6} = 57.33$	2. 344 [÷] 6 [=]	57.333333
3. Compute the standard deviation: $\sigma = \sqrt{\frac{\Sigma x^2}{N}} = \sqrt{\sigma^2}$	3. $\sqrt{57.33} = 7.57$	3. [√]	7.571878
		(round to 2 decimals)	7.57

As shown in Figure 5.4, the standard deviation σ for the sit-up data is 7.57. You should verify that the standard deviation for the Y-variable from Table 5.1 is zero, that is, no variability.

The range, mean deviation, variance, and standard deviation are four methods of describing and summarizing the degree of spread of a set of data. They are used only for data that are scaled on an interval or ratio level. When describing data in a summary form, the most common combination used is the mean (as a description of central tendency) and the standard deviation (as a description of dispersion).

COMPUTER FOCUS 5—Implementing an Algorithm for Computing the Variance and Standard Deviation

Purpose

The sum of squares is a widely used quantity, and the procedure for inserting the algorithm into programs for solving more complicated programs is presented later in the text. The twofold purpose of the present Computer Focus is to (1) acquaint you with important algorithms for computing the sum of squares of deviation scores, and (2) enable you to practice BASIC program modification by carefully following the set of instructions.

Problem

Building on the algorithm from the previous Computer Focus in which you verified a BASIC program for finding ΣX and $\overline{X}$, your task in this Computer Focus is to expand the program to compute the variance (V) and the standard deviation (SD) of a set of N values. This task requires that you compute ΣX^2 and Σx^2 before computing the variance.

Modifications for Program

Integrate the task of computing the standard deviation with your logic (flowchart) used in the previous Computer Focus. This can be accomplished by inserting the following segments as shown.

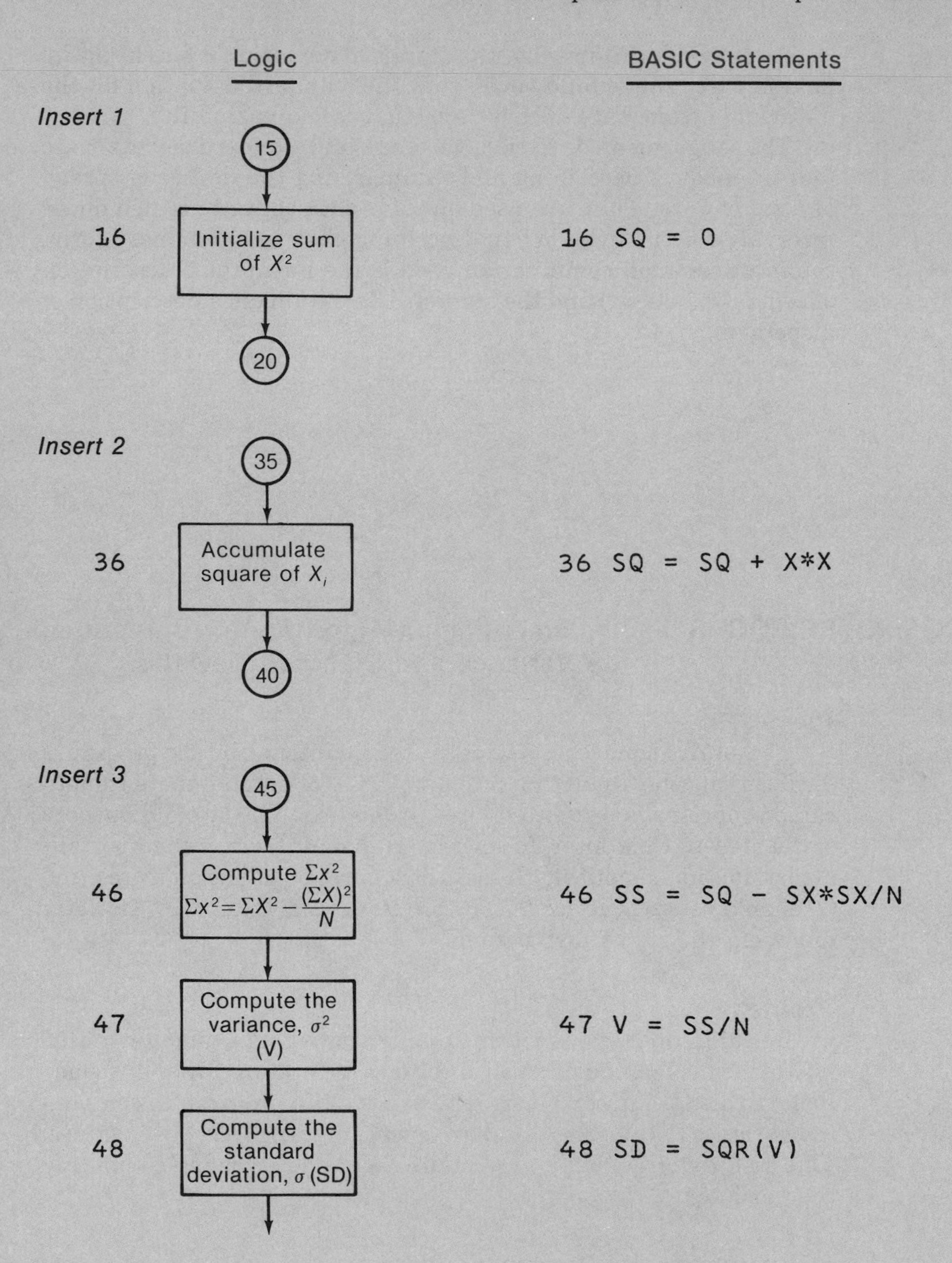

Note: The symbol * indicates multiplication in the BASIC language and SQR is a function that yields the square root.

Finally, write BASIC statements 53 and 54 that will PRINT the values of the variance (V) and standard deviation (SD), respectively.

Verify that the expanded version of the program that was started in the previous Computer Focus will now output N, ΣX^2, $\overline{X}$, σ^2, and σ.

KEY TERMS AND NOTATION

Absolute value	Σx^2
Deviation score	σ
Dispersion	σ^2
Mean deviation	Standard deviation
Range	Variance
Σ	$\Sigma(X - \overline{X})^2$
ΣX	$\lvert x \rvert$
Σx	

APPLICATION EXERCISES

5.1 Express the absolute value of each number that follows.

a. -2 c. $+150$

b. -9.35 d. 39

5.2 The variance of a well-known IQ test is 225. What is the standard deviation of the test?

5.3 If the standard deviation of scores on an achievement test is 10, what is the variance?

5.4 If, for 20 recorded weights of a group of young adults, ΣW is equal to 3,300 and ΣW^2 is equal to 544,570, what is the sum of squares of the W-variable; that is, $\Sigma w^2 = ?$

5.5 One distribution, say variable X, had a mean deviation of 40. Another distribution, variable Y, had a mean deviation of 60. In which distribution were the values more dispersed (scattered)?

5.6 For this list of values—0, 1, 2, 8, 4, 5, 6, 7, 9, 8—find the following values:

a. ΣX
b. ΣX^2
c. $(\Sigma X)^2$
d. Σx^2
e. Range
f. Mean deviation
g. Variance
h. Standard deviation

5.7 For the Y-variable distribution that follows, find the:

a. Mean
b. Median
c. Mode
d. Range
e. Mean deviation
f. Variance (σ^2)
g. Standard deviation (σ)

Y: 7, 8, 9, 9, 13, 12, 7, 11, 11, 13, 10, 10, 10, 10

5.8 Suppose we "curve" the scores in Application Exercise 5.7 and add 5 to each score. For the new distribution, find the mean and standard deviation. How did adding 5 to each value influence the mean? How did the newly created distribution compare to the original distribution with respect to variability?

5.9 Multiply each value in Application Exercise 5.7 by 2. Find the mean and standard deviation for the new distribution. How did multiplying each value by 2 affect the mean and standard deviation of the distribution?

5.10 From the results of Application Exercises 5.8 and 5.9, what generalization can be made regarding the mean and standard deviation of variables that are transformed by adding a constant or multiplying each value in a distribution by a constant?

5.11 From the frequency distribution that follows, find N, the mean, and the standard deviation.

Score = X	f
2	1
3	2
4	5
5	3
6	2

5.12 Given the frequency polygon that follows, find the value of N, the mode, the median, the mean, and the mean deviation.

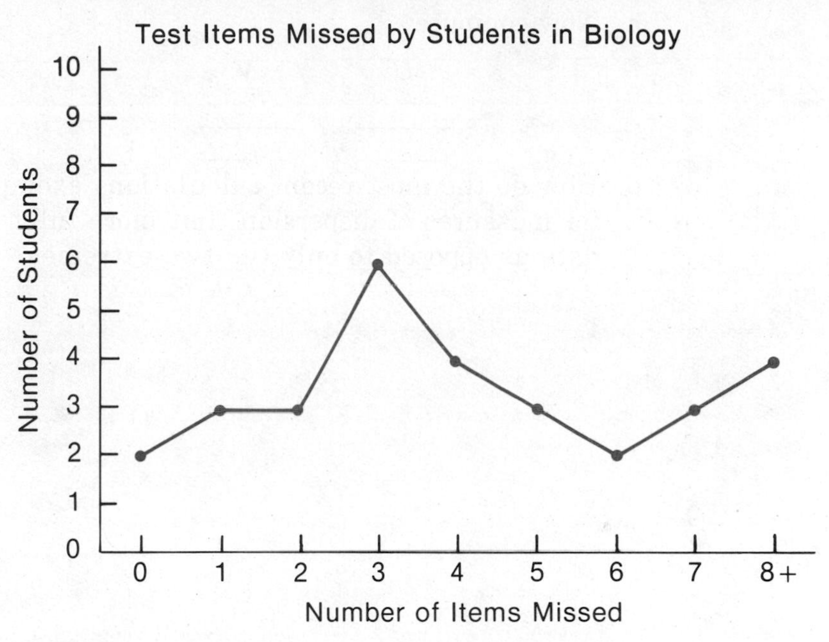

5.13 Given the values for two variables, U and V, perform the indicated operations and answer the questions that follow.

U	V
3	7
10	6
9	10
7	7
7	7
8	8
6	7
5	8
7	3
8	7

a. For each of the variables, find:

	U	V
N	______	______
Σ	______	______
Mean	______	______
Median	______	______
Mode	______	______
Range	______	______

b. Do the distributions appear to be identical based on the above measures of central tendency and dispersion?

c. Now compute:

	U	V
Σx^2	______	______
σ	______	______

d. How do the most recent calculations exemplify the need for measures of dispersion that more fully use all of the data as opposed to only the two extremes?

6

Describing Individual Performances

CHAPTER OBJECTIVES

Upon completion of the chapter, students will be able to:

1. Describe, in terms of the mean and standard deviation, *z*-score and *T*-score numerical distributions.
2. Given a set of raw scores, transform them into corresponding standard *z*-scores.
3. Convert a set of *z*-scores into a distribution of *T*-scores.
4. Convert a set of *z*-scores into a distribution of standard scores with any given mean and standard deviation.
5. Interpret the meaning of an individual standard score relative to the distribution of concern.
6. Describe characteristics of the normal curve.
7. Given a set of *z*-scores and using characteristics of the normal curve, convert the *z*-scores into percentile equivalents.
8. Given a percentile score and using the characteristics of the normal curve, transform the percentile to a standard *z*-score.
9. (Optional) Construct a box-and-whisker plot diagram from a set of numerical values.

Angela hurried home from school to tell her dad about her test results in natural science. When Angela reported that she had made 65 on the exam, her father was not favorably impressed. He couldn't understand why Angela was so excited. Angela then informed her father that 65 represented the number of correct items out of a possible 80. "Well," said her father, "that sounds a little better." Angela continued with additional information that the class mean was 48 and the standard deviation was 8.5. Now Angela's performance *relative* to the *typical performance* in her class began to clarify why she was so pleased with her score on the natural science test.

One way to view Angela's score is to specify how far it is from the mean. In Chapter 5 you learned that the difference between a score and the mean is referred to as a deviation score. Angela's relative position then could be expressed as $x = 17$, or 17 points above the class mean. Because σ (the standard deviation of the class scores) is a measure of spread, it could be used to help express how far Angela's score is "spread" from the mean. That is, how many standard deviations is Angela's score from the mean? Because $\sigma = 8.5$ points and Angela's score was 17 points from the mean, her performance could be expressed as 2 $(17 \div 8.5)$ standard deviations above the class mean.

It becomes clear that several methods are available for expressing Angela's score in a way that conveys information about her performance compared to the typical class member's performance. Scores such as the number correct, the percentage correct, and the number correct minus some correction factor are examples of **raw scores**. These numbers are direct results of scoring exams, timing an event, recording weights on a scale, and other familiar types of measures. They have not been transformed by mathematical procedures that permit an adequate description of the individual's performance relative to some defined group.

Raw scores by themselves are not very descriptive. In the previous example, Angela's score of 65 was relatively meaningless until additional information was provided about the average performance of her classmates. After a description of the overall group performance, Angela's performance became easier to interpret.

The previous chapters have been devoted to methods of describing and summarizing an *entire set* of numerical data. In this chapter you will learn the methods of describing the performance of *an individual* within the context of a group of performances. These methods are used to express the quality of an individual performance with numbers relative to the central tendency and dispersion of a group of numbers. The mean of a distribution will serve as the anchor and comparison point; the standard deviation will serve as the "yardstick" to indicate how far from the mean a score is located. More simply, the techniques described in this chapter are standard methods of expressing numerical scores.

Keith had a scaled *T*-score of 53 on his standardized social studies achievement test. Henry's standardized test score in language arts was reported as the 53rd percentile. The counselor told Joyce that she scored about one-half (0.50) of a standard deviation above the mean on her mathematics aptitude test. What do these scores mean? How are they derived? How are they interpreted? Did Keith, Henry, or Joyce score highest relative to the respective exams?

STANDARD SCORES

Standard scores are commonly used for the purpose of interpreting how well an individual performed relative to his or her peers. A standard score gives us at least two important pieces of information about a performance. First, it tells how far from the mean of a group of measures the score falls. Second, it indicates whether the score was above or below the typical (average) score. Scores called *z*-scores and *T*-scores are frequently used forms of standard scores and can be transformed to virtually any kind of standard score scale.

z-scores

To convert raw scores to ***z*-scores** (note the lowercase *z*), first find the mean and standard deviation of the entire set of scores as described in Chapters 4 and 5. Following is an equation for changing a raw score (X) to a standard score.

The formula for converting raw scores to *z*-scores is:

$$z = \frac{X - \overline{X}}{\sigma} \text{ or } \frac{x}{\sigma}$$

where X is the score, $\overline{X}$ is the mean, and σ is the standard deviation.

The equation states that to determine the corresponding deviation score (x), subtract the mean of the distribution from the score. Then z is equal to that deviation score divided by the standard deviation of the set of scores against which the score is being compared.

Suppose a distribution of raw scores represents the number of TV trivia items that are answered correctly by 30 students at a class party. The mean of the distribution of correct answers was 50, and the standard deviation of the distribution was 6. What would be the *z*-score that corresponds to Cheryl's raw score performance (number of correct trivia answers) of 53? To solve this problem, subtract the group mean (50) from the score (53) being considered. That difference is 3; that is, $x = 3$. This indicates that Cheryl scored 3 raw score points above the mean. Next, divide that difference by the standard deviation: $\frac{3}{6} = 0.5$. Therefore Cheryl's *z*-score was 0.5; the computation is summarized in Figure 6.1. The *z*-score shows that Cheryl scored 0.50 standard deviation from the class mean. Further, the positive algebraic sign of the *z*-score implies that her performance was *above* the class mean.

Figure 6.1—Calculator Keystrokes

PROBLEM:

Compute *z*-score

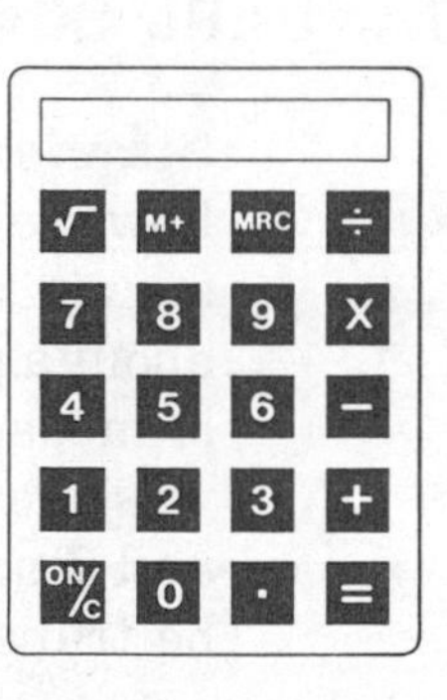

Procedure	Arithmetic Operation	Calculator Keystrokes	Display
1. Compute mean for entire set of data: $\overline{X} = \frac{\Sigma X}{N}$	**1.** $\overline{X} = 50$	**1.** See Figure 4.1	
2. Compute standard deviation: $\sigma = \sqrt{\frac{\Sigma x^2}{N}}$	**2.** $\sigma = 6$	**2.** See Figure 5.4	
3. Find the value of *x*, the deviation score	**3.** $x = X - \overline{X}$ $= 53 - 50$ $= 3$	**3.** 53 [−] 50 [=]	3
4. Divide the deviation score by σ	**4.** $z = \frac{3}{6} = 0.5$	**4.** [÷] 6 [=]	0.5

On the other hand, suppose Jerry got 41 trivia items correct at the same party. His z-score would be calculated as follows:

$$z = \frac{41 - 50}{6}$$

$$z = \frac{-9}{6}$$

$$z = -1.50$$

Jerry's z-score was 1.50 standard deviations from the mean. His score, as shown by the negative sign, was *below the average.*

Now consider the following characteristics of standard z-scores:

1. In a symmetric distribution, about half of the z-scores will be negative. This is because about half of the numerical values in a distribution will be below the mean.
2. Most of the z-scores in a distribution will need to be expressed to two-decimal-place accuracy to describe the location of a score sufficiently.
3. If every score in a distribution is converted to a standard z-score, the new distribution (z-scores) will have a mean of zero ($\bar{z} = 0$) and a standard deviation of one ($\sigma = 1$), within rounding error tolerances.

T-scores

Another standard score scale is denoted using an uppercase T. Some people prefer not to work with fractions (decimals) and negative numbers, both of which are prevalent when using z-scores. Even though a score of $z = 0$ would technically indicate average performance, such a score somehow conveys the impression that the person did not perform at all! This causes confusion for those not familiar with the z-score scaling.

***T*-scores** are derived from z-scores, but are scaled with a distribution mean of 50 and a standard deviation of 10. The arithmetic transformation is simple to perform after the z-scores are computed.

The formula for converting a z-score to a T-score is:

$$T = 10z + 50$$

where T is the T-score and z is the z-score.

Chapter 5 (see Application Exercise 5.10) illustrated that when a distribution of scores is multiplied by a constant, the standard deviation of the newly created distribution is also multiplied by that constant. Further, if a constant is added to each score in a distribution, that same constant is added to the original mean to form the mean of the new distribution.

The mechanics of these concepts can be further explained by transforming z-scores to T-scores. By examining the equation $T = 10z + 50$, we see that the z-scores are first multiplied by 10. This has the effect of making the T-score standard deviation 10 times the standard deviation of the set of z-scores. Because $\sigma = 1$ for a distribution of z-scores, the standard deviation of the corresponding T-score distribution will be 10×1 or 10. Second, 50 is added to each score in the z-score distribution. This implies that the new distribution of T-scores will have a mean of $0 + 50$, because $\bar{z} = 0$. Consequently, if an entire set of raw scores is converted to z-scores and the z-scores in turn are converted to T-scores, the T-score distribution will have a mean of 50 ($\overline{T} = 50$) and a standard deviation of 10 ($\sigma = 10$).

The computational procedures will be illustrated using the two previous examples. Recall that Cheryl's z-score on the trivia questions was 0.50 and Jerry's was -1.50. To convert each to a standard T-score, employ the T-score formula:

	Cheryl	Jerry
	$z = 0.50$	$z = -1.50$
therefore	$T = 10z + 50$	$T = 10z + 50$
	$T = (10)(0.5) + 50$	$T = (10)(-1.50) + 50$
	$T = 5 + 50$	$T = -15 + 50$
	$T = 55$	$T = 35$

When using a calculator to compute Cheryl's T-score, you can follow the sequence in Figure 6.2.

From the calculations just presented, Cheryl's T-score was 55 and Jerry's was 35. In a distribution of T-scores, because the mean is 50 and the standard deviation is 10, a close inspection of the T-scores shows Cheryl's performance is indeed 0.50 of a standard deviation above the mean (5 T-score points above), whereas Jerry's score of 35 is 1.50 standard deviations below 50. Consequently, T-scores give exactly the same information as z-scores, but they are scaled so that fractions can be rounded to whole numbers and negative numbers are eliminated.

To convert T-scores to z-scores, solve the equation $T = 10z + 50$ for z.

The formula for converting T-scores to z-scores is:

$$z = \frac{T - 50}{10}$$

where T is the T-score and z is the z-score.

Actually, the equation for converting T-scores to z-scores is exactly the same equation used for transforming any set of scores to z-scores, except that the mean of T is 50 and the standard deviation of T is 10. To check it out, let's use Cheryl's T-score of 55.

$$z = \frac{55 - 50}{10}$$

$$z = \frac{5}{10}$$

$$z = 0.5$$

This corresponds to her original z-score calculated earlier. Now use this equation to see if you can convert Jerry's T-score back to his original z-score.

In general, z-score distributions may be transformed into a scale with virtually any mean and standard deviation using the concept just illustrated. The procedure is this: Multiply each z-score in a distribution by whatever constant you would like to represent the standard deviation of your newly created scale. Then add whatever constant you wish to serve as the mean of the new distribution. Here are examples of scales with which you are probably

Figure 6.2—Calculator Keystrokes

PROBLEM:

Compute a *T*-score given a *z*-score

Procedure	Arithmetic Operation	Calculator Keystrokes	Display
1. Multiply the *z*-score by the value of the desired standard deviation: (10)(0.5)	**1.** (10)(0.5) = 5	**1.** 10 [×] 0.5 [=]	5
2. Add the results of (1) to desired mean: 5 + 50	**2.** 5 + 50 = 55	**2.** [+] 50 [=]	55

familiar: IQ scores have a mean of 100 and a standard deviation of 15 points for the general population; SAT scores have a mean of 500 and a standard deviation of 100. Thus you could create scales just as have already been done by the commercial test publishing companies.

For example, convert Cheryl's trivia score to a typical scale used for IQ tests. Because her *z*-score was 0.5, her performance on such a numerical scale would be found by using 15 as the new standard deviation and 100 as the new mean.

$$\begin{aligned} \text{Score} &= (\text{new standard deviation})(z) + (\text{new mean}) \\ &= 15z + 100 \\ &= (15)(0.5) + 100 \\ &= 7.5 + 100 \\ &= 107.5 \text{ (rounded to 108)} \end{aligned}$$

Obviously, this does not mean Cheryl's IQ is measured at 108. Aptitude, achievement, intelligence, personality, and so forth are functions of the content and problem-solving strategies, *not* the scaling of the scores. Performance on trivia items can be scaled on any common standard score scale; however, answering trivia questions can hardly be compared to answering items on an intelligence test.

One more illustration will help clarify this procedure. Suppose you wanted to scale Cheryl's *z*-score on a new distribution that had a mean of 50 (new mean) and a standard deviation of 6 (new standard deviation).

$$\begin{aligned} \text{New score} &= 6z + 50 \\ &= (6)(0.5) + 50 \\ &= 3 + 50 \\ &= 53 \end{aligned}$$

Did you note that 53 is her original raw score? Why? Because the original distribution of raw scores had a mean of 50 and a standard deviation of 6 points. Thus converting her *z*-score back to the scale of the raw score distribution yields her original raw score. How do standard scores fit into the scheme of a frequency distribution curve?

NORMAL CURVE

A frequency polygon that displays scores that are normally distributed has unique characteristics when smoothed into a frequency curve. Visually the graph resembles a **bell-shaped curve** because a relatively large number of values are located in the middle part of the distribution. Recall that the mode is one of the commonly used measures of central tendency. As you examine the number of occurrences of values that vary from the center of the distribution, the frequency (and relative frequency) tend to get smaller. As the scores depart more and more from the mean, median, and mode, the frequency (and relative frequency) become progressively smaller. Distributions such as "heights of American men," "hours per week spent on homework," "intelligence test scores," and "grade-point averages for a high school" are illustrative of many phenomena whose frequency polygons closely approximate the **normal curve**.

For more than a century, statisticians have made use of the theoretical normal curve, which was mathematically derived. Much of the data collected in our society, when graphed, will very closely approximate the theoretical normal curve. The approximation is accurate enough for the characteristics of the theoretical normal curve to be used in describing individual performance.

Figure 6.3 shows a frequency curve in the shape of the normal distribution. Notice that the frequency is large in the central portion of the distribution but declines as the values depart from the center. As a student of statistics, you must realize that when only a few cases (say, fewer than 30) are graphed, the resulting frequency curve will probably not appear as a bell-shaped normal distribution. Because of the discrete nature of measurement, the distribution will be displayed as a curve only after artificial "smoothing" of a histogram or frequency polygon. With small N's, random fluctuation in the location of only a few cases will distort the shape of the distribution, whereas if you collect data using a large N, the resulting distribution will approach normality.

The point is that "real" data resulting from measures with finite N's and less than perfect discrete measurement techniques cannot follow an exact normal curve. What is important is that the underlying trait, characteristic, or performance being measured is usually normally distributed in a large (say, over 100) group. In most of your work in this course, only small groups of individuals called samples will be studied, under the assumption that the actual phenomenon does conform to the normal distribution characteristics.

Several characteristics of the theoretical normal curve are important. First, the mean, median, and mode in a normal frequency curve are located at the same point. Second, the curve is **symmetric** around the mean. This means the right side of the

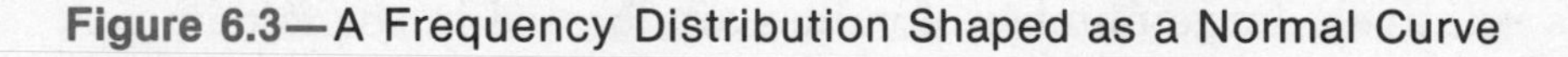
Figure 6.3—A Frequency Distribution Shaped as a Normal Curve

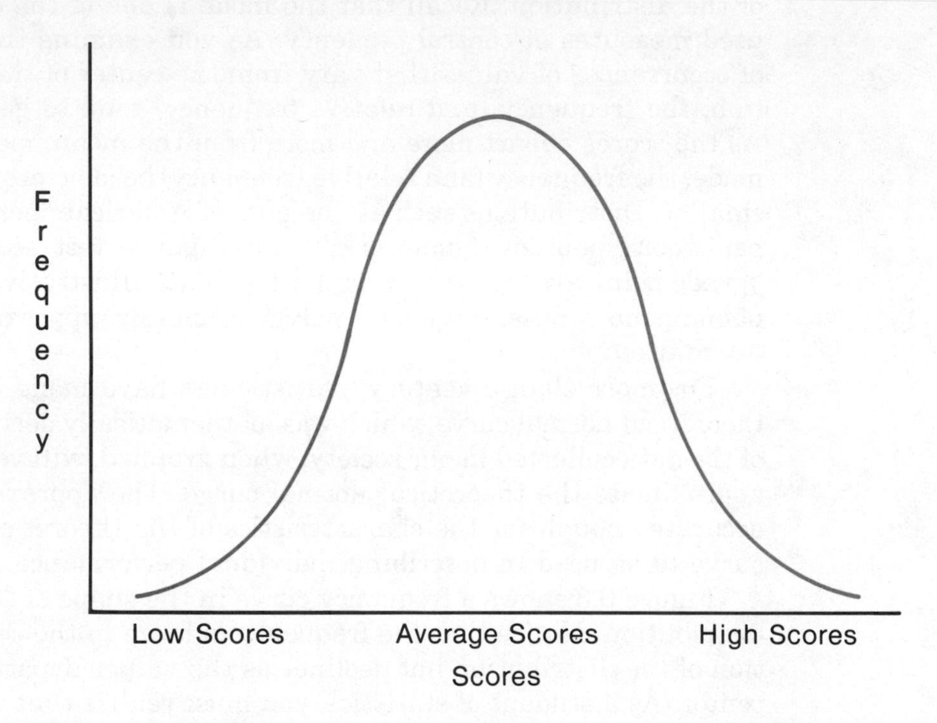

curve is a reflection or mirror image of the left side. Third, the "tails" of the theoretical normal curve approach the baseline, but never reach it. Notice that the central portion of the curve is curving down toward the baseline. This is known as "concave downward." However, on each side of the graph in the region of the tails, the graph begins to turn in an upward direction. This is known as "concave upward." The two points on each side where the curve changes from concave downward to concave upward are called **points of inflection**.

If you have trouble grasping the concept of a curve beginning to turn upward while still sloping downward, consider a roller coaster as a physical analogy. The roller coaster is going down at a steeper and steeper rate. Finally it begins to pull out of its dive, but it continues to drop at a rate that is not as steep. That spot at which the rate of drop begins to decrease is the point of inflection. The points on the baseline directly below the points of inflection are one standard deviation from the mean; the point to the right is one σ above the mean; and the point to the left is one σ below the mean. These characteristics of the normal curve are illustrated in Figure 6.4.

Figure 6.4—The Normal Curve and Points of Inflection Showing a Standard Deviation Unit (Distance)

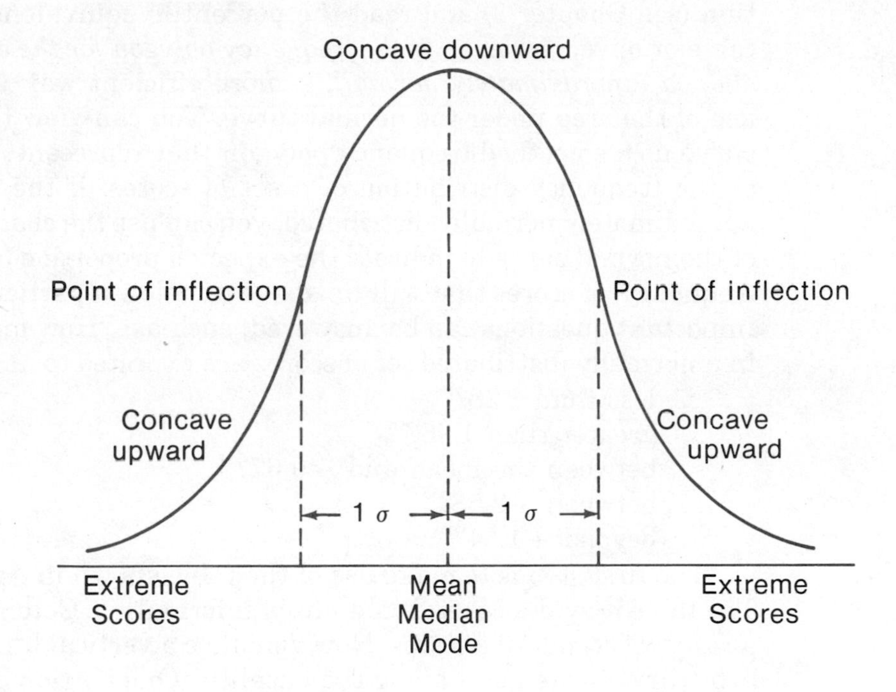

PERCENTILES

Fifty percent of the cases or values are below the median. The same interpretation from the graph of the normal curve in Figure 6.4 shows that 50% of the area under the curve is to the left of the median. Likewise, 50% of the area is above the median (mean and mode). It could be reasoned that the median is the 50th percentile, which means that 50% of the cases are equal to or less than that point on the baseline. A **percentile** equivalent of a particular point along the baseline is defined as the *area under the curve to the left (or below) that point*. Translated to test score terminology, a score reported as the 26th percentile means that 26% of the cases in the group received that score or a lower score. If at the point on the normal curve baseline that represents the 26th percentile a vertical line was drawn up to intersect the curve, then the area marked off to the left would be 26% (or 0.26) of the total area under the normal curve.

How are raw scores changed to percentile equivalents? One way is to construct an ojive or relative cumulative frequency distribution (see Chapter 3) and read the percentile equivalent from the table or ojive. However, *if the frequency polygon for the data is bell shaped (approximately normal)*, a more efficient way is to make use of the area under the normal curve. You can view the normal curve as a smoothed frequency polygon that represents the shape of the frequency distribution of a set of scores. If the scores are approximately normally distributed, you can use the characteristics of the normal curve to indicate the expected proportion (or relative frequency) of scores that will lie above or below a particular point. Important questions can be answered, such as: "How many scores in a normally distributed set of scores are expected to have z-scores

... less than 1.28?"
... greater than 1.96?"
... between the mean and -0.67?"
... between ± 2.58?"
... beyond $+1.54$?"

The first step is to make use of the table shown in Appendix C. The table provides several columns of information. Column 1 shows a range of standard z-scores. Now visualize a vertical line from any arbitrary z-score point along the baseline. One portion of the area under the curve would be between the mean and the vertical line. Further, this vertical line would generally divide the normal curve into two parts. Unless the line was drawn at the mean, it would define a larger portion and a smaller portion of the curve.

Column 2 shows the proportion of area (or relative frequency) between the mean and the corresponding z-score. Notice that negative z-scores are not listed in Appendix C. This is because the normal curve is symmetric, which permits use of the same values of z for proportions above and below the mean. For example, the proportion between the mean and a z-score of 0.79 is 0.2852. Thus the proportion between a z-score of -0.79 and the mean is also 0.2852. Figure 6.5 illustrates the area between the mean and two z-scores.

As you can see in Figure 6.5, column 2 in the table indicates that 0.4032 of the total area is between the mean and $z = 1.30$. The same is true, because of symmetry, for the proportion of area between the mean and $z = -1.30$. Using this information, the proportion to the left (which means below) of a z-score of 1.30 can be determined by adding 0.5000 (the area to the left of the mean) and 0.4032 (the area from the mean to the z-score). The results indicate that a 0.9032 proportion of the area is to the left of $z = 1.30$. Multiplying

Figure 6.5—The Proportion of Area Under the Normal Curve Between the Mean and *z*-Scores of +1.30 and −1.30

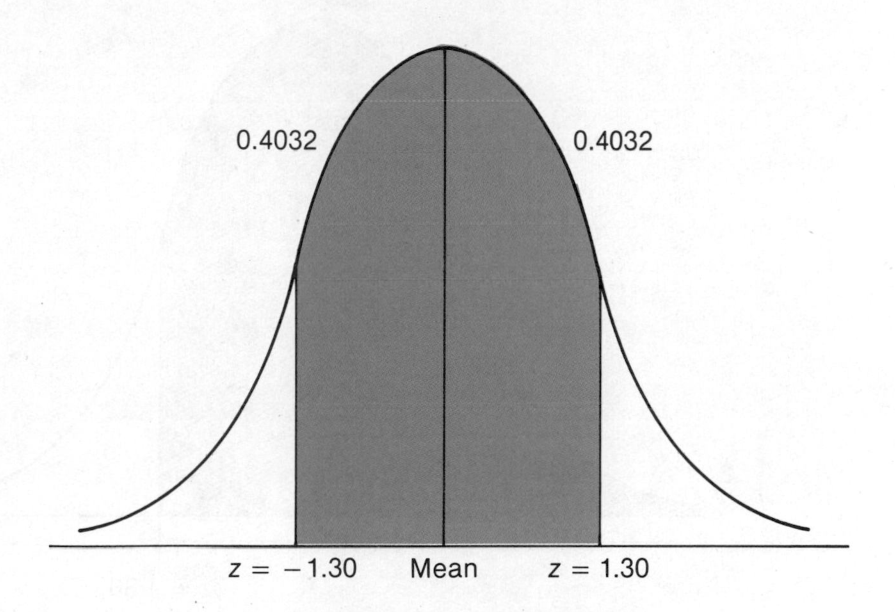

0.9032 by 100 and rounding to a whole number shows that 90% of the area under the normal curve is below a *z*-score of 1.30; that is, a *z*-score of 1.30 corresponds to the 90th percentile.

Again using the information from Figure 6.5 and Appendix C, the percentile equivalent of a *z*-score of −1.30 can be found. In this case you know that the proportion of the area below the mean is 0.5000. You also know that the proportion between a *z*-score of −1.30 and the mean is 0.4032. By subtracting 0.4032 from 0.5000, we find the percentage of area below the *z*-score is 10 (0.0968 times 100 rounded to the nearest whole number). The percentile equivalent of a *z*-score of −1.30 is 10, because 10% of the area of the normal curve is below that point.

Column 3 in Appendix C also may be used to determine percentile equivalents of *z*-scores. If the baseline is connected to the normal curve with a vertical line that is perpendicular to the baseline at a particular point, the normal curve is divided into two parts—a larger portion and a smaller portion as described previously. Column 3 indicates the proportion of area (or the relative frequency) under the curve in the *larger* part. This is illustrated in Figure 6.6.

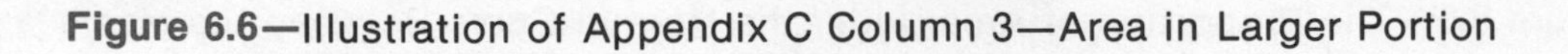

Figure 6.6—Illustration of Appendix C Column 3—Area in Larger Portion

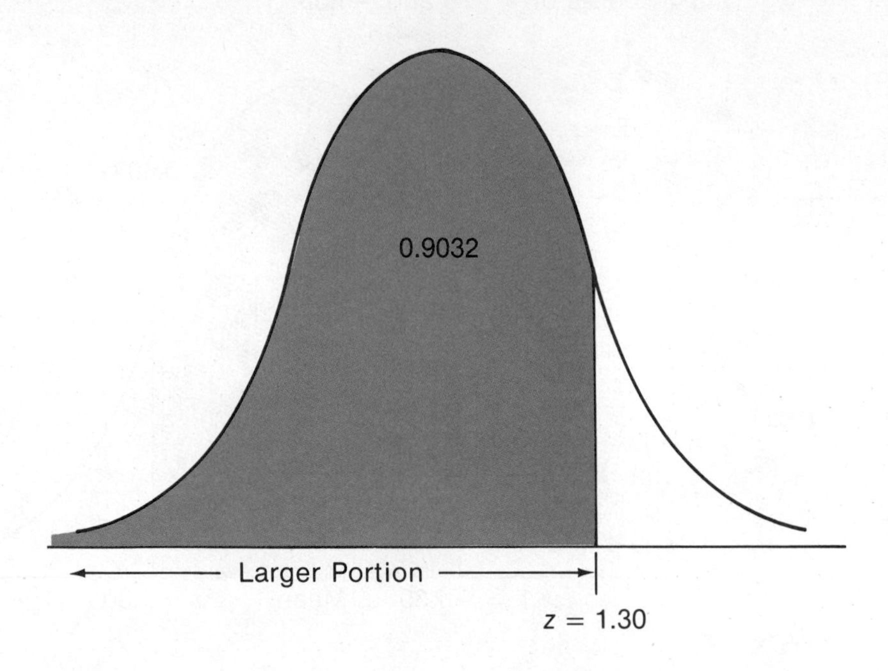

By locating a z-score of 1.30, you can see that the area to the left is the same area as the proportion in the larger section. Thus the percentile equivalent of $z = 1.30$ is 90 (when rounded to the nearest whole percent). This is exactly the same answer as was found when the area between the mean and $z = 1.30$ was added to 0.5000 earlier; however, using column 3 in this case (when the larger portion is to the left of the z-score) simplifies the procedure.

Finally, column 4 gives the proportion of the normal curve in the smaller portion when the curve is divided by a particular z-score. In Figure 6.6, the area to the right of a z-score of 1.30 is 0.0968. Again, column 4 could be used to find the percentile equivalent of $z = 1.30$ by subtracting 0.0968 from 1.0000 (because 1.0000 is the total area proportion under the curve). Column 4 would be the most efficient for finding the percentile equivalent of $z = -1.30$. Why?

As just described, the z-score may be used with any of the columns to determine a percentile. To find which column is the most efficient, sketch a normal curve and roughly locate the z-score as appropriate to the left or right of the mean. Then visually ascertain if the proportion of area in the larger or smaller portion should be used to determine the percentile.

Figure 6.7 provides a graphic summary of how standard scores and percentiles are related to the normal curve. As depicted in Figure 6.7, *z*-scores are distributed with a mean of zero and a standard deviation of one. By observation, the means and standard deviations of the other standard score scales can easily be determined. For example, as stated earlier in this chapter, deviation IQ scores have a mean of 100 and a standard deviation of 15; SAT scores have a mean of 500 and a standard deviation of 100.

Notice the regular and uniform spacing for intervals along the baseline between the standard scores. This does not hold true for the percentile scale, because there are large frequencies toward the middle of the distribution. The rate of increase in cumulative frequency in the "tails" of the curve is only slight.

Figure 6.7—Expressing Common Standard Scores and Percentiles on the Normal Curve

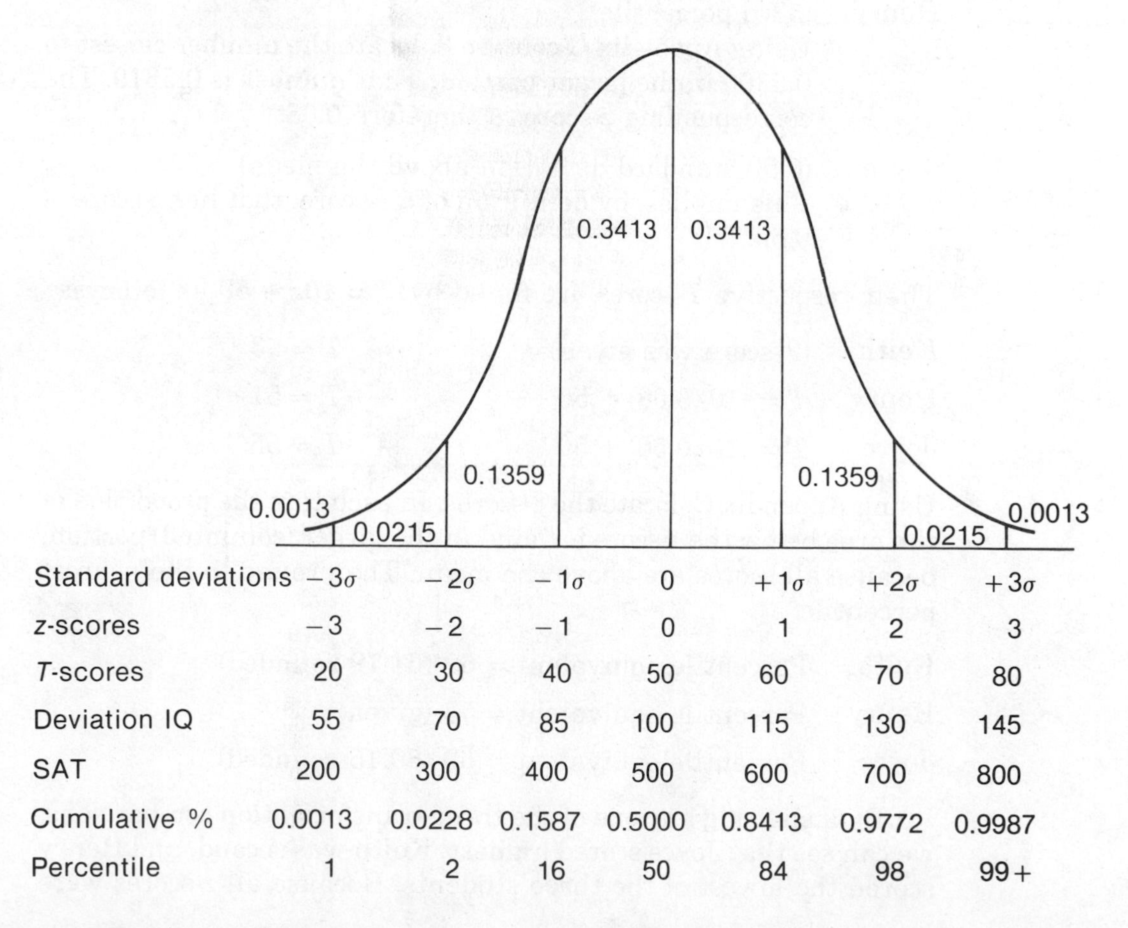

You now have the tools to answer the questions posed at the beginning of the chapter. To refresh your memory, Keith had a scaled T-score of 53, Henry's score was at the 53rd percentile, and Joyce scored 0.50 standard deviation above the mean on their respective exams. For the percentile equivalents to be accurate, the assumption must be made that the scores of the three students were part of a distribution that was bell shaped and was a reasonable approximation of the normal curve. Now, as a review, use this information to see how their respective performances compared.

First, describe their performance on a common scale. You can use z-scores, T-scores, percentiles, or any other common scale, as it doesn't matter which scaling is used. Convert them to three different scales for practice. First, the z-score equivalents for the three students' performances are determined as follows:

Keith (T-score = 53); $z = \dfrac{53 - 50}{10} = \dfrac{3}{10} = 0.30$

Henry (53rd percentile)
Using Appendix C column 3, locate the number closest to 0.5300 in the larger portion. That number is 0.5319. The corresponding z-score is therefore 0.08.

Joyce (0.50 standard deviation above the mean)
This implies, by definition of a z-score, that her z-score is 0.50.

Their respective T-scores are found by $T = 10z + 50$ as follows:

Keith	(T-score was given)	$T = 53$
Henry	$T = (10)(0.08) + 50$	$T = 51$
Joyce	$T = (10)(0.50) + 50$	$T = 55$

Using Appendix C, locate the z-scores. In each case the proportion of the area below the z-score is found in the larger (column 3) portion, because all scores are above the mean. Then round to the nearest percentile.

Keith	Percentile equivalent = 62 (61.79 rounded)
Henry	Percentile equivalent = 53 (given)
Joyce	Percentile equivalent = 69 (69.15 rounded)

By inspecting any one of the three transformation comparisons, we can see that Joyce scored highest, Keith was second, and Henry scored the lowest of the three students. Because all z-scores were positive, all T-scores were above 50, and all percentiles were greater than 50, all three students were above average for their respective reference groups.

The diagram in Figure 6.8 summarizes the various transformations presented in this chapter. Notice the importance of the z-score in the transformation process. In each case the z-score must first be determined either by computation or by the use of Appendix C.

Figure 6.8—Summary of the Relationship Between Raw Scores, Standard Scores, and Percentiles

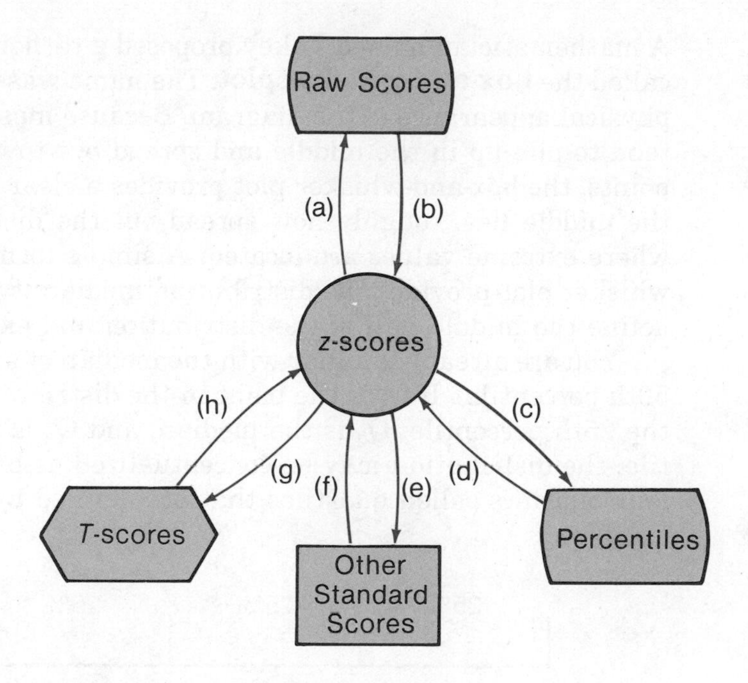

a. Raw score = $\sigma'(z) + \bar{X}$, where σ' and $\bar{X}$ are descriptors of the raw score distribution

b. $z = \dfrac{X - \bar{X}}{\sigma}$

c. Use Appendix C

d. Use Appendix C "in reverse"

e. $S = \sigma' z + \bar{S}$, where S = standard score; $\bar{S}$ and σ' are descriptors of the standard score distribution

f. $z = \dfrac{S - \bar{S}}{\sigma}$, where symbols are defined above

g. $T = 10z + 50$

h. $z = \dfrac{T - 50}{10}$

Note: (g) and (h) are special cases of (e) and (f).

Standard scores have important uses other than for interpreting individual performances. Percentiles are very popular as interpretive devices but are not particularly useful for other statistical manipulations. Both standard scores and percentiles have most of their application in educational and psychological testing, although they may be used to describe any individual performance that can be numerically measured on an interval or ratio scale.

BOX-AND-WHISKER PLOT—OPTIONAL

A mathematician named Tukey proposed a rather simple diagram called the **box-and-whisker plot**. The name was derived from the physical appearance of the diagram. Because most batches of data tend to pile up in the middle and spread out toward the two end points, the box-and-whisker plot provides a clear picture of where the middle lies, roughly how spread out the distribution is, and where extreme values are located. A simple form of the box-and-whisker plot provides the distribution median: two "hinges" that define the middle half of the distribution and extremes.

You are already familiar with the median of a distribution (the 50th percentile). If Q_1 is the point in the distribution representing the 25th percentile, Q_2 is the median, and Q_3 is the 75th percentile, the distribution may be conceptualized as being divided into four quarters called quartiles that are defined by Q_1, Q_2, and Q_3, as follows:

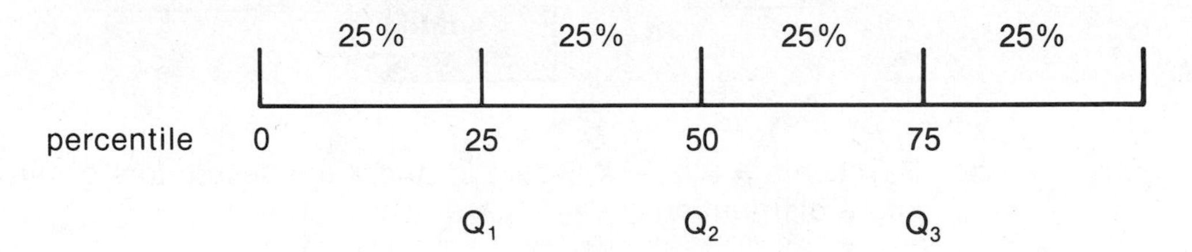

The "hinges" of a box-and-whisker plot are Q_1 and Q_3. Hence one half of the distribution lies between the hinges within the box. "Whiskers" extending from the box show variability from the central portion of the distribution. Finally, many sets of data include number values so small or so large that they stand apart from the rest of the distribution. These values are referred to as **outliers** or *strays*. Extremes represent the lowest and highest number values and may or may not be considered outliers depending on how far they are from the hinges. The example that follows is provided to help clarify the box-and-whisker plot. The data that follow represent the heights (in feet) of 40 buildings in selected U.S. cities.

First National Tower (Akron)	330
Erastus Corning II Tower (Albany)	589
Winston Peachtree Plaza (Atlanta)	723
Austin National Bank (Austin)	328
Merritt Tower (Baltimore)	625
State Capitol (Baton Rouge)	460
First National, Southern National (Birmingham)	390
John Hancock Tower (Boston)	790
Marine Midland Center (Buffalo)	529
NCNB Plaza (Charlotte)	503
Sears Tower (Chicago)	1,454
Carew Tower (Cincinnati)	568
Terminal Tower (Cleveland)	708
Main Centre (Dallas)	939
Kettering Tower (Dayton)	405
Republic Plaza (Denver)	714
Detroit Plaza Hotel (Detroit)	720
Center Tower II (Fort Worth)	646
State Office Tower #2 (Harrisburg)	334
City Place (Hartford)	535
Texas Commerce Tower (Houston)	1,002
American United Life Insurance (Indianapolis)	533
Independent Life & Accident Ins. (Jacksonville)	535
Power & Light Building (Kansas City)	476
Sundance Hotel (Las Vegas)	400
First National Bank (Little Rock)	454
First Interstate Bank (Los Angeles)	858
American General Center (Nashville)	452
World Trade Center (New York)	1,350
Woodmen Tower (Omaha)	469
Valley National Bank (Phoenix)	483
Gateway Arch (St. Louis)	630
California First Bank (San Diego)	388
Transamerica Pyramid (San Francisco)	853
Wachovia Building (Winston-Salem)	410
Vehicle Assembly Building (Cape Canaveral)	552
Bank of Oklahoma Tower (Tulsa)	667
Northwest Center (Minneapolis)	950
Southeast Financial Center (Miami)	764
First Interstate Tower (Portland)	546

Figure 6.9 displays a box plot for the building heights.

Figure 6.9—Box Plot of the Heights of 40 Selected Prominent Buildings in the U.S.

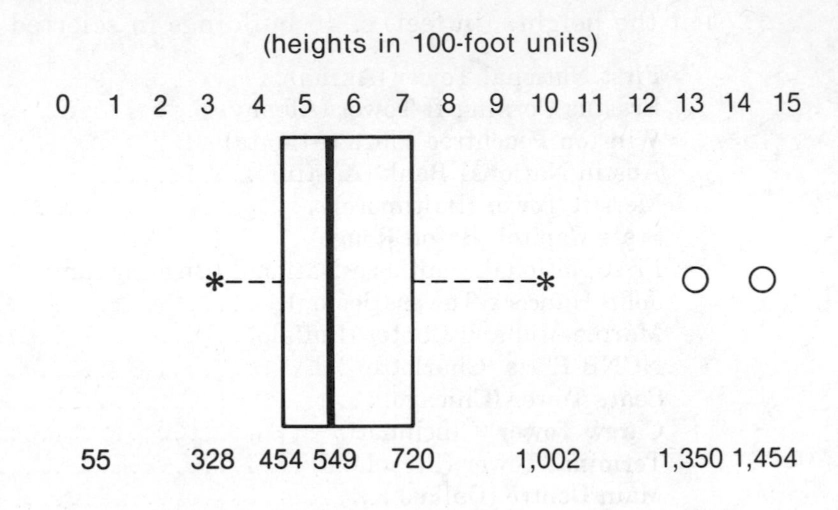

To find Q_1, take 25% of 40 ($0.25 \times 40 = 10$) to find the number of buildings in the first quartile. Alternatively, use the number (10 in this case) and count down from the highest building to find Q_3. The 25th percentile (Q_1) for the building data is 454. The 75th percentile (Q_3) is 720. The width of the box between the hinges ($Q_3 - Q_1 = 720 - 454 = 266$) is called the *interquartile range* (I). The dark vertical line through the box marks the median (549) of the distribution. Extending from each side of the box are whiskers. The dashed line extends to a point called the *adjacent value.* The lower adjacent value is $Q_1 - 1.5(I)$; or in the present case $454 - 1.5(266) = 454 - 399 = 55$. Because the shortest building in this set is 328 feet, the dashed line (whisker) stops at that point. The upper adjacent value is the most extreme value between Q_3 and $Q_3 + 1.5(I)$; or $720 + 1.5(266) = 720 + 399 = 1{,}119$. The tallest building between 720 and 1,119 feet in the data is 1,002 feet tall, which marks the upper adjacent value. Outliers are marked with O's and lie outside the adjacent values. Thus the Sears Tower (1,454 feet) and the World Trade Center (1,350 feet) are marked as outliers in the box-and-whisker plot.

The box-and-whisker plot provides a great deal of information about the central tendency and dispersion of a distribution. Further, hints about the skewness can be detected at a glance. In the example just cited, notice that the median is to the left of the center of the box. All the values lie well above the lowest adjacent value, and some are outliers above the largest adjacent value. Half the buildings in the set are between 454 and 720 feet tall, and the distribution probably has a positive skew.

COMPUTER FOCUS 6—Writing a Program to Transform Raw Scores to Standard Scores

Purpose

The purpose of this activity is to provide practice in following a structured sequence of events and translating the procedures into BASIC code. The Problem, Program Specifications, and Flowchart provide the structure; you are to provide the program.

Problem

A flowchart is provided to illustrate an algorithm for changing raw scores to *z*-scores and *T*-scores. Your task is to write a BASIC computer program that follows the strategy shown in the flowchart.

Program Specifications

Input:
1. Enter a value of the mean, $\overline{X}$.
2. Enter a value of the standard deviation, σ.
3. Enter raw score values, X_i.

Process:
1. Transform X to z by:

$$z = \frac{X - \overline{X}}{\sigma}$$

2. Transform z to T by:

$$T = 10z + 50$$

3. Enter a negative value of X to terminate execution of the program.

Output:
1. raw score, X_i.
2. *z*-score, and
3. *T*-score

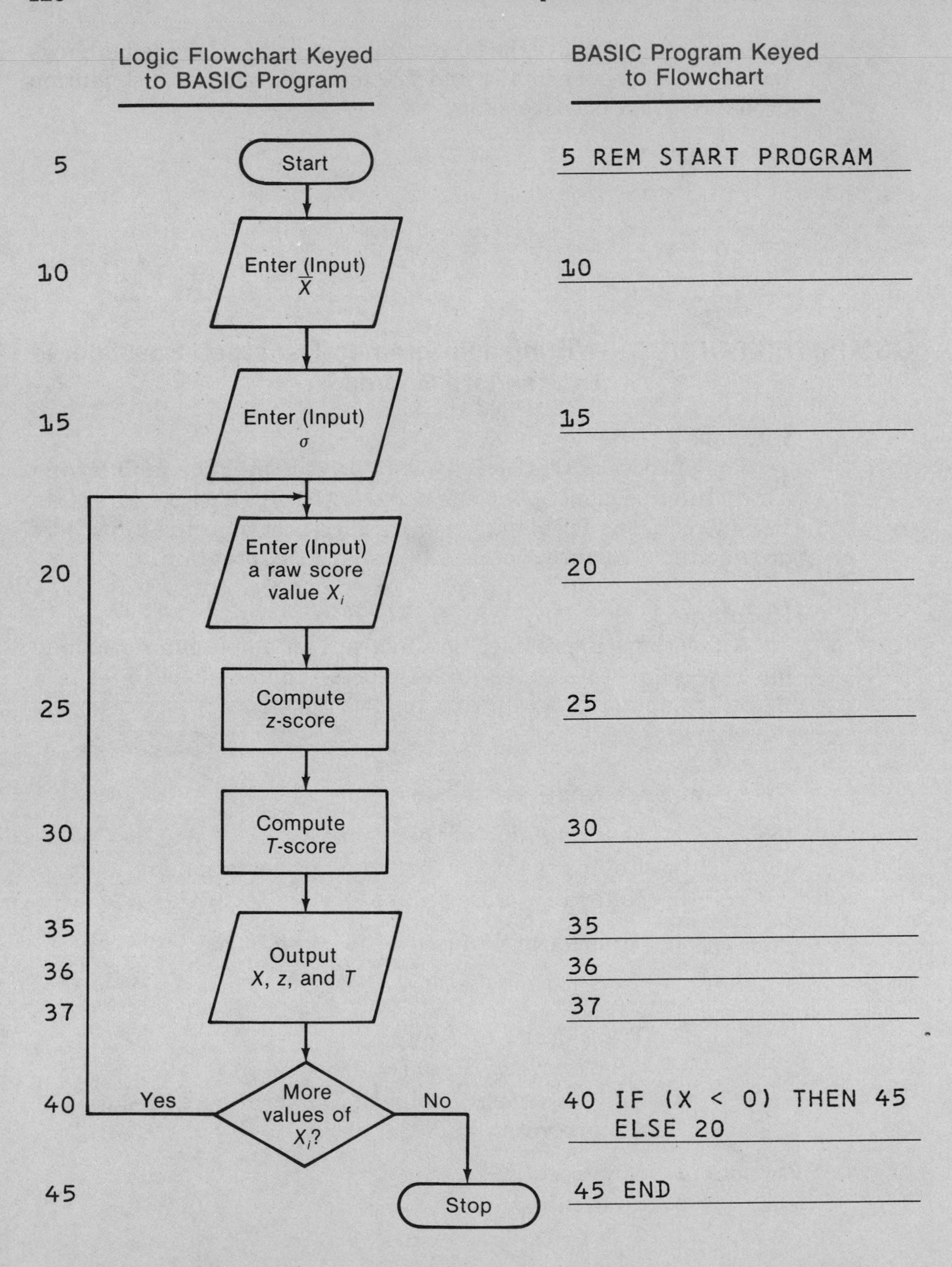
Logic Flowchart Keyed to BASIC Program
BASIC Program Keyed to Flowchart
5
Start
5 REM START PROGRAM
10
Enter (Input) X̄
10
15
Enter (Input) σ
15
20
Enter (Input) a raw score value Xi
20
25
Compute z-score
25
30
Compute T-score
30
35
36
37
Output X, z, and T
35
36
37
40
Yes
More values of Xi?
No
40 IF (X < 0) THEN 45 ELSE 20
45
Stop
45 END

KEY TERMS AND NOTATION

Bell-shaped curve
Box-and-whisker plot (optional)
Normal curve
Outlier (optional)
Percentile
Points of inflection
Raw score
Standard scores
Symmetric
T-score
$\frac{x}{\sigma}$ or $\frac{X - \overline{X}}{\sigma}$
z-score

APPLICATION EXERCISES

6.1 Given a score point located 1.42 standard deviations above the mean in a normal distribution, determine the:
a. proportion of area between the mean and the score
b. proportion of area to the left of (below) the score
c. proportion of area to the right of (above) the score
d. percentile equivalent

6.2 Given a score point located 1.42 standard deviations below the mean in a normal distribution, determine the:
a. proportion of area between the mean and the score
b. proportion of area to the left of the score
c. proportion of area to the right of the score
d. percentile equivalent

6.3 Given the following *z*-scores, find the corresponding percentile equivalents using Appendix C.
a. 1.00
b. 0.00
c. −1.96
d. 0.57
e. −1.58
f. −2.58

6.4 Compute *T*-score equivalents for each of the following *z*-scores:
a. 0.40
b. 1.20
c. −0.80
d. −1.00
e. 2.10
f. −0.89

6.5 Given the following *T*-scores, determine the corresponding *z*-scores.
a. 50
b. 20
c. 65
d. 72
e. 33
f. 16

6.6 Identify the proportional area (relative frequency) in a normal curve that would fall:
a. below a z-score of -1.25
b. above a z-score of 2.38
c. between the mean and $z = 1.45$
d. between $z = -1.00$ and $z = +1.00$
e. above a T-score of 65
f. below a T-score 32
g. between $T = 30$ and $T = 54$

6.7 A normal distribution of 700 scores on a school-wide mathematics placement test has a mean of 48 and a standard deviation of 4. How many scores would you expect to be:
a. above a raw score of 50?
b. below a raw score of 45?
c. between the mean and 49?
d. between 49 and 52?

6.8 Given a percentile of 20 in a normal distribution, find the corresponding z-score.

6.9 What T-score in a normal distribution corresponds to the 93rd percentile?

6.10 What proportion of the scores in a normal distribution lies between z-scores of $+1.96$ and -1.96?

6.11 Suppose a counselor had a z-score on each person known to be in a group of students with normally distributed scores. Further, suppose the counselor wanted to identify the middle 50% of the group. What z-score range would define the middle 50% of the distribution?

6.12 In a normally distributed set of raw scores with a mean of 60 and a standard deviation of 5, a raw score of 72 would convert to a:
a. standard z-score of __________
b. standard T-score of __________
c. percentile of __________

6.13 If a senior student scored 470 on the SAT, what would be the percentile equivalent assuming the SAT is scaled with a mean of 500 and a standard deviation of 100?

6.14 A deviation IQ score of 130 has a percentile equivalent of __________.

6.15 If Eugene had a raw score of 40 on a test with $\mu = 35$ and $\sigma = 10$ while Sally scored 62 on an exam with $\mu = 55$ and $\sigma = 8$, which student had the higher score relative to his or her group? (Remember: μ is the group mean.)

6.16 For each raw score that follows, compute z- and T-scores and use Appendix C to determine the corresponding normal curve percentile equivalents.

X	z	T	Percentiles
3	______	______	______
4	______	______	______
5	______	______	______
11	______	______	______
7	______	______	______
8	______	______	______
9	______	______	______
12	______	______	______
11	______	______	______
10	______	______	______

6.17 In Chapter 3, Pamela, with help from Officer Mahaffey, constructed a frequency distribution of auto speeds on a new boulevard. The mean of her distribution was 52 miles per hour, and the standard deviation was 4.00 mph. Assuming that her sample was representative of velocities for the entire day and that automobile speed is normally distributed, what proportion of the vehicles would be traveling over the 55 mph speed limit?

6.18 Suppose an algebra teacher found the total points earned by her students on homework, classwork, and exams for one grade-reporting period were normally distributed with a class $\mu = 85$ and $\sigma = 10$. Suppose any student scoring above 1.50 standard deviations above the mean received an A for the period.

a. What is the minimum number of total points a student would need to receive an A?

b. If there were 30 students in the class, about how many would receive an A?

6.19 Measure and record the foot sizes of the male class members to the nearest whole centimeter.
a. Compute μ and σ.
b. Convert each size to a standard z-score.
c. Transform each z-score in b. to a T-score.
d. Use Appendix C to convert each size to a percentile equivalent.

6.20 Assume a normal distribution has a standard deviation of 13 points. If a raw score of 35 corresponds to the 74th percentile, find the mean of the distribution.

6.21 *Optional* Following are 50 scores on a business aptitude test of students in Clearview High School who had expressed an interest in enrolling in a business curriculum. Construct a box-and-whisker plot.

90	85	120	110	95
96	92	92	85	90
110	100	105	110	90
93	85	115	112	125
99	86	100	115	95
100	85	105	110	135
60	90	100	102	98
112	143	88	90	100
89	95	75	95	105
96	109	100	92	99

6.22 *Optional* Construct a box-and-whisker plot from the following data:

12	14	15	15	18	18
18	20	20	23	24	25
25	27	27	27	28	30
35	35	35	35	35	36
38	40	40	44	45	48
49	52	54	59	60	60

PART TWO
Probability

7

Elementary Probability

CHAPTER OBJECTIVES

Upon completion of the chapter, students will be able to:

1. Relate a statement of probability to a relative frequency distribution.
2. Use elementary rules to determine the probability of the occurrence of an event.
3. Define the numerical limits (range) of probability statements.
4. Recognize mutually exclusive events.
5. Recognize independent events.
6. Use ratios of geometric measures to determine the probabilty of success on an experimental trial.
7. Illustrate the addition rule and the multiplication rule of probability with Venn diagrams.
8. Compute conditional probability when provided appropriate data.
9. Translate probability of occurrence to odds.
10. Construct a tree diagram to represent probability outcomes.
11. Use Bayes' theorem to compute conditional probability.
12. Distinguish between combinations and permutations.
13. Calculate the number of combinations and permutations of r things taken N at a time.
14. (Optional) Calculate the probability that the value of a random variable will deviate from the distribution mean by some specified amount using Tchebycheff's Inequality.

The Coronado High School science club has scheduled a model rocket launch for Saturday morning. Denise Jones, club president, is considering inviting the local press to cover the launch and to write a story about the science curriculum at Coronado High. She would like to show them a successful rocket flight so that the science curriculum will get a favorable write-up.

Past experience has indicated that there will be a successful flight on about 92% of the launches. But a launch can take place only if the wind speed is less than 15 miles per hour and it is not raining. The weather report for Saturday calls for calm winds and a 40% chance of rain. What are the chances that the rocket will have a successful flight? What determines whether or not there is a launch? What are the chances the weather will cooperate? If there is a launch, can Denise be sure the flight will be successful?

Questions dealing with uncertainties are posed daily and can be addressed using concepts of modern *probability theory*. Many fields of science, notably genetics, use the concept of probability. Even physics, once believed to be almost an entirely deterministic scientific discipline, now relies on probability distributions to describe basic particles of matter.

INTRODUCTION TO PROBABILITY

The inferential branch of statistics includes the study of randomly occurring events and outcomes of experiments, or, more formally, the science of **probability**. Probability is the basis for statistical decisions, and there are rules governing chance or random occurrences.

The probability that a particular event will occur is expressed as a fraction between zero (0) and one (1). It is usually in the form of a decimal or percentage. A 40% chance of rain threatened the launch of the model rocket. In statistical terminology, an equivalent statement would be: "The probability of rain is 40% (or 0.4 or $\frac{4}{10}$) for Saturday." A probability of zero (0) means that the event under consideration will not or cannot occur, whereas a probability of one (1) means that the event is certain to occur. For example, the probability that the Sahara desert will flood tomorrow is close to zero, whereas the probability that the Los Angeles freeways will be crowded on Wednesday at 9:00 a.m. is close to one.

The probabilities between 0 and 1 express the expected proportion or percentage of the time an event will occur. The set or collection of all possible outcomes of a probability experiment is called a **sample space**. The sample space for the science club experiment consists of three possible outcomes: (1) no launch; (2) launch and successful flight; and (3) launch and unsuccessful flight. A particular outcome, which is an element in the sample space, is called a **simple event**. The three simple events just listed collectively form the sample space.

In some cases the probability of occurrence of an event can be determined from the information known about the phenomenon. For example, the flip of a coin can result in two possible outcomes: one is a head and the other is a tail. The roll of a single die can likewise result in six possible outcomes with the face showing 1, 2, 3, 4, 5, or 6. Similarly, the act of drawing a playing card from a regular deck of 52 cards has 52 possible outcomes.

Selecting a particular card at random occurs within a set of circumstances that permits the computation of certain probabilities based simply on the physical aspects of the event. The sample space is known from the physical characteristics of the experiment, and the probability of the occurrence of any particular event can be determined from these physical characteristics. When the probability can be determined from known characteristics of the sample space before an experiment is performed, it is referred to as ***a priori* probability**.

However, the problem requires a different approach when we are considering the probability that a thumbtack will land point up when tossed on a table. In this case an experiment must be conducted to estimate the probability of the tack landing with the point up. You could toss out 500 tacks and count the number with points turned up. Only after such an experiment would there be sufficient information to determine the probability of the tack coming to rest with the point up. This technique determines the **relative frequency** of the tacks with points turned upward and is called **empirical probability**. It was only after a number of *previous flight attempts* that the science club was able to determine that 92% of the flights after launch are successful. That is, the empirical probability of having a successful flight, given that a launch has taken place, is 0.92 or 92%.

Another common example of probability based on relative frequency is the chance of rain reported for Saturday morning. Recall that the chance of rain was 40%, which means that over many observations under similar weather conditions, rain has been the outcome in 40% of the cases. *The probability that an event will occur is equal to the relative frequency of occurrence of that event for a large number of trials under the same circumstances.* If the event does in fact occur, that outcome is called a **success**. There are seven basic rules of probability.

BASIC RULES OF PROBABILITY

As an illustration of the seven fundamental rules of probability, consider this example. Suppose the student council in your high school wanted one student appointed at random to serve as an at-large representative. The president of the student body selects one name at random from the school roster. Suppose there are 150 sophomores, 120 juniors, and 180 seniors enrolled in the school (a total enrollment of $N = 450$).

Rule 1. The probability that an event will occur when the event is one of a set of equally likely events is one divided by the total number of events in the set. The equation for the probability is: $p(S) = \frac{1}{N}$, which is read, "The probability of a success, or of the occurrence of a particular event, is equal to the proportion 1 divided by N, the total number of elements in the sample space."

What is the probability that the president will select the name of the student body vice-president in a random selection of a single name? In this case each of the 450 names has the same chance of being selected, which means that each of the possible simple events is *equiprobable*. Because there are 450 names ($N = 450$) and the student body vice-president is the one name under consideration as a "success,"

$$p(\text{vice-president}) = \frac{1}{450} \text{ or about } 0.002.$$

Rule 2. The probability of an event cannot exceed one or be less than zero; $0 \leq p(S) \leq 1$, *where* $p(S)$ *stands for the probability of a "success."* This rule basically defines the range of the value of probabilities. From the student body in the example just cited, the probability of randomly drawing an eighth-grade student from the set is zero because there are no eighth graders in the sample space. On the other hand, the probability of drawing a high school student is one, a sure bet because all elements of the sample are high school students.

Rule 3. If there is more than one event that could be considered a success, the probability of occurrence of any one of the successful events is the frequency of the successful events divided by the total number of events. Symbolically, $p(S) = \frac{f}{N}$ where f is the frequency of the "successful" events and N is the total number of possible events. What is the probability that the new appointee to the student council will be a junior? In this case the number of juniors is the number of successes (f). Consequently, the probability that a junior would be randomly selected would be:

$$p(\text{junior}) = \frac{120}{450} \text{ or approximately } 0.27.$$

Rule 4. The sum of the probabilities of every possible event in a set of events is equal to one; that is, $\Sigma p(E) = 1$, *where* E *represents a simple event.* To illustrate, the probability that the new randomly appointed member of the student council is a:

$$\begin{aligned} \text{sophomore is } \tfrac{150}{450} &= 0.33 \\ \text{junior is } \tfrac{120}{450} &= 0.27 \\ \text{senior is } \tfrac{180}{450} &= \underline{0.40} \\ \Sigma p(E) &= 1.00 \end{aligned}$$

Rule 5. The probability that an event will occur is equal to one minus the probability that the event will not occur. The *complement* of an event includes all possibilities in the sample space other than that event. Hence Rule 5 states that the probability of the occurrence of an event is equal to one minus the probability of the complement. If the symbol p stands for the probability of a "success" and q denotes the probability that the event was not a success (the complement), then: $p = 1 - q$. The probability that the randomly selected student is not a junior is equal to one minus the probability that the student is a junior, or symbolically: $p(\text{not a junior}) = 1 - p(\text{junior})$. The prime symbol $'$ is also used to denote the complement of an event. Therefore:

$$\begin{aligned} p(\text{junior}') &= 1 - p(\text{junior}) \\ &= 1 - \tfrac{120}{450} \\ &= \tfrac{330}{450} \\ &= \tfrac{11}{15} \text{ or about } 0.73. \end{aligned}$$

Frequently statisticians combine the probability that an event will occur (p) with the probability that the event will not occur (q) by stating the **odds**. The *odds in favor* of the occurrence of an event are $p{:}q$ (read "p to q"); the *odds against* its happening are $q{:}p$. Thus the odds in favor of a junior being selected are $\frac{120}{450}{:}\frac{330}{450}$, or in a reduced form $\frac{4}{15}{:}\frac{11}{15}$, which is frequently read: "The *odds in favor* of selecting a junior are 4 to 11. The *odds against* selecting a junior would be 11 to 4. This is a shorthand way of saying that of every 15 selections over a large number of experiments, 4 would result in a junior being selected and 11 of the 15 would result in a sophomore or senior (not a junior) being selected.

Rule 6. The so-called **addition rule of probability** *is used in situations in which statisticians must compute the probability that at least one of two events will occur.* If J and K are two events and their respective probabilities are $p(\text{J})$ and $p(\text{K})$, then the probability of either J or K (or both) symbolically stated $p(\text{J or K})$ is equal to $p(\text{J}) + p(\text{K}) - p(\text{J and K})$, where $p(\text{J and K})$ is the probability that *both* J and K will occur. Suppose that J is the event "selection of a member of the junior class" and K is the event "selection of a member of the senior class." What is the probability that the student selected at random to serve on the student council is a junior or a senior?

$$\begin{aligned} p(\text{J or K}) &= p(\text{J}) + p(\text{K}) - p(\text{J and K}) \\ &= p(\text{junior}) + p(\text{senior}) - p(\text{junior and senior}) \\ &= \tfrac{120}{450} + \tfrac{180}{450} - 0 \\ &= \tfrac{300}{450} \\ &= \tfrac{2}{3} \text{ or } 0.67 \end{aligned}$$

Notice that $p(\text{J and K})$ was zero, because the student *cannot be both a junior and a senior*. The events J and K in this case are known as *mutually exclusive* events. **Mutually exclusive** means that no more than one of the events can occur in a single trial. In the case where J and K are mutually exclusive events, the addition rule of probability reduces to $p(\text{J or K}) = p(\text{J}) + p(\text{K})$.

To illustrate that the rule can be expanded to several mutually exclusive events, the probability of drawing an ace from a regular 4-suit, 52-card deck of cards is $\frac{4}{52}$ or $\frac{1}{13}$. Also, the probability of drawing a jack is $\frac{4}{52}$ or $\frac{1}{13}$, and the probability of drawing a king on a single draw is $\frac{1}{13}$. What is the probability of randomly selecting an ace or a king or a jack in a single trial?

$$\begin{aligned} p(\text{ace or jack or king}) &= p(\text{ace}) + p(\text{jack}) + p(\text{king}) \\ &= \tfrac{1}{13} + \tfrac{1}{13} + \tfrac{1}{13} \\ &= \tfrac{3}{13} \text{ or } 0.23 \end{aligned}$$

Again notice that the events are mutually exclusive, because no more than one of the "successful" outcomes can occur on a single draw (trial). The results indicate that over a large number of trials (experiments or draws) one would expect to draw either an ace, a jack, or a king about 23% of the time.

Consider an experiment in which two coins are tossed. Such an experiment illustrates two outcomes that are *not mutually exclusive.*

The sample space showing the possible outcomes is:

1st Coin	2nd Coin
H	H
H	T
T	H
T	T

Thus there are four possible outcomes from the toss of two coins. Let J be defined as the event that the first coin comes up heads (H) and let K be the event that the second coin turns up heads. Then $p(\text{J}) = \frac{1}{2}$ and $p(\text{K}) = \frac{1}{2}$. Because *one of the four* outcomes in the sample space is both J and K (both coins showing heads), $p(\text{J and K}) = \frac{1}{4}$. Thus the probability of *at least one head* resulting from the experiment is found using rule 6 as follows:

$$\begin{aligned} p(\text{J or K}) &= p(\text{J}) + p(\text{K}) - p(\text{J and K}) \\ &= \tfrac{1}{2} + \tfrac{1}{2} - \tfrac{1}{4} \\ &= \tfrac{3}{4} \text{ or } 0.75 \end{aligned}$$

Hence, in such an experiment, at least one head is expected in 75% of the trials. In this case J and K are not mutually exclusive, because both J and K can occur in a single trial; therefore $p(\text{J and K})$ does not reduce to zero as with the previous illustrations of the addition rule.

Rule 7. Whereas the addition rule is useful when you are interested in computing the probability that *at least one* of two or more events will occur, many situations call for the probability that *all* of several events will take place. For example, when two coins are tossed, what is the probability that both coins will turn up heads? What is the probability that the student selected to serve as an at-large representative on the student council will be female and will be a sophomore? Or what is the probability that it will not rain on Saturday and the model rocket flight will be successful? In cases such as these, the *multiplication rule of probability*, not the addition rule, is appropriate for calculating the probability. However to understand the multiplication rule, one first must understand the concepts of joint probability and conditional probability.

Joint Probability

The multiplication rule is used to determine $p(\text{J and K})$, the probability of *both* J and K occurring. The expression $p(\text{J and K})$ is an example of a joint probability. As you may have surmised, a **joint probability** is the probability of the joint occurrence of two or more events.

The **intersection of two sets** of events is a subset of events that are common to both of the original sets. A **joint event** is any event that is an element of the intersection of two or more events. The sample space formed by the intersection of two classes of events is a set of joint events. The sample space for the outcome of the flip of a coin is H (heads) or T (tails). The sample space for the outcome of rolling a single die is {1, 2, 3, 4, 5, 6}. The joint sample space formed when a coin is tossed and a die is rolled is:

Coin Toss						
H	H1	H2	H3	H4	H5	H6
T	T1	T2	T3	T4	T5	T6
	1	2	3	4	5	6

Outcome of the Roll of a Die

As you can see, the joint sample space consists of 12 possible joint outcomes of the two events. Further, because the outcome of the roll of the die is not influenced by the result of the coin toss, these two sets are said to be independent events. In general, two or more events are **independent** if the probability of occurrence of each of them is not influenced by the occurrence of another. In the joint sample space, the probability of any 1 of the 12 joint events occurring is $\frac{1}{12}$. For example, the probability of tossing a head with the coin and a 3 with the die is $p(\text{H and } 3) = p(\text{H3}) = \frac{1}{12}$. Also note that the sum of the probabilities of the top row [that is, $p(\text{H1}) + p(\text{H2}) + \ldots + p(\text{H6})$], which is the intersection of the event "head on the coin toss" with the event "roll of the die," is $\frac{6}{12}$ or $\frac{1}{2}$, which is the same probability as getting a head on the single toss of a coin with or without the roll of a die. This is a consequence of independence.

Conditional Probability

Let H be the event that the coin turns up heads and let D be the event of an even number on the die. The joint events, or the intersection of H and D, denoted $\text{H} \cap \text{D}$, are the events H *and* D. Consider the probability that H (head) will occur after D (an even number) occurs. In statistical language, this situation is referred to as the **conditional probability** of H given D, written $p(\text{H}|\text{D})$.

The formula for conditional probability of two independent events H and D is:

$$\boldsymbol{p(\text{H}|\text{D}) = \frac{p(\text{H} \cap \text{D})}{p(\text{D})}}$$

where | means "given" and $\cap$ means "and."

This equation assumes that $p(\text{H} \cap \text{D})$ and $p(\text{D})$ are known. When these probabilities are unknown, the sample space can be used as an alternate but equivalent way of conceptualizing the conditional probability. Two alternate formulas for determining conditional probability are:

$$p(\text{H}|\text{D}) = \frac{\text{number of elements or events in H} \cap \text{D}}{\text{number of elements or events in D}}$$

and

$$p(\text{H}|\text{D}) = \frac{\text{number of ways H and D can occur}}{\text{number of ways D can occur}}$$

These conditional probability formulas can be used to address the question asked earlier: What is the probability of tossing a head (H) given an even number (event D) has resulted from the roll of the die? The sample space for the possibilities in H ∩ D is smaller than the original 12 possibilities and is therefore called a *reduced sample space.* It consists of the two possible results of a coin flip and the even die rolls (D) and is:

Coin Toss	2	4	6
H	H2	H4	H6
T	T2	T4	T6

Event D

The number of ways H can occur is 3. The total number of possible results of the coin flip and an even number on the die roll is 6. The number of elements in H ∩ D is 3; hence $p(\text{H}|\text{D}) = \frac{3}{6} = 0.50$. From this example, notice that $p(\text{H}) = \frac{1}{2}$ and that $p(\text{H}|\text{D}) = \frac{3}{6} = \frac{1}{2}$. The fact that an even number resulted from the roll of the die did not alter the probability of getting a head (H), illustrating again that H and D are independent events.

As another example, suppose that in a certain high school, 10% of the students fail algebra, 15% of the students fail chemistry, and 5% of the students fail both algebra and chemistry. If a student is selected at random, what is the probability that the student failed chemistry if he or she failed algebra? Let A be the set of students who fail algebra and C be the set of students who failed chemistry. Then $p(\text{A}) = 0.10$ and $p(\text{C}) = 0.15$ and $p(\text{A} \cap \text{C}) = 0.05$. The conditional probability that the student failed chemistry given that she or he failed algebra is:

$$p(\text{C}|\text{A}) = \frac{p(\text{C} \cap \text{A})}{p(\text{A})} = \frac{0.05}{0.10} = \frac{1}{2} \text{ or } 0.5.$$

As a result of multiplying both sides of the equation for conditional probability by $p(\mathrm{A})$, rule 7, the **multiplication rule**, can now be stated:

If A and C are two events, the joint probability p(A *and* C) *that both will occur is equal to the conditional probability of* A *and* C *times the probability of* A. Symbolically, this is denoted:

$$p(\mathrm{A\ and\ C}) = p(\mathrm{C}\,|\,\mathrm{A})p(\mathrm{A})$$

Or, taking this one step further, the probability that several *independent events* will occur in a single trial is the product of the separate probabilities of the events.

Returning to the model rocket flight example, the probability that the weather will cooperate and the flight will be successful is the product of the probabilities of the two events. Assume that because the winds are predicted to be calm for Saturday's launch attempt, the probability of no rain (complement of rain) is 0.60; or $p(\mathrm{rain}') = 0.60$. The probability of a successful flight after launch is 0.92. Consequently, Denise figures the probability of a successful flight for the news media is $(0.60)(0.92) = 0.55$, or there is a 55% chance for a successful flight. Or, stated in a slightly different form, the odds are 55 to 45 in favor of a successful flight.

As a further illustration of the multiplication rule, consider the student enrollment discussed earlier. There were 150 sophomores, 120 juniors, and 180 seniors. Assume that one half of the student body is female and then determine the probability that the random selection from the student roster will be a female sophomore. The solution of the problem is as follows:

$p(\mathrm{female}) = \frac{1}{2}$ (assumed)

$p(\mathrm{sophomore}) = \frac{150}{450}$ (rule 3)

Then by the multiplication rule (rule 7):

$$\begin{aligned} p(\text{female and sophomore}) &= p(\text{female}) \text{ times } p(\text{sophomore}) \\ &= (\tfrac{1}{2})(\tfrac{150}{450}) \\ &= \tfrac{150}{900} \\ &= \tfrac{1}{6} \text{ or } 0.17 \end{aligned}$$

Geometric Probability

Geometric probability is a branch of probability in which procedures for determining the probability of success on a single experimental trial are analyzed using points selected at random in a defined geometric region that represents the sample space of the experiment. Suppose U is the set of points in a geometric region that represents the sample space or Universe for an experiment. A particular event represented by a subregion (a portion of the sample space) is symbolized as u. In general, the probability of a geometric point in U also being in u is

$$p(u) = \frac{\text{measure of } u}{\text{measure of } U}$$

If the measure is one of units of length, then U is one dimensional. If the measure is in square units of area, U is two dimensional. The four examples that follow illustrate the concept of geometric probability. The first two exemplify one-dimensional measures of U (length), and the latter two examples illustrate two-dimensional measures (area).

For the first illustration (see Figure 7.1), suppose a barnyard is fenced in a rectangular shape 140 feet long and 80 feet wide. In response to a loud clap of thunder, a wild mustang breaks out of the barnyard at a random location. Find the probability that the mustang breaks out of one of the smaller sides (one of the ends) of the rectangular barnyard.

Figure 7.1—Geometric Reproduction of Barnyard Probability Problem

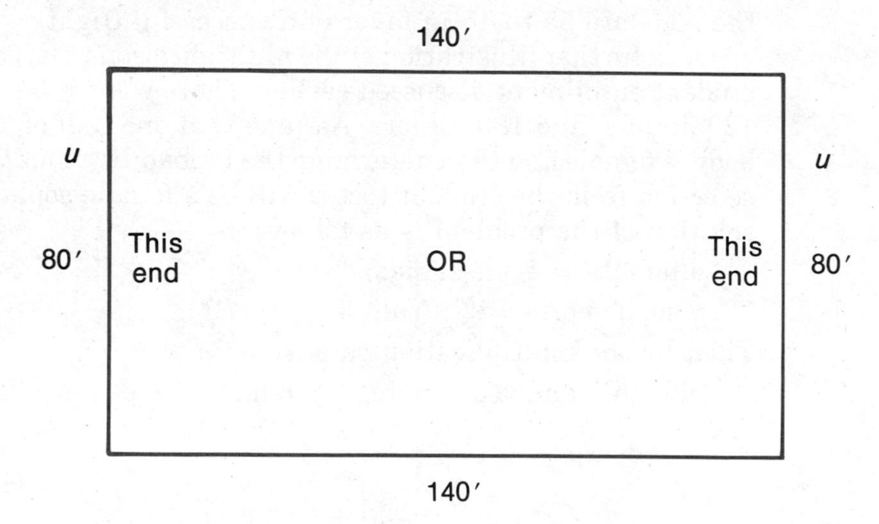

The geometric probability can be determined by:

$$p(\text{ends}) = p(u) = \frac{\text{length of } u}{\text{length of } U}$$

where length of u = combined length of the two ends

$$= 80 + 80$$

$$= 160$$

and length of U = perimeter of the barnyard

$$= 2(80) + 2(140)$$
$$= 160 + 280$$
$$= 440.$$

Hence $p(u) = \frac{160}{440} = 0.36$; there is about a 36% probability that the mustang breaks out of one of the ends of the rectangle.

Similarly, use geometric properties of lengths to find the probability of randomly selecting a point on the subregion u (in the arc) shown on the circle in Figure 7.2.

Figure 7.2—Circle with Circumference *U* and Subregion *u*

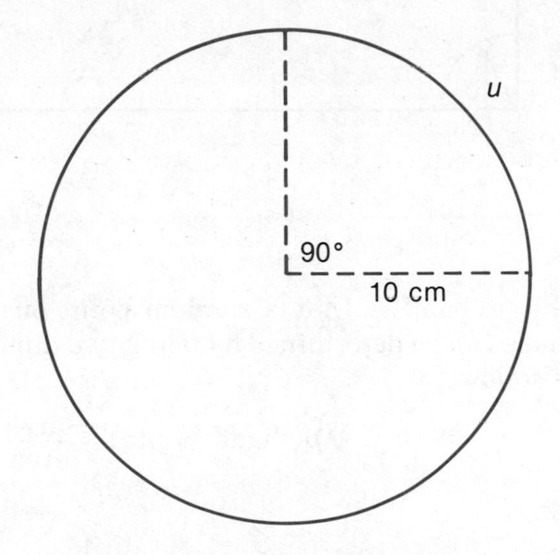

The measurement can be one of length or of angle in this case. Using measures of length,

$$p(u) = \frac{\text{length of } u}{\text{length of } U}$$

where length of u = length of the arc

$$= (\tfrac{1}{4})(2)(\pi)(r)$$
$$= 5\pi$$

and length of U = circumference of the circle

$$= (2)(\pi)(r)$$
$$= 20\pi$$

Therefore, $p(u) = \frac{5\pi}{20\pi} = \frac{5}{20} = \frac{1}{4} = 0.25.$

For the third geometric probability example, assume floodlights shine down on a rectangular floor from the four corners as shown in Figure 7.3.

Figure 7.3—Geometric Representation of Rectangular Floor

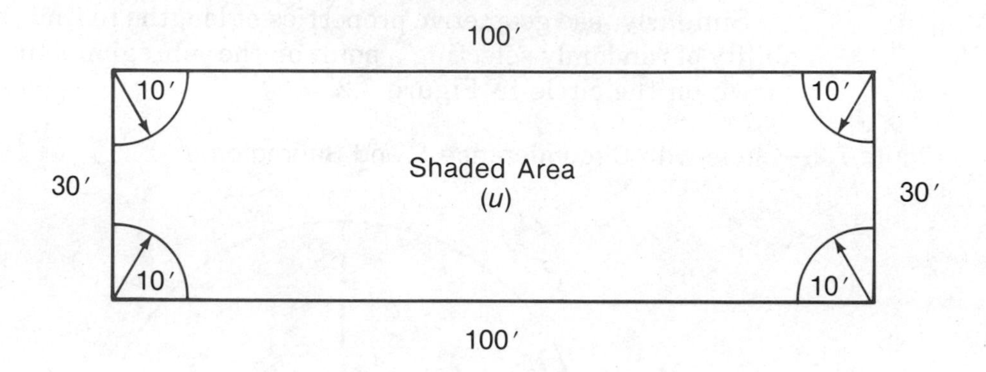

The probability that a random point on the floor will be in the shade can be determined by using two-dimensional measures (area) as follows:

$$p(\text{shade}) = p(u) = \frac{\text{area of } u}{\text{area of } U}$$

where

$$\begin{aligned} \text{area of } U &= \text{length} \times \text{width} \\ &= 100 \times 30 \\ &= 3{,}000 \text{ square units} \end{aligned}$$

and

$$\begin{aligned} \text{area of } u &= [\text{length} \times \text{width}] - [4(\tfrac{1}{4})(\pi)r^2] \\ &= [\text{length} \times \text{width}] - (\pi)r^2 \\ &= 3{,}000 - 100\pi \\ &= 3{,}000 - 314 \\ &= 2{,}686 \text{ square units} \end{aligned}$$

Thus $p(u) = \frac{2{,}686}{3{,}000} = 90\%$ probability that a randomly selected point would be in the shaded area.

Finally, area as a measure of U and u is appropriate for finding the probability that a random point on the target will be located in the center circle (bull's eye) given that the dart hits the target (conditional probability) as shown in Figure 7.4.

Figure 7.4—Target

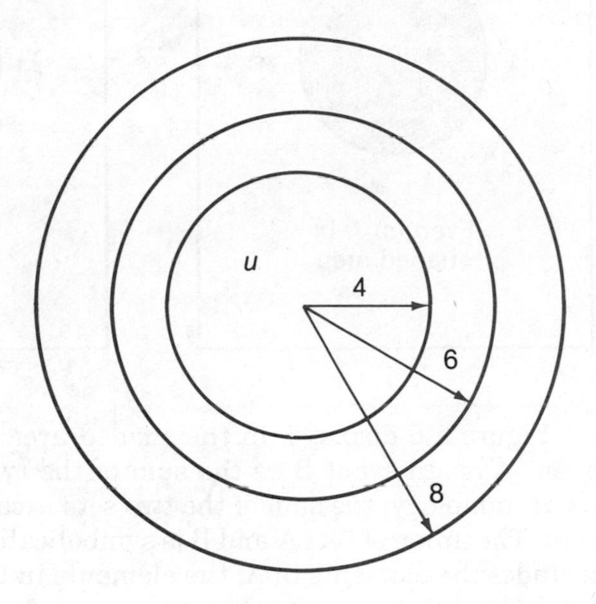

$$p(u) = \frac{\text{area of } u}{\text{area of } U} = \frac{\pi r^2}{\pi r^2} = \frac{\pi 4^2}{\pi 8^2} = \frac{1}{4} = 0.25.$$

The probability that the dart hits the bull's eye when it hits the target is one fourth or 25%. As demonstrated, length and area can be used to represent sample spaces. Other aids such as Venn diagrams and tree diagrams, topics discussed in subsequent sections, are also useful for helping to determine the probability of occurrence of certain events.

Venn Diagrams

Venn diagrams, named after a ninteenth-century English logician, can be used to help clarify set concepts associated with the basic rules of probability. Assume that A and B are either single outcomes or classes of "successful" outcomes. These outcomes are shown in Figure 7.5 in Venn diagram circles that represent the respective outcomes. In set theory terminology, the sets of outcomes A and B may each contain a single element or more than one element.

Figure 7.5—Venn Diagram Circles Illustrating Elements (Outcomes) of Two Sets

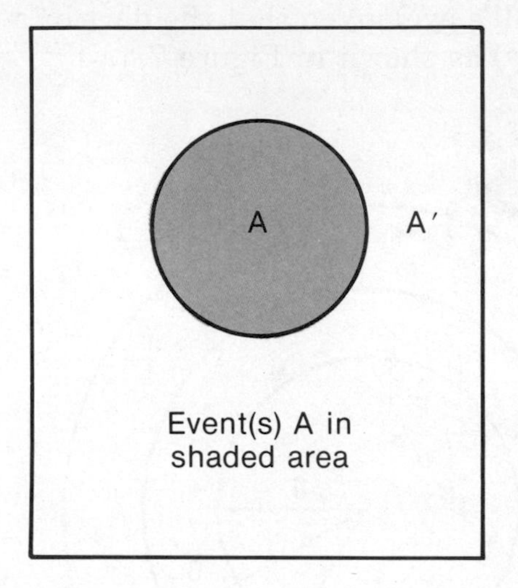

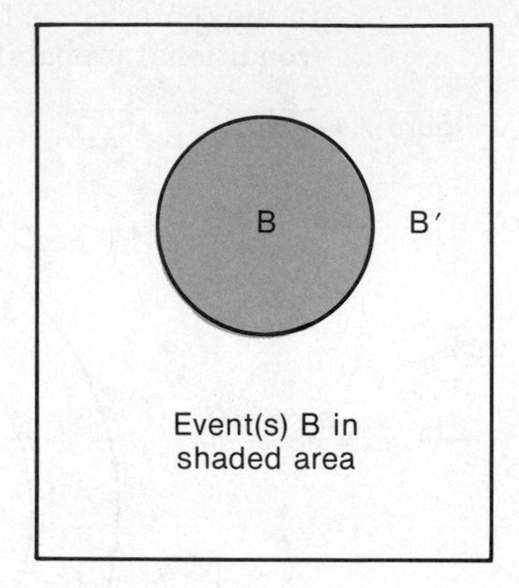

Figure 7.6 displays, in the shaded area, the occurrence of an event A *or* an event B as the sum of the two sets of A and B. In set terminology, the sum of the two sets is called the **union of the sets**. The union of sets A and B is symbolically shown as $A \cup B$ and includes the elements in A, the elements in B, and the elements in both A and B if any exist. In such a case, if event A occurs or event B occurs, the experimental outcome will constitute a "success." Thus rule 6 (addition rule) states that $p(\text{A or B}) = p(A \cup B) = p(A) + p(B)$ if A and B are mutually exclusive. The mutually exclusive aspect is shown as nonoverlapping circles in the Venn diagram (Figure 7.6) because there are no elements (outcomes) that belong to both A and B.

Rule 7, the multiplication rule, can be illustrated as the overlapping portion of sets A and B in Figure 7.7. The shaded area graphically depicts the events that belong to both A and B, such as female *and* sophomore.

In set theory, Figure 7.7 displays the intersection of two sets and is denoted $A \cap B$ or A *and* B. The elements that are in both A and B form the *intersection* of A and B.

Consequently, the union of mutually exclusive sets is associated with the addition rule (rule 6) for finding the probability that at least one successful outcome will result from an experiment. The intersection of two sets is associated with the occurrence of events in both sets and therefore represents the multiplication rule (rule 7).

Figure 7.6—Venn Diagram Illustrating Mutually Exclusive Events A and B and $p(\text{A or B}) = p(\text{A}) + p(\text{B})$

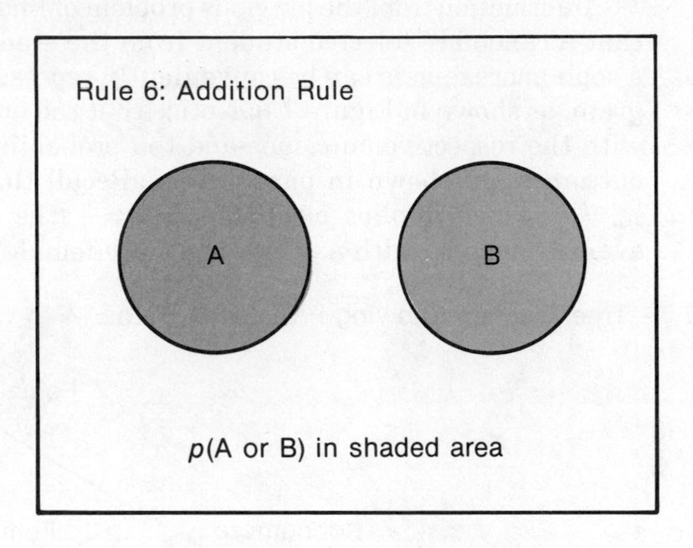

Figure 7.7—Venn Diagram illustrating Events A and B and $p(\text{A and B}) = p(\text{A})p(\text{B})$ when A and B Are Independent Events

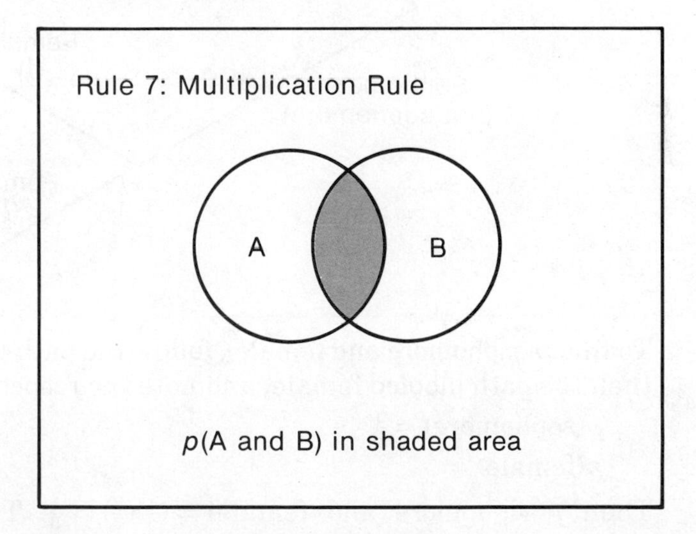

Tree Diagrams

An extremely useful technique for representing and describing probability information on the stages of an experiment is a **tree diagram**. The multiplication rule (rule 7) of the earlier section is used to calculate the probability of the occurrence of an event represented by any given path of the tree.

Information from the previous problem of finding the probability that a randomly selected student from the student body would be a sophomore female can be conveniently represented by a tree diagram, as shown in Figure 7.8. Notice that the branches are labeled with the respective outcomes and the probabilities of the various outcomes are shown in parentheses. Recall that $p(\text{sophomore}) = \frac{150}{450} = \frac{1}{3}$, which implies $p(\text{sophomore}') = \frac{2}{3}$ (the complement of an event is denoted with a prime $'$). And $p(\text{female}) = p(\text{female}') = \frac{1}{2}$.

Figure 7.8—Tree Diagram Showing Probability Paths

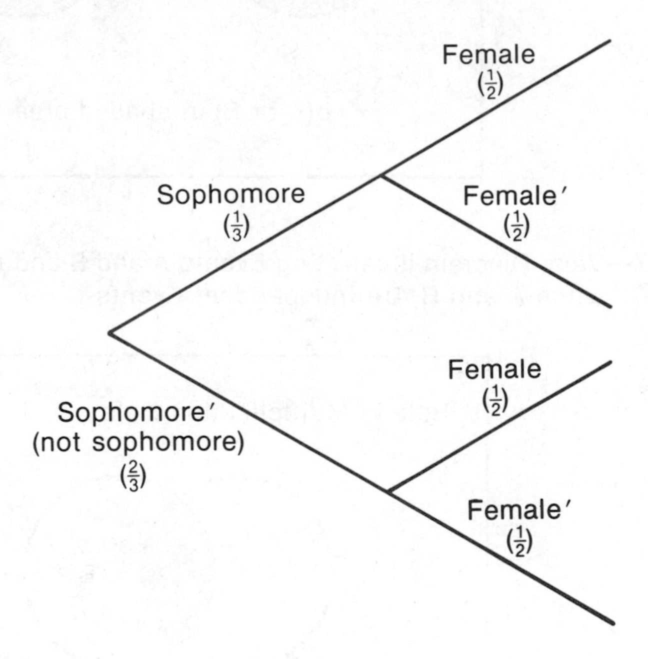

To find $p(\text{sophomore and female})$, follow the path labeled sophomore, then the path labeled female, and note the respective probabilities:

$p(\text{sophomore}) = \frac{1}{3}$

$p(\text{female}) = \frac{1}{2}$

Then $p(\text{sophomore and female}) = (\frac{1}{3})(\frac{1}{2}) = \frac{1}{6}$. This tree diagram represents all four of the possible outcomes:

$$p(\text{sophomore and female}) = (\tfrac{1}{3})(\tfrac{1}{2}) = \tfrac{1}{6}$$
$$p(\text{sophomore and female}') = (\tfrac{1}{3})(\tfrac{1}{2}) = \tfrac{1}{6}$$
$$p(\text{sophomore}' \text{ and female}) = (\tfrac{2}{3})(\tfrac{1}{2}) = \tfrac{2}{6}$$
$$p(\text{sophomore}' \text{ and female}') = (\tfrac{2}{3})(\tfrac{1}{2}) = \tfrac{2}{6}$$

Notice that the four probabilities add to 1.00 ($\frac{1}{6} + \frac{1}{6} + \frac{2}{6} + \frac{2}{6} = \frac{6}{6} =$ 1.00). Can you determine why these sum to 1.00?

Bayes' Theorem

Bayes' theorem was first stated as a proposition in Thomas Bayes' 1764 publication entitled "An Essay Towards Solving a Problem in the Doctrine of Chance." Subsequently refined by later mathematicians, the theorem provided the foundation for an entire branch of modern statistics called Bayesian statistics.

A rather complex formula from **Bayes' theorem** for solving conditional probability problems is:

$$p(A|B) = \frac{p(B|A)p(A)}{p(B|A)p(A) + p(B|A')p(A')}$$

where:

$p(A|B)$ is the probability of event A, given event B;
$p(B|A)$ is the probability of event B, given event A;
$p(A)$ is the probability of event A (unconditionally);
$p(A')$ is the probability of event "not A" (or the probability of the complement of A); and
$p(B|A')$ is the probability of event B, given the event "not A."

As can be seen, Bayes' theorem shows the relation between $p(A|B)$ and $p(B|A)$.

As an illustration of Bayes' theorem, suppose that your student body consists of 47% male students and 53% female students. Thus for a randomly selected student:

$p(M) = 0.47$ (probability of selecting a male) and
$p(M') = 0.53$ (probability of selecting a "not male," that is, the complement of male or female student)

Let C be the event that a student supports a proposed change in the school's constitution. Further, suppose it is known that 40% of the female (M′) and 30% of the male (M) students support the change; that is,

$p(C|M') = 0.40$ and
$p(C|M) = 0.30.$

An editorial that favors the constitutional change will be written anonymously and will appear in the school newspaper. What is the probability that the phantom editor is a male? Using Bayes' formula and the probability values already determined, the probability that the editorial will be written by a male, given that the editorial will be in favor of the change, is solved as follows. (Notice that C's and M's have been substituted for the A's and B's in the original formula.)

$$\begin{aligned} p(\text{M}\,|\,\text{C}) &= \frac{p(\text{C}\,|\,\text{M})p(\text{M})}{p(\text{C}\,|\,\text{M})p(\text{M}) + p(\text{C}\,|\,\text{M}')p(\text{M}')} \\ &= \frac{(0.30)(0.47)}{(0.30)(0.47) + (0.40)(0.53)} \\ &= \frac{0.141}{0.141 + 0.212} \\ &= \frac{0.141}{0.353} \\ &= 0.40 \end{aligned}$$

Therefore, given that the editorial is favorable toward a constitutional change, the probability is about 0.40 that the writer is male. As just illustrated, Bayes' theorem is an alternative procedure for solving conditional probability problems.

Consider another example of how Bayes' theorem can be applied. Suppose that results from a recent Census Bureau study have classified 30% of the families in the Brookside School District as having upper-class incomes and 70% as having lower-class incomes. Further, the study found that 60% of the families with upper-class incomes owned two or more cars, while only 20% of the lower-class income earners owned two or more cars. What is the probability that a family living in the Brookside School District will have an upper-class income, given that the family owns two or more cars?

Define the events as follows: U is the event "upper-class income" and U′ is the event "lower-class income" or not upper-class income. Then

$$p(\text{U}) = 0.30 \text{ and}$$
$$p(\text{U}') = 0.70.$$

If T represents the event "owning two or more cars," then

$$p(\text{T}\,|\,\text{U}) = 0.60 \text{ and}$$
$$p(\text{T}\,|\,\text{U}') = 0.20.$$

By making the appropriate substitutions in Bayes' formula:

$$\begin{aligned} p(\text{U}\,|\,\text{T}) &= \frac{p(\text{T}\,|\,\text{U})p(\text{U})}{p(\text{T}\,|\,\text{U})p(\text{U}) + p(\text{T}\,|\,\text{U}')p(\text{U}')} \\ &= \frac{(0.60)(0.30)}{(0.60)(0.30) + (0.20)(0.70)} \\ &= 0.56. \end{aligned}$$

Thus, given that a family in the Brookside School District has two or more cars, the probability is about 0.56 or 56% that it is an upper-class income family.

Notice that the information used by Bayes' formula can be represented by a tree diagram. Using the problem of the Brookside School District as an example, the diagram is constructed as shown in Figure 7.9.

Figure 7.9—Tree Diagram Showing the Source of Each of Four Possible Outcomes

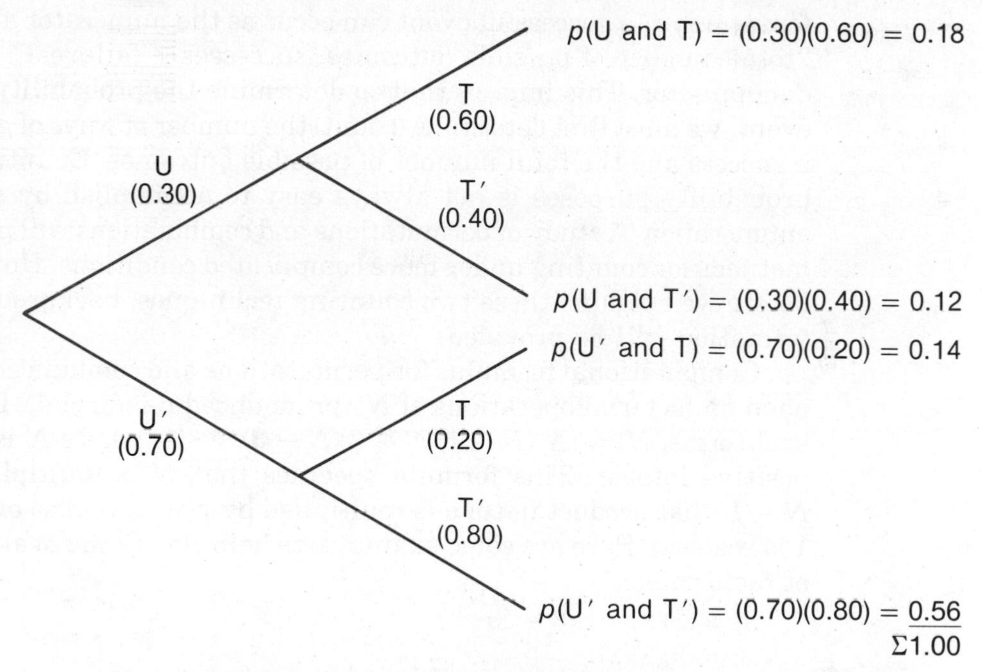

The question posed earlier was: What is the probability that a family living in the Brookside School District will have an upper-class income, given that the family owns two or more cars? Notice how the tree diagram represents the following probabilities:

$$p(\mathrm{U} \cap \mathrm{T}) = (0.30)(0.60) = 0.18 \quad \text{(rule 7)}$$

$$\begin{aligned} p(\mathrm{T}) &= p(\mathrm{U \text{ and } T}) + p(\mathrm{U' \text{ and } T}) \\ &= (0.30)(0.60) + (0.70)(0.20) \\ &= 0.18 + 0.14 \\ &= 0.32 \quad \text{(rule 7 and rule 6)} \end{aligned}$$

But the formula for conditional probability presented earlier implied the relationship:

$$p(\mathrm{U}|\mathrm{T}) = \frac{p(\mathrm{U} \cap \mathrm{T})}{p(\mathrm{T})}$$

which in this case gives:

$$p(\mathrm{U}|\mathrm{T}) = \frac{0.18}{0.32} = 0.56$$

which is precisely the same result given by Bayes' theorem earlier.

METHODS OF COUNTING

As pointed out earlier in this chapter, probability is expressed as the relative frequency of the class of events classified as "successes." An equivalent way of conceptualizing relative frequency as described earlier is in the form of a common fraction with the "number of different ways" a successful event can occur as the numerator and the "total number of possible outcomes (successes + failures)" as the denominator. This implies that to determine the probability of an event, we must first determine (count) the number of ways of getting a success and the total number of possible outcomes. Counting for probability purposes is not always easy to accomplish by simple enumeration. A study of permutations and combinations will provide methods for counting under more complicated conditions. However, before we examine these two counting techniques, background information will be provided.

Computational formulas for permutations and combinations depend on factorial operations of $N!$ (pronounced n-factorial). In general terms, $N! = (N)(N-1)(N-2)(N-3) \cdots (1)$ where N is some positive integer. This formula specifies that N is multiplied by $N-1$, that product in turn is multiplied by $N-2$, and so on until 1 is reached. Here are some examples to help clarify the evaluation of factorials:

$$\begin{aligned} 3! &= (3)(3-1)(3-2) \\ &= (3)(2)(1) \\ &= 6 \end{aligned}$$

$$\begin{aligned} 5! &= (5)(4)(3)(2)(1) \\ &= 120 \end{aligned}$$

$$\begin{aligned} 8! &= (8)(7)(6)(5)(4)(3)(2)(1) \\ &= 40{,}320 \end{aligned}$$

By definition, 0! and 1! are both equal to 1; $0! = 1$; $1! = 1$. The reason for such a definition will become obvious as you work with factorials in the formulas for permutations and combinations.

Permutations

Suppose you have three distinct events: W, X, and Y. *The number of ways that these three events can be ordered in sequence or arranged in some particular order is known as the number of* **permutations**. The listing that follows shows all possible orderings for W, X, and Y, or the sample space for the experiment.

W X Y	X Y W	Y W X
W Y X	X W Y	Y X W

There are six ways to order the three events, or there are six permutations possible for the three events. There are $N!$ permutations or arrangements for N events. In the previous example where $N = 3$, the number of permutations for W, X, and Y can be calculated: $3! = 3 \times 2 \times 1 = 6$, which is precisely the number of arrangements shown in the listing.

More generally, how many permutations of N events will result if we are considering r events at a time? In the case of W, X, and Y, consider the number of permutations of these three events when taken two ($r = 2$) at a time.

The equation for determining the number of permutations of N events taken r at a time is:

$$_N\mathbf{P}_r = \frac{N!}{(N - r)!}$$

Applying this equation to the example:

$$\begin{aligned} {}_3P_2 &= \frac{3!}{(3 - 2)!} \\ &= \frac{(3)(2)(1)}{(3 - 2)!} \\ &= \frac{6}{1} \\ &= 6. \end{aligned}$$

The six permutations in the sample space resulting from this computation can be verified by examining the following listing:

W X	X Y	Y W
W Y	X W	Y X

These six permutations exhaust the possibilities of the three events when the events are considered two at a time.

Assume that the basketball coach was asked to rank order the top three players from the starting five players. How many possible ways could the coach accomplish this task?

$$N = 5$$
$$r = 3$$

Therefore the solution calls for finding the number of permutations of five players considered three at a time.

$$
\begin{aligned}
{}_5P_3 &= \frac{5!}{(5-3)!} \\
&= \frac{(5)(4)(3)(2)(1)}{2!} \\
&= \frac{120}{(2)(1)} \\
&= \frac{120}{2} \\
&= 60.
\end{aligned}
$$

There are 60 possible ways the coach could rank order three players out of a total number of five.

Perhaps you have noticed some computational "short-cuts." Usually both the numerator and denominator contain some of the same factors. These factors can be divided out of the numerator and denominator (in somewhat less than precise mathematical language, some refer to the arithmetic operation of division as "cancelling").

Combinations

If the earlier problem of dealing with five starting basketball players is modified slightly by not requiring a rank ordering of the three best players, the problem becomes one of determining how many ways three players can be selected from five when the order is not important. The order of arrangement of objects is a necessary condition for determining the number of permutations. For example, for two objects A and B, A B is one permutation and B A is another permutation because of different arrangements of the two objects. *The number of ways* r *objects can be selected from a set of* N *objects when the order of arrangement is ignored is called the number of* **combinations** *of the* N *objects taken* r *at a time.*

The equation showing the symbolic notation and the computational procedures for determining the number of combinations of N objects taken r at a time is:

$$
{}_N C_r = \frac{N!}{r!(N-r)!}
$$

This equation can be used to determine how many possible ways the coach could select three players (while not ordering them) from five.

$$
\begin{aligned}
{}_5C_3 &= \frac{5!}{3!(5-3)!} \\
&= \frac{5!}{3!2!} \\
&= \frac{(5)(4)(3)(2)(1)}{(3)(2)(1)(2)(1)} \\
&= \frac{20}{2} \\
&= 10.
\end{aligned}
$$

The coach could select ten combinations of three players from the starting five.

How many combinations of the six letters A, B, C, D, E, F can be formed by selecting two at a time without regard to ordering?

$N = 6$

$r = 2$

Therefore

$$
\begin{aligned}
{}_6C_2 &= \frac{6!}{2!(6-2)!} \\
&= \frac{6!}{2!4!} \\
&= \frac{30}{2} \\
&= 15 \text{ possible combinations of six objects taken two at a time.}
\end{aligned}
$$

In this example, the arrangement C D would not be considered a different combination from D C because order is not a consideration when determining the number of combinations.

The following example illustrates how the counting techniques can be used to solve a probability problem. A subcommittee consisting of three students is to be randomly selected from a larger set of eight students including Bob, Dorothy, and Kelvin. What is the probability that these three students will be selected?

Because order of selection can be ignored, the total number of ways three students can be selected from eight is:

$$
\begin{aligned}
{}_8C_3 &= \frac{8!}{3!(8-3)!} \\
&= \frac{8!}{3!5!} \\
&= \frac{(8)(7)(6)(5)(4)(3)(2)(1)}{(3)(2)(1)(5)(4)(3)(2)(1)} \\
&= (8)(7) \\
&= 56 \text{ possible combinations. (See Figure 7.10.)}
\end{aligned}
$$

Figure 7.10—Calculator Keystrokes

PROBLEM:
Evaluate ${}_8C_3$

Procedure	Arithmetic Operation	Calculator Keystrokes	Display
1. Set up expression for ${}_8C_3$	**1.** $\frac{8!}{3!(8-3)!}$	**1.** (If your calculator does not have a factorial function key, start at step 3.)	
2. Simplify expression	**2.** $\frac{8!}{3!5!}$		
3. Evaluate denominator and store in memory	**3.** (3)(2)(5)(4)(3)(2)	**3.** 3 [×]	3
		2 [×]	6
		5 [×]	30
		4 [×]	120
		3 [×]	360
		2 [=]	720
		[M+]	720
4. Evaluate numerator	**4.** (8)(7)(6)(5)(4)(3)(2)	**4.** 8 [×]	8
		7 [×]	56
		6 [×]	336
		5 [×]	1,680
		4 [×]	6,720
		3 [×]	20,160
		2 [=]	40,320
5. Divide numerator (on display) by denominator (in memory)	**5.** 40,320 ÷ 720 = 56	**5.** [÷]	40,320
		[MRC]	720
		[=]	56

Because Bob, Dorothy, and Kelvin make up only 1 of the 56 possible combinations in the sample space, p(Bob and Dorothy and Kelvin) = $\frac{1}{56}$ or about 0.02 (approximately 2 out of 100 chances).

If, on the other hand, consideration is given to some particular order, the problem just cited changes drastically. Assume the members of the subcommittee of three people were designated by position: chair, vice-chair, and secretary. What is the probability that Bob would be selected as chair, Dorothy as vice-chair, and Kelvin as secretary (note the particular arrangement)?

Because order is important, the total number of outcomes in the sample space is given by:

$$
\begin{aligned}
{}_8P_3 &= \frac{8!}{(8-3)!} \\
&= \frac{8!}{5!} \\
&= \frac{(8)(7)(6)(5)(4)(3)(2)(1)}{(5)(4)(3)(2)(1)} \\
&= (8)(7)(6) \\
&= 336 \text{ possible arrangements according to position}
\end{aligned}
$$

The probability of selecting Bob, Dorothy, and Kelvin as chair, vice-chair, and secretary, respectively is $\frac{1}{336}$ or about 0.003. Notice that requiring the order according to office creates a large number of outcomes for the sample space compared to the number of elements in the sample space when order or sequence of arrangement is not a requirement.

TCHEBYCHEFF'S INEQUALITY—OPTIONAL

Using the elementary rules of probability and the methods of counting described thus far, probabilities of occurrence of specified events can usually be determined. However, to this point the applications have relied on enumerating or counting outcomes from rather straightforward circumstances. The utility of inferential statistical methods lies in their ability to provide probabilities about the occurrence of certain events in numerical distributions in general. Indeed, statistical inference depends on such probability statements. With only the knowledge of the mean and variance of a numerical distribution, some probability statements can be determined. The Russian mathematician Tchebycheff discovered and verified a formula that gives the likelihood of a particular event deviating from the mean by a certain amount. This result is foundational for probability statements about certain occurrences within a numerical distribution.

The following relation, known as **Tchebycheff's Inequality**, is named after the mathematician who was the first to prove this principle.

The formula for Tchebycheff's Inequality is:

$$p(|X - \mu| \geq \epsilon) \leq \frac{\sigma^2}{\epsilon^2}$$

where μ is the mean, σ is the standard deviation of the distribution, X is a value of a random variable, and ϵ is a small positive value.

The inequality means that the probability that any random value of a variable X (called a random variable) will differ absolutely from the distribution mean by ϵ or more units ($\epsilon > 0$) is *always* less than or equal to the ratio of σ^2 to ϵ^2. In other words, any deviation from the mean of a distribution of ϵ or more units can be no more probable than $\frac{\sigma^2}{\epsilon^2}$.

As an illustration of this inequality, suppose a distribution has a mean (μ) of 100 and a variance (σ^2) of 25. The probability of any value (X) deviating *10 or more units from the mean is less than* $\frac{\sigma^2}{\epsilon^2} = \frac{25}{100} = 0.25$. The probability of randomly selecting a value from the same distribution that deviates 15 or more units is at most $\frac{25}{225} = \frac{1}{9}$ or about 0.11.

If each value in the numerical distribution is transformed to a standardized z-score with a distribution mean of zero and a standard deviation of one, Tchebycheff's Inequality is simplified to:

$$p(|z| \geq \epsilon) \leq \frac{1}{\epsilon^2}$$

This inequality implies that the probability of a randomly drawn standardized z-score having an absolute value greater than or equal to some positive number ϵ is always less than $\frac{1}{\epsilon^2}$. Thus the probability of randomly drawing a standardized z-score of 2 or more from a distribution is less than or equal to $\frac{1}{4} = 0.25$. The probability of drawing a z-score of 3 or more standardized units from the mean is $\frac{1}{9}$. The probability that $|z| \geq 5$ is at most $\frac{1}{25} = 0.04$, and so on.

Tchebycheff's Inequality is used in the derivation of other probability theorems that are beyond the scope of this text. Therefore it is considered an important theorem in inferential statistics. For the present purposes, Tchebycheff's Inequality suggests several important concepts. The inequality demonstrates the key roles of the mean and variance in probability. However, Tchebycheff's Inequality, although quite important theoretically, is not an extremely powerful method of estimating probabilities in applied situations. As more is known about the characteristics of a distribution (in addition to the mean and variance), stronger probability statements can be made about departures from the mean. Throughout the remainder of the text, you will be dealing with distributions of random variables that are assumed to conform to certain characteristics (called "assumptions"). Therefore in applied and practical situations you will usually be able to use other techniques to make "tighter" probability statements than Tchebycheff's Inequality would permit.

What is the point of this discussion? Statisticians are concerned with the probability that some phenomena will occur. They deal with questions such as: "What is the probability that what I observe in my descriptive statistics is a departure from what I would expect to happen just by random chance?" If such a probability is "small" (or, as a popular saying has it, "The chances are slim and none"), then the statistician concludes that the departure is probably the result of something other than chance or random occurrence. Depending on the experimental context, the statistician is often able to scientifically infer certain alternative reasons for the deviation between what was observed and what would be expected on the basis of random chance. Such a decision may confirm an existing theory that predicts how the event should have occurred, or it may lead to a discovery of factors that relate the phenomena under consideration. In brief, probability provides techniques to assist the statistical worker in dealing with uncertainty. When compared to probability statements derived from Tchebycheff's Inequality, more refined and accurate means of dealing with probability will be discussed in the subsequent chapters.

COMPUTER FOCUS 7—Permutations

Purpose

Up to this point the Computer Focus sections have concentrated on computers, planning, and programming. The purpose of Computer Focus 7 is to use a relatively short program to stimulate your thinking about a statistical topic, permutations. Many who have prior experience in programming will recognize the use of FOR-NEXT loops. So, although the section is not devoid of computer science topics, the activities will require a shift from a development mode to dealing with output of data that stimulate "statistical thinking."

Problem

Key the following BASIC computer program into a computer.

```
5 FOR I = 1 TO 3
10        FOR J = 1 TO 3
15                FOR K = 1 TO 3
20                        PRINT I;J;K
25                NEXT K
30        NEXT J
35 NEXT I
40 END
```

Run the program and respond to the following items.

1. How many times is statement 20 executed?
2. How many lines of print are repeated (same numbers in the same order) before the end of the program run?
3. Write out (or have your printer print) the 27 ordered triples and circle the six permutations of 1, 2, and 3, which are: 1 2 3, 1 3 2, 2 1 3, 2 3 1, 3 1 2, and 3 2 1.
4. Insert these BASIC statements into the program as lines 16, 17, and 18.

   ```
   16 IF J = K THEN 25
   17 IF I = J THEN 30
   18 IF I = K THEN 25
   ```

   Then run the program. What function do these three statements perform?
5. How does the output of the program with lines 16, 17, and 18 compare with the list of permutations?
6. Write a program that will list the 24 permutations of the numbers 1, 2, 3, and 4.

KEY TERMS AND NOTATION

A priori probability
Addition rule of probability
Bayes' theorem
Combination
Conditional probability
Empirical probability
Geometric probability
Independent events
Intersection of two sets
Joint events
Joint probability
Multiplication rule of probability
Mutually exclusive
Odds
Permutation
Probability
Relative frequency
Sample space
"Success"
Tchebycheff's Inequality (optional)
Tree diagram
Union of two sets
Venn diagram

APPLICATION EXERCISES

7.1 If a single playing card is randomly selected from a standard deck of 52 cards, what is the probability that the card will be the queen of hearts?

7.2 Using the experiment in Application Exercise 7.1, what is the probability of selecting a queen (any of the four suits)?

7.3 One student is randomly selected from a roster of 1,200 students to serve as "Principal for a Day." What is the probability that the selection will be the president of the school's computer club?

7.4 With reference to Application Exercise 7.3, what is the probability that the student selected will be one of the tri-captains of the football team?

7.5 In a group of 25 students interested in speech and drama, 8 specialized in debate, 12 specialized in dramatic acting, and the remaining 5 were emphasizing public speaking. If one student is randomly selected to participate in the local talent show, what is the probability that he or she will be a debate specialist?

7.6 Again with reference to Application Exercise 7.5:
a. What is the probability that the student selected will *not* be an acting specialist?
b. What are the odds against the student being an acting specialist?

7.7 From the 25 students described in Application Exercise 7.5, what is the probability that the random selection will be either a public speaking or an acting specialist?

7.8 Mrs. Sanchez had five high-ability, seven average-ability, and three low-ability students in her special gymnastics class. One student is randomly selected to perform at the next student body assembly. What are the chances that the student selected from Mrs. Sanchez's class will be either a high-ability or an average-ability performer?

7.9 The probability that Don left his industrial arts text in Judy's locker is $\frac{1}{4}$. The probability that Judy did not lock her locker door is $\frac{1}{2}$. If Don does not have a key and Judy is in class, what is the probability that he will be able to get his book from the locker?

7.10 The astronomy club consists of 20 male and 30 female students. The business club is made up of 25 male and 20 female students. Assume that there are no students who belong to both clubs. Further, suppose the two clubs have a jointly sponsored party with all members in attendance. A door prize is randomly awarded to one student at the party. What is the probability that the door prize will go to a female student?

7.11 Refer to Application Exercise 7.10.
a. What is the probability that the door prize will be given to a member of the astronomy club?
b. What are the odds in favor of a member in the astronomy club being selected?

7.12 Once again refer to Application Exercise 7.10 and determine the probability that the door prize winner will be a male member of the business club.

7.13 Of 40 students enrolled in an auto mechanics class, 8 own their own cars. Twenty-one of the students are classified as seniors.

If one auto mechanics student is randomly selected to represent the group on a school executive council, what is the probability that the student will be a senior who owns a car?

7.14 Phil, a member of the varsity golf team, has 10 clubs in his golf bag.

a. For a particular shot, Phil has a 0.7 probability of hitting the ball onto the green with one particular club. If his choice is wrong, he has a 0.4 chance of landing on the green. Construct a tree diagram representing the probability data in this problem.

b. If Phil picks a club at random from his bag, what is the probability that he will be on the green with his next shot?

c. What is the probability that Phil will miss the green on his next shot?

7.15 In a group of 100 scouts who took the physical exam for summer camp, 20% had type A blood. Six percent had both blond hair and type A blood. Find the probability that one scout selected at random will have blond hair, given that the blood test reveals type A.

7.16 A single die is tossed. Given that an odd number resulted, list the sample space and determine the probability that the result is also a prime number. Recall that a prime number is an integer that has no integral factor other than 1 and itself.

7.17 A pair of fair dice is thrown. What is the probability that the sum shown on the dice is 9 or greater if a 6 appears on the first die? (*Hint:* The reduced sample space is: {(6, 1), (6, 2), (6, 3), (6, 4), (6, 5), (6, 6)}.)

7.18 From a sample group of students who were asked their political party preference, 65 indicated Democrat and 15 indicated Republican. Further, it was found that 30 of those indicating Democrat were female and 10 preferring Republican were female. If one student is drawn from this group at random:

a. What is the probability that the student will be a male or a Democrat?

b. What is the probability that the student will be a male Democrat?

c. What is the probability that the student will be a Democrat or a Republican?

7.19 Stuart and Ellie are students in the archery club. Stuart hits the center of the target 50% of the time whereas Ellie hits it 25% of the time. If Stuart and Ellie each shoot at the target, what is the probability that:
a. Both Stuart and Ellie will hit the center of the target?
b. Both will miss the center of the target?
c. Stuart or Ellie will hit the center of the target?

7.20 If N = the total number of objects and r = the number of objects to be taken at a time, find the number of combinations and permutations for the following values:
a. $N = 5$; $r = 2$
b. $N = 4$; $r = 1$
c. $N = 8$; $r = 4$
d. $N = 10$; $r = 7$

7.21 From 12 students, how many *unique* 4-member committees could be formed?

7.22 Renda has five objects on her desk and concentrates on two of them. Kyle tries to guess on which two objects she is concentrating. Assuming that he cannot read Renda's mind, what is the probability that Kyle will guess the two objects on which Renda is concentrating?

7.23 From 10 competing orchestra students in a particular section, how many ways could the teacher select a first chair and a second chair?

7.24 Although 3519 and 1953 are different permutations, they represent a single ________.

7.25 Why should a "combination" lock such as those used on lockers and bikes technically be called a "permutation" lock?

7.26 In a particular senior class, 6% of the male and 2% of the female students are over 6 feet tall. This class is made up of 55% female students. If a senior selected at random is over 6 feet tall, use Bayes' theorem to find the probability that the student is a female. Construct a tree diagram showing the four outcomes of the probability sample space for this problem.

7.27 Of the high school marching band members, 40% have blond hair and 30% are male. Further, 15% of the band members are male with blond hair. If a member selected at random is blond, what is the probability that the member is a male?

7.28 *Optional* IQ scores are nationally distributed with a mean of 100 and a variance of 225. Using Tchebycheff's Inequality, determine the probability of randomly selecting a person from the population with an IQ of 130 or more.

7.29 *Optional* Most standardized achievement tests provide results on a *T*-score distribution with a mean of 50 and a standard deviation of 10 on national norms. According to Tchebycheff's Inequality, what is the maximum probability of a randomly selected student scoring below 30 on the natural science portion of a standardized test?

8

Probability Distributions

CHAPTER OBJECTIVES

Upon completion of the chapter, students will be able to:

1. Compute binomial probabilities by enumeration of sample spaces and by examination of terms of a binomial expansion.
2. Determine binomial coefficients using a formula or using Pascal's triangle when appropriate.
3. Calculate the mean and standard deviation of a binomial distribution.
4. Apply the characteristics of the normal curve to determine probabilities of classes of the random variable.
5. (Optional) Use the Poisson distribution to compute probabilities of events or classes of events when given the distribution mean.
6. (Optional) Approximate binomial and Poisson probabilities with the normal curve when n is large and p is close to $\frac{1}{2}$.
7. (Optional) Use the Poisson distribution to approximate the binomial distribution when p is small.

Lynn, Barry, and Donna were taking a break from their fifth-period course entitled "Home and Family Living," a senior-level elective. They were discussing their plans as future parents and making predictions about the gender of their first child. "I wonder what the chances are that we'll each have a son as our first child," said Barry. "I doubt we'll see that," responded Donna, "but I guess it could happen." Lynn took charge of the conversation at that point, saying, "Let's list all of the possibilties for the first child for the three of us." They listed all elements of the sample space, as follows:

Lynn's First Child	Barry's First Child	Donna's First Child
son	son	son
son	son	daughter
son	daughter	daughter
son	daughter	son
daughter	daughter	son
daughter	son	son
daughter	son	daughter
daughter	daughter	daughter

Assuming that $p(\text{son}) = p(\text{daughter}) = \frac{1}{2}$, each of the eight possibilities would have an equal chance of occurring. In only one of the eight possibilities would the students all have sons. Similarly, one of the eight possible events would be three daughters. In three of the eight cases their first born would be two sons and one daughter. Similarly, three of the eight possibilities show two daughters and one son. Consequently, from information in the sample space, the students were able to determine that:

$$p(3\text{ sons}) = p(3\text{ daughters}) = \tfrac{1}{8}$$
$$p(2\text{ sons}) = p(2\text{ daughters}) = \tfrac{3}{8}$$
$$p(1\text{ son}) = p(1\text{ daughter}) = \tfrac{3}{8}$$
$$p(0\text{ sons}) = p(0\text{ daughters}) = \tfrac{1}{8}.$$

The students understood the chances of the various patterns of gender of their first born. However, five other students joined Lynn, Barry, and Donna and were also interested in the topic of the sex of their first child. They tried to list the entire sample space but could not complete the task before the class break had ended. In fact, they doubted whether they could list all of the possibilities (there are 256 of them, as you will find out) even if they were given an entire hour. The problem, which is everyday fare for statisticians, can be solved using a particular probability distribution.

To make inferences or generalizations about a population from sample statistics, it is often necessary to understand the random behavior of the statistics. Associated with such random occurrences are *probability* or *relative frequency* distributions. In the short history of statistical analysis, experience with data analysis shows that a few distributions are very frequently encountered in practical situations. Three of the most common probability distributions are the (1) binomial, (2) normal, and (3) Poisson distributions. Because characteristics of these distributions are known, precise probability statements may be determined about certain occurrences or classes of occurrences.

BINOMIAL DISTRIBUTION

A large number of experiments have these three properties in common:

1. a sequence of independent trials is conducted;
2. for each trial only two outcomes are possible; and
3. a particular outcome has the same probability of occurrence for each trial.

These characteristics imply that: (1) any trial event does not influence the outcome of any other trial; (2) the population is divided into two groups (for example, male-female, right-wrong, Democrat-Republican, heads-tails, etc.); and (3) either the population is very large or samples are drawn and replaced in the population before the next sample is chosen. Virtually any experiment with these three properties is referred to as a **binomial experiment** or **Bernoulli* experiment**. The **binomial distribution** is the probability distribution that describes the possible outcomes of a binomial experiment. In a binomial experiment, the two potential outcomes are generally called successes (s) and failures (f). Although it does not matter which outcome is referred to as "success" and which is called "failure," be consistent within a given problem. The symbol p denotes the probability of success on a given trial, q is the probability of failure, and n refers to the number of trials in a particular experiment.

**The well-known Bach family produced no fewer than two dozen eminent musicians in eight generations. Though less well known, what the Bachs were to music, the Bernoulli family was to science. Within the course of a century, eight of the Bernoullis attained prominence in various branches of mathematics and science. Jakob (also known as James and Jacques) initially studied theology but became a mathematician during the latter part of the seventeenth century. His book on the theory of probabilities and the binomial distribution was published posthumously in 1713.*

In a binomial experiment, s (the number of successes) is a random variable, because the outcome of any particular trial is random and s is determined by the individual outcomes. Hence the probability distribution of s provides the probabilities of s for various values. Another name for the probability distribution of s is the *binomial distribution*. The parameters that define a binomial distribution are n and p. The following example illustrates the binomial distribution.

A fair (unbiased) coin is tossed three times. This constitutes a binomial experiment. Notice that such an experiment conforms to the criteria for a binomial experiment listed earlier. First, each flip of the coin is an independent trial, because the outcome of one coin flip does not influence the outcome of another. Second, only two outcomes (heads or tails) are possible for each trial (flip of a coin). Finally, the probability does not change from trial to trial. It always remains 50% or $\frac{1}{2}$. Each toss corresponds to a trial, and the two possible outcomes of a trial are heads and tails. Heads and tails may arbitrarily be designated as "success" and "failure" (not a success), respectively. The number of trials is $n = 3$, and $p = q = \frac{1}{2}$ under the assumption that the coin is unbiased (fair).

The experiment is set up so that a single coin is tossed three times: the first toss is called trial 1, the second toss is labeled trial 2, and the third is referred to as trial 3. Specifically, when a coin comes up heads, the trial is arbitrarily defined as a "success" and denoted p. This experiment can have eight possible results. That is, there are eight unique patterns that the three simple events can yield. The number of successes (number of heads) can range from zero (no heads) to three (all three tosses turn up heads). The sample space of all possible results of tossing the coin three times is as follows:

Trial 1	Trial 2	Trial 3	Possible Outcomes of s
tail (q)	tail (q)	tail (q)	$s = 0$
head (p)	tail (q)	tail (q)	$s = 1$
tail (q)	head (p)	tail (q)	$s = 1$
tail (q)	tail (q)	head (p)	$s = 1$
head (p)	tail (q)	head (p)	$s = 2$
tail (q)	head (p)	head (p)	$s = 2$
head (p)	head (p)	tail (q)	$s = 2$
head (p)	head (p)	head (p)	$s = 3$

If the probability of obtaining exactly two heads (successes) is desired, an examination of the preceding experiment outcomes reveals that $s = 2$ on three of the eight possibilities. Thus $p(s = 2) = \frac{3}{8}$. When n is small as in the present example, the probability of each value of s can easily be determined by inspection. As shown in the list that follows, the probabilities are simply the number of successes (s) divided by the total number of possibilities:

$p(s = 0) = \frac{1}{8}$
$p(s = 1) = \frac{3}{8}$
$p(s = 2) = \frac{3}{8}$
$p(s = 3) = \frac{1}{8}$.

Notice that the distribution of s is the same as determined by the sample space used for determining the probability of 0 sons, 1 son, 2 sons, and 3 sons as first-born children of Lynn, Barry, and Donna. Further notice that in each case the denominator (total number of possible outcomes for an experiment) is given by 2^n (the nth power of 2), or in this case, $2^3 = 8$. The numerators are the number of combinations of n items (three in this instance) taken s at a time. Recall that 0! and 1! equal 1 by definition. Thus:

$${}_3C_0 = \frac{3!}{0!(3-0)!} = \frac{3!}{(1)(3!)} = 1$$

$${}_3C_1 = \frac{3!}{1!(3-1)!} = \frac{3!}{(1)(2!)} = 3$$

$${}_3C_2 = \frac{3!}{2!(3-2)!} = \frac{3!}{(2)(1!)} = 3$$

$${}_3C_3 = \frac{3!}{3!(3-3)!} = \frac{3!}{(6)(0!)} = 1$$

The point illustrated by this example is that the probability of s successes in a binomial experiment with n trials can be calculated.

The formula for binomial probability is:

$$\boldsymbol{p(s) = \frac{{}_nC_s}{2^n}}$$

where ${}_nC_s$ is the number of combinations of n objects taken s at a time.

A concept closely related to the theory of probability is that of a *random variable*, generally called a *variate* for brevity. In the previous example, if $s =$ the number of heads showing when three coins are tossed, then $s = 0, 1, 2,$ or 3. Assuming that the coins are fair (unbiased), so that $p = \frac{1}{2}$ for each coin, the following information is known with regard to the random variable:

s:	0	1	2	3	other value
p:	$\frac{1}{8}$	$\frac{3}{8}$	$\frac{3}{8}$	$\frac{1}{8}$	0 (zero)

As shown in the earlier example, such probabilities for the random variable can be determined directly by listing all elements of the sample space. However, in practice, mathematical functions are generally used to determine the probabilities of occurrence of the various values of the random variable s, or the occurrence of an s-value within a specified class of outcomes. A frequency distribution curve of a random variable s is called the probability density of s and is extremely useful in determining probabilities. One such probability distribution is the binomial distribution.

The binomial distribution is obtained by expanding the binomial $(p + q)^n$. Thus

$$(p + q)^n = p^n + np^{n-1}q + \frac{n(n-1)}{(1)(2)}p^{n-2}q^2 + \frac{n(n-1)(n-2)}{(1)(2)(3)}p^{n-3}q^3 + \ldots + q^n.$$

The terms of the binomial expansion of $(p + q)^n$ may be obtained by actual multiplication of the binomial expressions. Or, without performing the multiplication, the following laws (known as the **binomial theorem**) can be used to expand the binomial. For the expansion of $(p + q)^n$:

1. The *number of terms* in the expansion is $n + 1$. The first term contains only p, and the last term contains only q. The other terms contain pq and are positive.
2. The *exponent of p* in the first term is n, and it decreases by 1 in each succeeding term until it becomes 1 in the next to last term. The *exponent of q* in the second term is 1, and it increases by 1 in each succeeding term until it becomes n in the last term.
3. The *coefficient* of the first term is 1; the coefficient of the second term is n. The coefficient of any term farther on may be computed from the previous term by multiplying the term's coefficient by the exponent of p and dividing by 1 more than the exponent of q.

For the previous example of the three coin tosses, the expansion of $(\frac{1}{2} + \frac{1}{2})^3$ would be:

$$(\tfrac{1}{2})^3 + 3(\tfrac{1}{2})^2(\tfrac{1}{2}) + \frac{(3)(2)}{(1)(2)}(\tfrac{1}{2})(\tfrac{1}{2})^2 + (\tfrac{1}{2})^3$$

$$= (\tfrac{1}{2})^3 + 3(\tfrac{1}{2})^2(\tfrac{1}{2}) + 3(\tfrac{1}{2})(\tfrac{1}{2})^2 + (\tfrac{1}{2})^3$$

(Verify the *terms*, *exponents*, and *coefficients* in this expression with the three statements of the binomial theorem.)

$$= \tfrac{1}{8} + \tfrac{3}{8} + \tfrac{3}{8} + \tfrac{1}{8}.$$

Notice that the terms of the binomial expansion give the probabilities of $s = 0$, $s = 1$, $s = 2$, and $s = 3$, respectively. Thus expansion of the binomial provides a third alternative for solving the probability distribution problem raised by Lynn, Barry, and Donna. Again, if the probability of obtaining a sample with $s = 2$ is desired, this is given by the third term in the binomial expansion. Or it is the term in which p (the probability of success) is raised to the power of 2 (because $s = 2$).

Note that the *coefficients* of p and q in the binomial expansion provide the number of ways a sample with s successes can be obtained, that is, ${}_nC_s$. For any binomial experiment, the probability that X successes will occur is determined by:

$$p(X) = p(s = X) = \frac{n!}{X!(n - X)!} p^X q^{n-X}$$

Consequently, for more complicated experiments, the binomial distribution efficiently defines the probabilities of the values of the random variable s. Figure 8.1 gives an example of evaluating this equation.

Shape of *s*

Suppose a binomial experiment is conducted on a population with $p = \frac{1}{2}$ and $n = 8$. The possible number of successes (s) for the experiment has a range of integers with values 0 through 8. The probabilities of s are shown in Table 8.1.

Table 8.1 Binomial Expansion for $n = 8$

Value of s	Binomial Terms	p(s)
0	$(1)(\frac{1}{2})^0(\frac{1}{2})^8$	$\frac{1}{256}$
1	$(8)(\frac{1}{2})^1(\frac{1}{2})^7$	$\frac{8}{256}$
2	$(28)(\frac{1}{2})^2(\frac{1}{2})^6$	$\frac{28}{256}$
3	$(56)(\frac{1}{2})^3(\frac{1}{2})^5$	$\frac{56}{256}$
4	$(70)(\frac{1}{2})^4(\frac{1}{2})^4$	$\frac{70}{256}$
5	$(56)(\frac{1}{2})^5(\frac{1}{2})^3$	$\frac{56}{256}$
6	$(28)(\frac{1}{2})^6(\frac{1}{2})^2$	$\frac{28}{256}$
7	$(8)(\frac{1}{2})^7(\frac{1}{2})^1$	$\frac{8}{256}$
8	$(1)(\frac{1}{2})^8(\frac{1}{2})^0$	$\frac{1}{256}$

Only when $p = q = \frac{1}{2}$ *will the shape of the binomial be symmetrical about the midpoint as shown in the probability histogram* in Figure 8.2.

Figure 8.1—Calculator Keystrokes

PROBLEM:

Evaluate: $p(s = 2) = \dfrac{n!}{X!(n - X)!} p^X q^{n-X}$

where: $n = 5$, $X = 2$, $p = \frac{1}{4}$, $q = \frac{3}{4}$

or

$p(s = 2) = \dfrac{5!}{X!(n - X)!} (\frac{1}{4})^2 (\frac{3}{4})^3$

Procedure	Arithmetic Operation	Calculator Keystrokes	Display
1. Evaluate denominator	1. $X! = (2)(1)$	1. 2 [×]	2
	$(n - X)! = (3)(2)(1)$	3 [×]	6
	$(4^2)(4^3) = 4^5$	2 [×]	12
	$= (4)(4)(4)(4)(4)$	4 [×]	48
	$\therefore (2)(3)(2)(4)(4)(4)(4)(4)$	4 [×]	192
	$= 12{,}288$	4 [×]	776
		4 [×]	3,072
		4 [=]	12,288
		[M+]	12,288
2. Evaluate numerator	2. $n! = (5)(4)(3)(2)(1)$	2. 5 [×]	5
	$3^3 = (3)(3)(3)$	4 [×]	20
	$\therefore (5)(4)(3)(2)(3)(3)(3)$	3 [×]	60
	$= 3{,}240$	2 [×]	120
		3 [×]	360
		3 [×]	1,080
		3 [=]	3,240
3. Divide numerator by denominator	3. $\dfrac{3{,}240}{12{,}288} = 0.2636718$	3. [÷]	3,240
	$= 0.26$	[MRC]	12,288
		[=]	0.2636718 or 0.26

Figure 8.2—Probability (Relative Frequency) Histogram of the Binomial Distribution

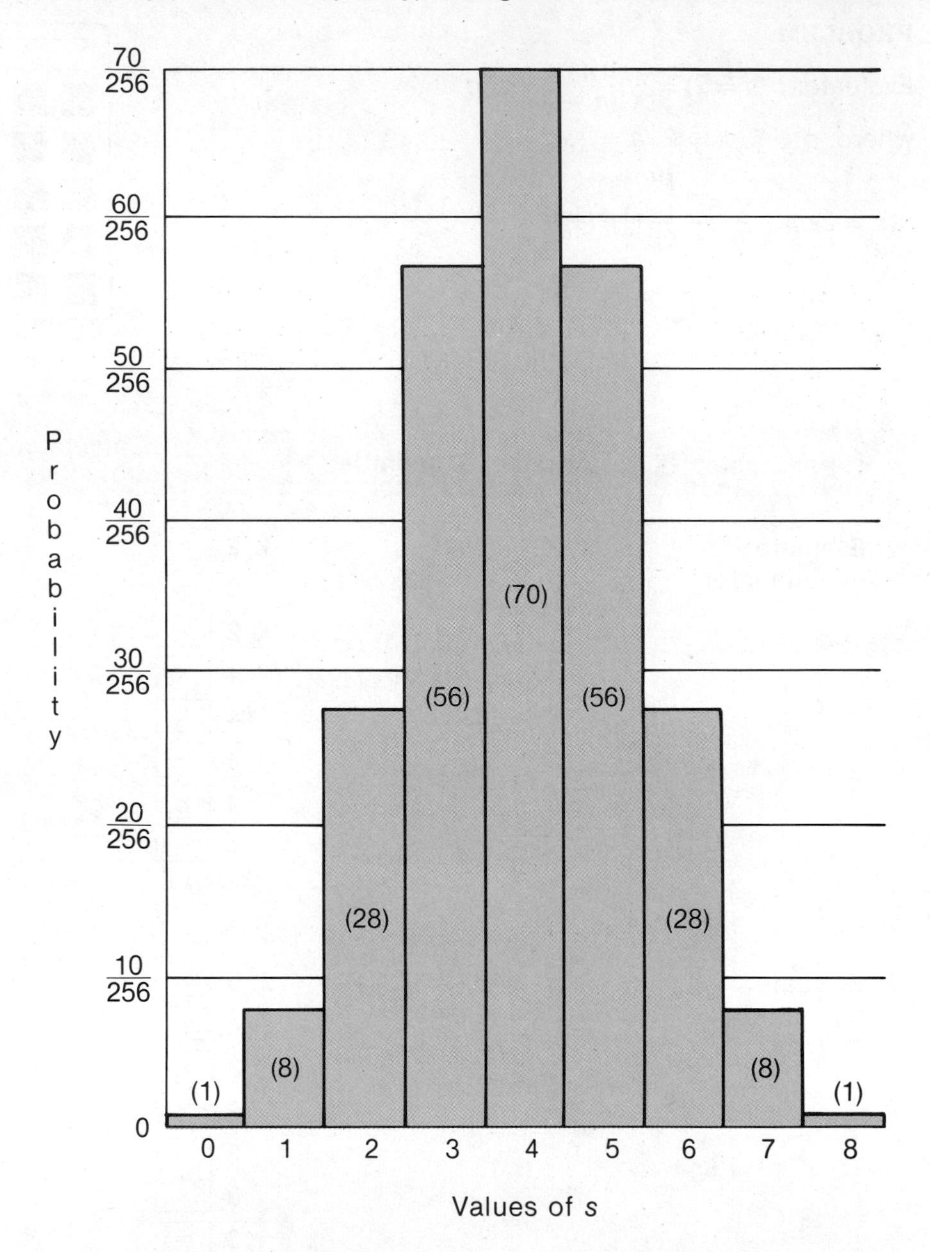

If n is relatively small and p is not equal to q, the binomial distribution of s will be skewed. You may wish to refer to Chapter 4 to review the shapes of skewed distributions. If $p > q$, the distribution will be negatively skewed, whereas if $p < q$, the skew will be positive.

Pascal's Triangle

So far, three methods have been described for computing the probability of s in a binomial experiment: (1) listing the sample space and counting the relative frequency of s in the sample space; (2) dividing the number of combinations of n objects taken s at a time by 2 to the nth power; and (3) using the terms of the binomial expansion of $(p + q)^n$. If $p = q$, **Pascal's triangle** may be used to determine the binomial coefficients. These coefficients are shown for the first 11 rows of Pascal's triangle in Figure 8.3. As can be seen, each row is formed with 1's on each end and each intermediate value is the sum of the two numbers above it in the previous row.

Figure 8.3—Pascal's Triangle of Binomial Coefficients

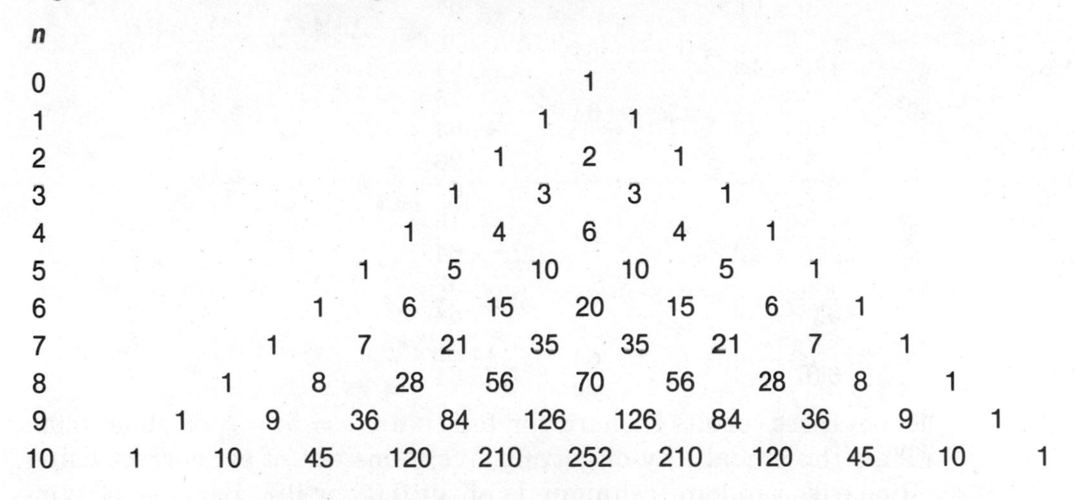

Again, to illustrate for $n = 8$, Pascal's triangle may be used to expand the binomial assuming $p = q = \frac{1}{2}$ as follows:

$$(\tfrac{1}{2} + \tfrac{1}{2})^8 = (\tfrac{1}{2})^8 + 8(\tfrac{1}{2})^7(\tfrac{1}{2}) + 28(\tfrac{1}{2})^6(\tfrac{1}{2})^2 + 56(\tfrac{1}{2})^5(\tfrac{1}{2})^3 + 70(\tfrac{1}{2})^4(\tfrac{1}{2})^4$$
$$+ 56(\tfrac{1}{2})^3(\tfrac{1}{2})^5 + 28(\tfrac{1}{2})^2(\tfrac{1}{2})^6 + 8(\tfrac{1}{2})(\tfrac{1}{2})^7 + (\tfrac{1}{2})^2$$
$$= \tfrac{1}{256} + \tfrac{8}{256} + \tfrac{28}{256} + \tfrac{56}{256} + \tfrac{70}{256} + \tfrac{56}{256} + \tfrac{28}{256} + \tfrac{8}{256} + \tfrac{1}{256}.$$

Notice that the terms are the same as the ones derived in Table 8.1 and shown in Figure 8.2. At the beginning of the chapter the eight students in the "Home and Family Living" class were unable to complete a listing of the sample space for their prospective first-born children with regard to gender. With the information just provided, if $p(s)$ is the probability of s males out of the eight trials (first child), what could you tell the class members about the probability distribution of the males?

Application 1

Suppose Richard Hall, a social studies student, randomly answers a six-item true-false exam by flipping a coin. In this case, $p = \frac{1}{2}$ and $n = 6$. He wants to determine the probability that he will get exactly five items correct with this method.

Richard's chore consists of finding the sixth term of the binomial expansion of the sixth power of $(p + q)$, or finding $_6C_5$ and the sixth power of 2. The entire sample space for values of s from 0 (none correct) to 6 (all correct) is given by:

s	$_6C_5$	2^6	$p(s) = \frac{_nC_s}{2^6}$
0	$\frac{6!}{0!(6-0)!} = 1$	64	$\frac{1}{64}$
1	$\frac{6!}{1!(6-1)!} = 6$	64	$\frac{6}{64}$
2	$\frac{6!}{2!(6-2)!} = 15$	64	$\frac{15}{64}$
3	$\frac{6!}{3!(6-3)!} = 20$	64	$\frac{20}{64}$
4	$\frac{6!}{4!(6-4)!} = 15$	64	$\frac{15}{64}$
5	$\frac{6!}{5!(6-5)!} = 6$	64	$\frac{6}{64}$
6	$\frac{6!}{6!(6-6)!} = 1$	64	$\frac{1}{64}$

From these results Richard can tell that $p(s = 5) = \frac{6}{64}$ or about 0.09. Thus the probability of getting five items out of six correct using Richard's random technique is about 0.09 or 9%. Because of symmetry, notice that $p(s = 1)$ has the same probability as $p(s = 5)$. This is not generally true except when $p = q = \frac{1}{2}$, because that is a requirement for distribution symmetry. Consider the following application.

Application 2

Instead of a true-false test, suppose that Richard was administered a six-item multiple-choice exam with four alternative responses for each item. What is the probability that Richard will correctly answer five questions by randomly choosing an answer to each item? In this case, $p = \frac{1}{4}$, $q = \frac{3}{4}$, and $n = 6$. The $p(s = 5)$ is given by the second term of the binomial expansion of $(\frac{1}{4} + \frac{3}{4})^6$. The only term (because the distribution is not symmetrical) that will yield the correct probability is the term with p raised to the fifth power (second term).

$$
\begin{aligned}
np^5q^1 &= (6)(\tfrac{1}{4})^5(\tfrac{3}{4})^1 \\
&= (6)(\tfrac{1}{1{,}024})(\tfrac{3}{4}) \\
&= 0.004.
\end{aligned}
$$

The chances are very small that Richard will correctly answer five questions by random methods. How many items would Richard be expected to get right just by random selection on the multiple-choice exam if p is $\frac{1}{4}$ on each item? The expected value in this case is also the "typical" value based on the hypothetical experiment being carried out over many thousands of trials. As you will recall, "typical" is also known as the "average" or the "mean." Because s represents a random variable, it should come as no surprise that the distribution has a mean and a standard deviation.

Mean and Variance of the Binomial Distribution

Central tendency and variability can be easily computed for a binomial distribution. If p is the probability of success, q is the probability of failure, and n is the number of trials, then the mean of the binomial distribution is found by: $\mu = np$. That is, the mean is equal to the product of the number of trials (n) times the probability of success on a single trial (p). Thus the mean of a binomial for which $p = \frac{1}{2}$ and $n = 10$ is $(10)(\frac{1}{2}) = 5$. So, for example, an average of five heads would result over many experimental trials when 10 fair coins were tossed. On the average, Richard Hall would be expected to get one-fourth of the six multiple-choice items correct, or $\mu = 1.5$.

The variance of a binomial distribution is ***npq***, which implies that the standard deviation (σ) is $\sqrt{npq}$. Again, for the experiment of tossing 10 fair coins, the standard deviation of the binomial would be found by:

$$
\begin{aligned}
\sigma &= \sqrt{(10)(\tfrac{1}{2})(\tfrac{1}{2})} \\
&= \sqrt{\tfrac{10}{4}} \\
&= \sqrt{2.5} \\
&= 1.58.
\end{aligned}
$$

NORMAL CURVE

In Chapter 6 you learned to find the proportion of the total area under the normal curve between any two points scaled as z-scores with the assistance of the table of values in Appendix C. You also learned that the area under a particular part of the curve was the proportion of cases whose scores fell on the corresponding section of the baseline. Recall that: (1) the proportion of the cases of a large set of individuals falling within a particular interval on the baseline is nothing more than the *relative frequency* for that class; and (2) the probability that an event will occur is the *relative frequency* of that event for a large number of trials.

It should therefore come as no surprise that what you have learned about the normal curve can be used to obtain probabilities of certain classes of events. As discussed in Chapter 6, many variables are distributed approximately as a normal distribution frequency curve. For such variables, the area segments under the normal curve can be treated directly as probabilities of occurrence. As you will see, *the* **normal curve**, *although continuous, may be used as a very close approximation of the discrete binomial distribution under certain conditions.* It may be beneficial at this point to review the normal curve characteristics discussed in Chapter 6.

Consider a set containing a large number of normally distributed z-scores. What is the probability of selecting a z-score of 1.37 or greater in a single random draw? To solve this problem it is necessary to find the area under the normal curve to the right (in the smaller portion) of $z = 1.37$, as shown in Figure 8.4. With the assistance of Appendix C we see that the area to the right of $z = 1.37$ is 0.0853, or $p(z > 1.37) = 0.0853$ or 0.09 (rounded to two decimal places).

Figure 8.4—Normal Curve Showing Probability of the Random Occurrence of z Greater than 1.37

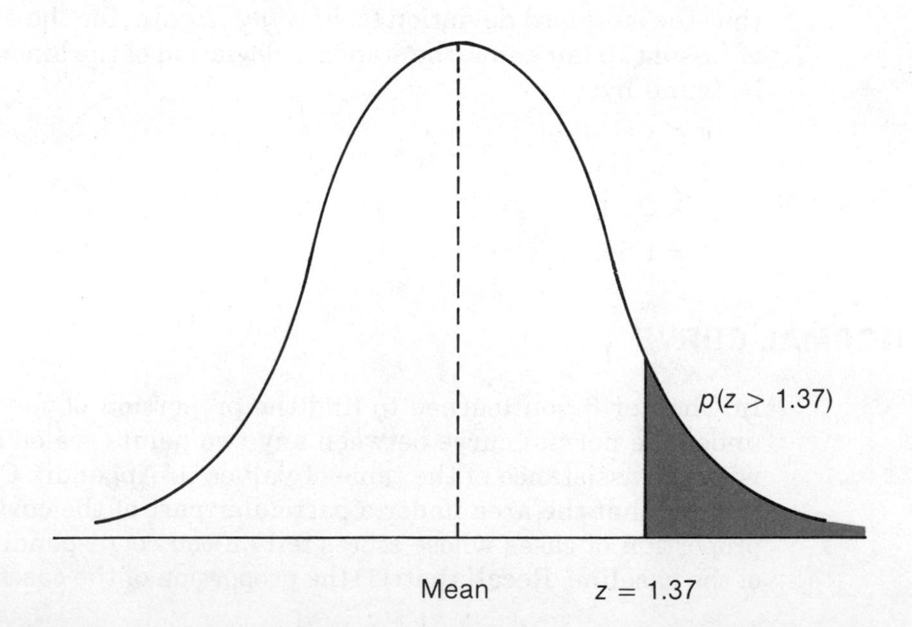

IQ (intelligence quotient) scores are distributed with a mean of 100 and a standard deviation of 15 in the general population. If one person is selected at random from the general public, what is the probability that the person will have an IQ score greater than 115 *or* an IQ of less than 95? The first step in the solution is to find the z-scores corresponding to the two IQ scores (remember that z-scores indicate how many standard deviation units the scores are departing from the mean). Thus for an IQ = 115, $z = \frac{115 - 100}{15} = 1.00$. For an IQ = 95, $z = \frac{95 - 100}{15} = -0.33$. Second, using Appendix C, determine the areas under the curve to the right of 1.00 and to the left of $z = -0.33$, as shown in Figure 8.5.

Figure 8.5—Normal Curve Showing Probability of the Random Occurrence of z Greater than 1.00 or Less than −0.33

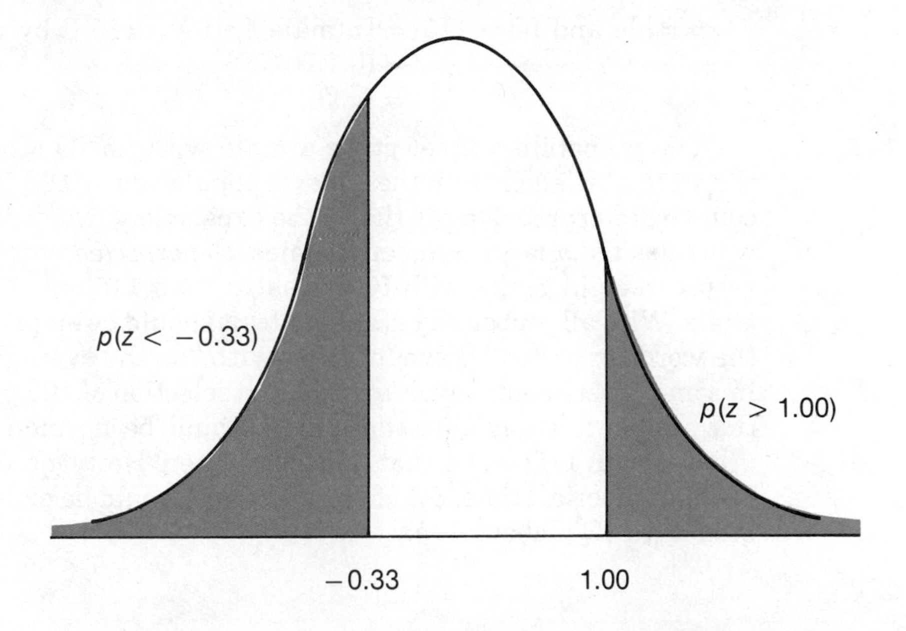

The area to the right of $z = 1.00$ is 0.16 (rounded), and the area to the left of $z = -0.33$ is 0.37 (rounded). Therefore $p(z > 1.00) = 0.16$ and $p(z < -0.33) = 0.37$. Applying the addition rule or rule 6 (yes, the rules still apply!) results in:

$$\begin{aligned} p(z < -0.33 \text{ or } z > 1.00) &= p(z < -0.33) + p(z > 1.00) \\ &= 0.37 + 0.16 \\ &= 0.53. \end{aligned}$$

Hence the probability is slightly over one-half (0.53) that the person would have an IQ score greater than 115 or less than 95.

You may have noticed that there is no distinguishing between the area to the right of $z = 1.00$ and the area to the right of or equal to $z = 1.00$. This is because the normal distribution is a continuous curve representing a continuous random variable. Theoretically, the probability of the occurrence of any particular point on the normal curve is zero; therefore no information is lost by not distinguishing between ($\geq$) and ($>$). If a discrete variable such as the binomial was being used, such a distinction would be necessary.

As a further illustration of the use of the normal curve as a probability distribution, determine the probability that a person drawn from the general public is a male with an IQ score greater than 115. In the previous example you found $p(\text{IQ} > 115) = 0.16$. Assuming an equal number of men and women in the population, $p(\text{male}) = 0.50$, the solution is:

$$\begin{aligned} p(\text{male and IQ} > 115) &= [p(\text{male})][p(\text{IQ} > 115)] \quad \text{by rule 7} \\ &= (0.16)(0.50) \\ &= 0.08. \end{aligned}$$

The probability of selecting a male with an IQ score greater than 115 at random from the general population is 0.08. This result could be interpreted to say that if the experiment (random selection) was repeated a large number of times, the *expected* outcome would result in eight males with IQ's greater than 115 out of every 100 trials. With all probability statements, it should be emphasized that the word "expected" is synonymous with "on the average." In fact, in some experiments involving random selection of 100 people from the public, it is possible that there would be no men with IQ's greater than 115; or, for that matter, no men! However, over a very large number of trials, 8% of the outcomes would be expected to be men with IQ's above 115.

POISSON DISTRIBUTION—OPTIONAL

The binomial and normal probability distributions are appropriate for a wide variety of probability problems. However, there is frequently a need to study phenomena that occur rarely over a specified unit of time, space, or distance. Some examples are the number of deaths per year in a given region caused by a rare disease, the number of midair collisions per year, the number of worms observed per head of cabbage from a harvest, and the number of "prank" phone calls per minute received by a community help hot line.

The characteristic that is common to such events is that the probability of a "success" is small and the number of trials is large. Such occurrences can best by modeled by a *Poisson distribution.* Like the binomial distribution, the **Poisson distribution** is a discrete random variable. If simple events occur at random within a time frame, the number of such events occurring in a fixed interval of time has a Poisson distribution. Generally, if events are occurring at a rate of λ (**lambda**) events per unit of time, a Poisson distribution provides the probability that X will occur k times within a unit of time.

The formula for the Poisson distribution is:

$$p(X) = \frac{\lambda^k e^{-\lambda}}{k!} \quad \text{or} \quad \frac{\lambda^k}{k!}e^{-\lambda}$$

where: $k = 0, 1, 2, 3, \ldots$
e is the base of the natural logarithm and is approximately 2.71828
λ is both the mean and the variance of the Poisson distribution
$\sqrt{\lambda}$ is the standard deviation.

The Poisson distribution is applicable to many natural phenomena, such as the number of cars on a highway passing a certain point per minute, the number of alpha particles emitted by a radioactive substance during a period of time, and the number of emergency cases at a hospital within a one-month period. Actually, the time variable does not have to be present. For example, the Poisson distribution is also applicable to situations such as the number of misspelled words in a large textbook, the number of raisins per slice in many loaves of raisin bread, and the number of defective lightbulbs from a certain manufacturer. Again, notice that p is relatively small and n is large.

The Poisson distribution may be used to compute probabilities of random variables from a binomial experiment when the normal approximation is inadequate, that is, when $np < 5$ or $nq < 5$. More specifically, if n is large (more than 20) and if p (the probability of success) is small (less than 0.05), then for practical purposes the Poisson distribution can be used to closely approximate the number of successes (X) in a binomial experiment. In such a case $\lambda = np$. The variance and standard deviation are still λ and $\sqrt{\lambda}$, respectively.

The following example will help clarify the utility of the Poisson distribution. Suppose that between 7:30 and 9:30 a.m. the school switchboard receives an average of three outside calls per minute. What is the probability that during a particular minute the school will receive:

1. exactly one call?
2. four or fewer calls?
3. more than four calls?

Given that the mean $\lambda = 3$, let X represent the random variable (number of outside calls). Then, by the Poisson distribution, the solutions are as follows:

a. $$p(X = 1) = \frac{(3)^1(2.7183^{-3})}{1} = (3)(0.04979) = 0.14937$$

b. $$\begin{aligned} p(X \leq 4) &= p(X = 1) + p(X = 2) + p(X = 3) + p(X = 4) \\ &= \frac{3(e^{-3})}{1!} + \frac{3^2(e^{-3})}{2!} + \frac{3^3(e^{-3})}{3!} + \frac{3^4(e^{-3})}{4!} \\ &= (2.7183^{-3})(\tfrac{3}{1} + \tfrac{9}{2} + \tfrac{27}{6} + \tfrac{81}{24}) \\ &= 0.7655 \end{aligned}$$

c. $$\begin{aligned} p(X > 4) &= 1 - p(X \leq 4) \\ &= 1 - (0.7655) \\ &= 0.2345 \end{aligned}$$

Generally, λ is relatively small, that is, less than 10, and the computation of the desired probabilities usually requires only a few terms. In such cases Appendix B provides values for the $e^{-\lambda}$ factor in the formula. However, when λ is large, the normal curve is a good approximation with $\mu = \lambda$ and $\sigma = \sqrt{\lambda}$. This means that the normal curve probability characteristics will introduce only very small errors when used as an approximation to the Poisson distribution when n is large.

COMPUTER FOCUS 8—The Binomial Theorem

Purpose

The task required in this Computer Focus, if successfully completed, will result in a program that can be used to solve "real-life" probability problems involving binomial experiments. The purpose of this Computer Focus is to illustrate how a theory (the binomial theorem in this case) can be translated into practice. Statements of the theorem provide specific guides about algorithms that in turn are programmed and coded in BASIC language to provide a practical tool—a lesson that theory and practice can be directly related.

Problem

Using the logic specified in the three statements of the binomial theorem, your task is to write a BASIC program that will output the probability of s successes out of n binomial trials. If $p =$ the probability of success for any trial, then the probability of s successes in n trials is computed by:

$$p(s) = {}_{n}C_{s}p^{s}q^{n-s}$$

where $q = 1 - p$.

As specified in the binomial theorem, write your program by relying on the algorithm of the recurrence formula, which relates the probability of s successes to the probability of success on the $(s - 1)$st trial. If $p(s)$ is the probability of success on the sth trial, then $p(s - 1)$ denotes the probability of success on the previous trial; or

$$p(s) = \frac{(n - s + 1)}{s}\left(\frac{p}{q}\right)[p(s - 1)]$$

The following flowchart represents one approach to solving the programming problem:

Flowchart

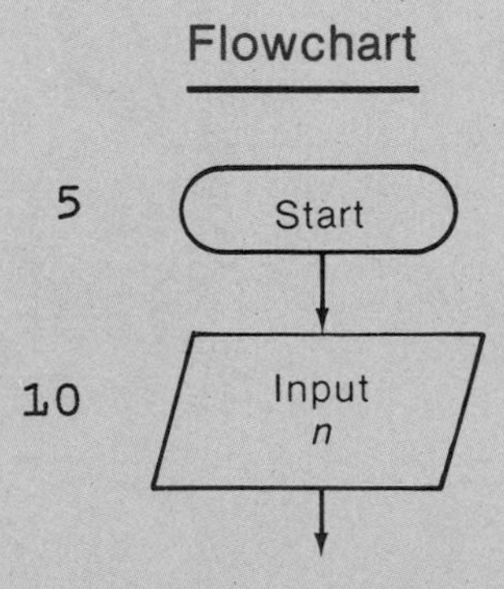

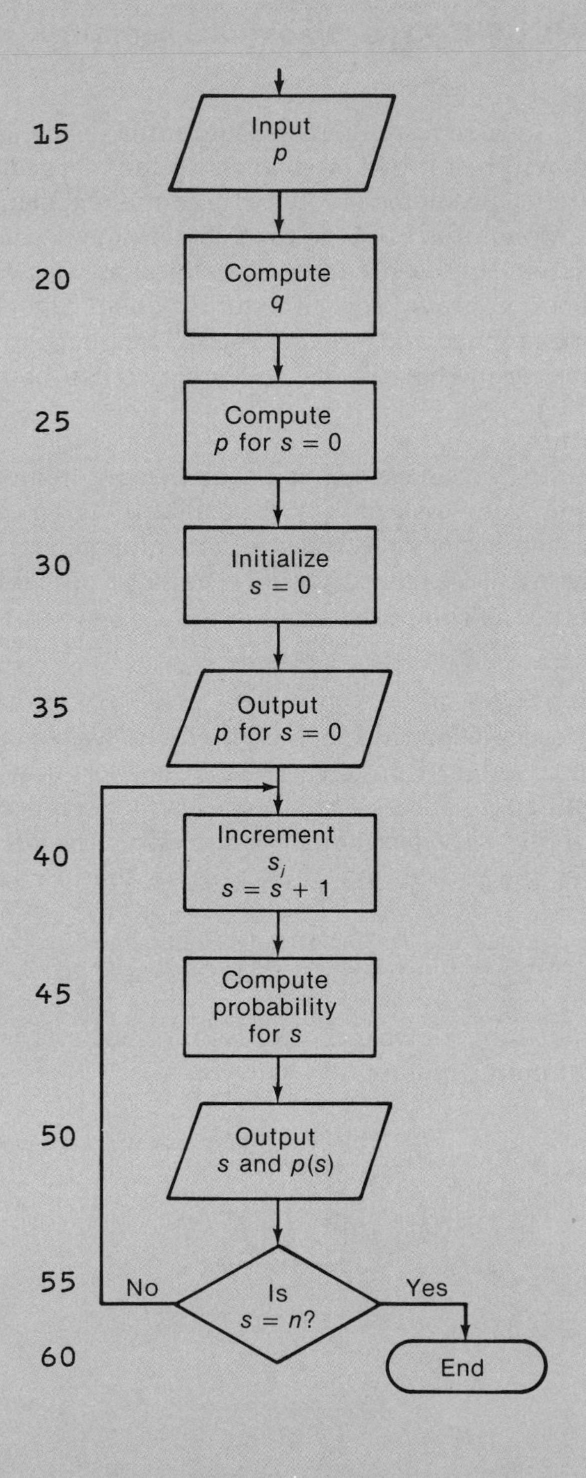
15
Input
p
20
Compute
q
25
Compute
p for s = 0
30
Initialize
s = 0
35
Output
p for s = 0
40
Increment
s;
s = s + 1
45
Compute
probability
for s
50
Output
s and p(s)
55
No
Is
s = n?
Yes
60
End

KEY TERMS AND NOTATION

Binomial distribution
Binomial experiment (Bernoulli experiment)
Binomial theorem
λ (lambda) and $\sqrt{\lambda}$
${}_nC_r$
np
npq and $\sqrt{npq}$
Normal curve
Pascal's triangle
Poisson distribution (optional)
z-score

APPLICATION EXERCISES

8.1 For a binomial experiment, assume that $p = \frac{1}{2}$, $q = \frac{1}{2}$, and $n = 4$. Write the terms for the expanded binomial $(\frac{1}{2} + \frac{1}{2})^4$.

8.2 If $p = \frac{1}{3}$ and $q = \frac{2}{3}$ in a binomial experiment of 5 trials, find the following probabilities for the random variable X:

a. $p(X = 2)$
b. $p(X = 0)$
c. $p(X = 5)$
d. $p(X \leq 3)$

8.3 If $p = \frac{1}{2}$, $q = \frac{1}{2}$, and $n = 10$ in a binomial experiment, find the probability that the random variable X will equal or exceed 8; that is, $p(X \geq 8)$.

8.4 For a binomial experiment with $n = 6$ and $p = \frac{1}{2}$:

a. Compute the probability distribution; that is, list the possible values of the random variable s and the probability associated with each value of s.
b. Construct a probability bar graph such as that shown in Figure 8.2 of your text.
c. Find the mean and standard deviation of the binomial distribution.

8.5 Rogers High will play Memorial High in baseball six times during the upcoming season. Assume the teams are of equal ability; that is, $p = \frac{1}{2}$. Within the context of a binomial experiment, what is the probability that:

a. Rogers will win 4 games and lose 2?
b. Rogers will win *at least* 4 games?

8.6 From a population of high school juniors and seniors, 8 students are selected at random to serve on the student parking committee. Assume that the class sizes are equal and let s = the number of juniors in the sample. Use the binomial distribution to find:

a. $p(s = 4)$
b. $p(s = 0)$
c. $p(s \leq 3)$
d. $p(s > 6)$

8.7 Assuming a normal distribution of a set of standard z-scores, find:

a. $p(z > 1.96)$
b. $p(z < 2.58)$
c. $p(0 < z < 1.58)$
d. $p(1.0 < z < 2.0)$

8.8 The mean height of the male sophomore class members at a certain school is 70 inches, and the standard deviation is 3 inches. If H is the height of an individual male selected at random, what is:

a. $p(H > 70)$?
b. $p(H < 67)$?
c. $p(67 < H < 73)$?
d. $p(H > 75)$?

8.9 Over the years, the average income from the student-sponsored Halloween Carnival is $1,200. The standard deviation of the income distribution is $150. Assume a normal distribution and assume the upcoming carnival is fairly typical of carnivals in the past. Let I represent the income for the upcoming event and find:

a. $p(I > \$1,300)$
b. $p(I < \$900)$
c. $p(\$1,000 < I < \$1,500)$
d. $p(\$1,000 < I < \$1,200)$

8.10 In a normal curve distribution with a mean of 5 and a standard deviation of 1.58:

a. Find the probability that a random selected case will have a value equal to or greater than 7.5.
b. How does the value in a. compare with the result of Application Exercise 8.3?

8.11 *Optional* In a Poisson distribution with a mean of 2, what is the probability that a random variable X equals:

a. 1?
b. 2?
c. 3?

8.12 *Optional* If the mean of a Poisson random variable is 4, what is its:

a. standard deviation?
b. variance?

8.13 *Optional* If $n = 20$ and $p = 0.05$, what is the probability that $X = 2$ when computed on a:
a. Poisson distribution? (*Hint:* $\lambda = np$.)
b. binomial distribution?

8.14 *Optional* The number of motor vehicle accidents per week involving students in a particular school system conforms to a Poisson distribution. Assuming that the distribution mean is 2 per week, what is the probability that no motor vehicle accidents involving students will occur during a particular week?

8.15 *Optional* For the senior class of 100 students, it was found that the probability was 0.04 that each student would be absent because of illness. Assume a Poisson distribution.
a. What is the mean number of absences per day caused by illness in the senior class?
b. What is the standard deviation?

8.16 *Optional* For Application Exercise 8.15, answer the two questions assuming a binomial distribution.

8.17 *Optional* In a senior English class, the records revealed that the number of students who failed was a Poisson distribution with a mean of 4 per semester. What is the probability that the number of failures for a semester will:
a. equal 0?
b. be less than or equal to 3?
c. equal 4?
d. be greater than 3? [*Hint:* $p(f > 3) = 1 - p(f \leq 3)$.]

PART THREE
Inferential Statistical Concepts

9

Applied Sampling

CHAPTER OBJECTIVES

Upon completion of the chapter, students will be able to:

1. Distinguish between a population and a sample.
2. Describe the role of sampling in inferential statistics.
3. Discuss four common sampling techniques.
4. Explain the importance of ''representativeness'' in forming a sample.
5. Determine an appropriate sample size n for a given population of size N.
6. Explain the general relationship between sample size and size of population.
7. Identify biased samples.
8. Identify an important characteristic of random sampling.
9. Discuss the role of computed statistics in inferential statistical work.
10. Contrast a statistic and a parameter.

As part of a physical examination required for admission to a mountain-climbing summer camp, Jerry needed a blood test performed. The procedure involved pricking his index finger and collecting a few drops of blood. This blood was subjected to thorough analysis by laboratory technicians. From the small amount of blood taken, the technicians were able to describe red and white cell counts, hemoglobin concentration, sedimentation rate, blood sugar, and levels of cholesterol and triglycerides (fat-like substances) in Jerry's blood. These findings described the blood throughout his system, not just the blood in his finger.

From the findings in a very small portion of Jerry's blood, **inferences** can be made about all the blood in his body. The technicians did not need to examine all of Jerry's blood to determine its characteristics. Accurate inferences about the total volume of Jerry's blood were possible only because the small portion of extracted blood had the same characteristics as the rest of the blood in his system. To use statistical terminology, the few drops of blood used in the extensive tests were **representative** of the remainder of Jerry's blood.

Although this example is from the medical field, the principles used can also be applied to work performed in the field of inferential statistics. For example, responses to questions asked of only a small group of people can be inferred from that small group to a larger group. Responses from a small group can tell what residents of Kansas City think of a city-wide bond referendum, the most popular TV programs during the week for the entire nation, and how political leaders are perceived. Other information about large segments of society can be obtained by studying only a small subset of the total universe of concern. **Sampling** is the operation that makes such activities possible.

POPULATIONS AND SAMPLES

Populations

In inferential statistics there is much discussion about *populations* and *samples*. In ordinary conversation you speak of a population of people in a city, state, nation, school district, and so on. Usually the term is tied to some kind of geographical boundaries or locations. Although this meaning is quite consistent with the term "population" used in statistics, statisticians do not restrict populations to locations, nor do they limit the elements of their populations to humans. *A* **population** *in a statistical context is any well-defined set of elements that have at least one characteristic in common.* Population data refer to the group of observations made on a defined set of elements about which the statistician wishes to draw conclusions. Therefore a population may consist of a large number of elements such as all taxpayers in the nation; or a population may be fairly small, as in the ages of United States senators. In some cases infinite populations are studied, such as the set of all positive integers. You could define a population as the set of automobiles in the student parking lot at your school. You could also define a population as the class officers in your school. Notice that in each case the elements or objects that make up the population have some specific trait or quality that identifies them as members of the population under consideration. In your mathematics classes, you may recall dealing with "universal" sets. If so, it may help to think of a population in a statistical setting as a well-defined "universal set" of elements.

Samples

A **sample** is a portion of a defined population or a subset of a population. If you define a population as the full-time students in your high school, one of the many possible samples that could be part of the population is the full-time members of the senior class. Samples are important, because statistical workers study sample data. Just as the few drops of Jerry's blood were analyzed in detail, samples from populations can provide the data for intense analysis. Samples are used because populations are often too large or too inaccessible to be efficiently studied in their entirety. The procedures for drawing conclusions about a population from a sample taken from that population are fairly straightforward.

1. Determine what information is desired.
2. Identify the population about which conclusions are to be made.
3. Select a sample of subjects or elements from the entire population and extract the necessary information.
4. Statistically analyze the sample data.
5. From the results of the statistical analysis, make inferences or generalizations about the entire population.

Figure 9.1 schematically illustrates the process of sampling from a population, obtaining information about the sample, and making one or more inferences about the population.

Figure 9.1—Sampling and Inferential Statistics

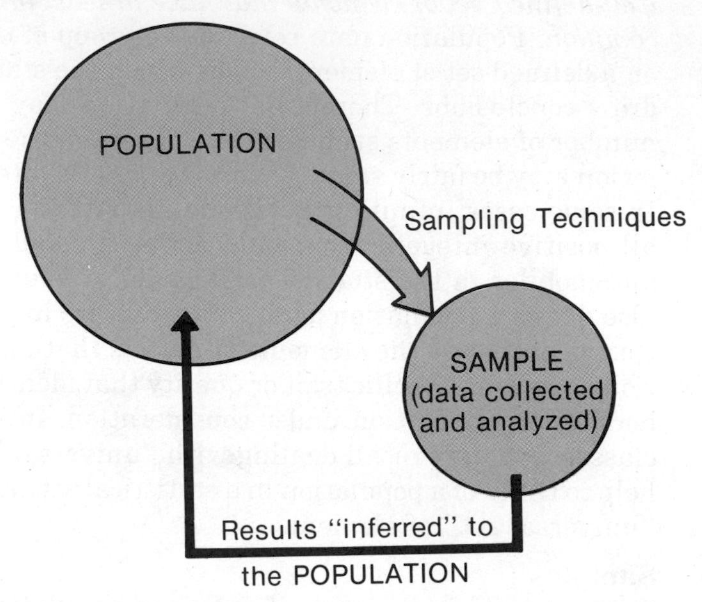

Inferences will be valid only if certain steps are taken to make the sample as *representative* of the population as possible. This means that the sample would have the same characteristics in the same proportion as the population. Suppose one of your school counselors wanted to assess the attitudes toward studying of the student body. A commercial attitude inventory could be selected, or a questionnaire could be constructed for students to complete. Then, if assessing the entire student body was not practical because of expense, time requirements, or other problems, a smaller subset of students (a sample) would be selected to complete the attitude assessment instrument.

If the attribute under consideration is attitude toward studying, do you think the sample would provide valid inferences about attitude for the entire student body if it consisted of:

1. the first 50 students showing up at the building on a Monday morning?
2. the students found in the library 2 hours after the end of school on a Friday afternoon?
3. all male students?
4. all seniors applying for summer-school classes to complete graduation requirements?
5. members of the varsity basketball team?

With a little thought, you can probably conclude that all of these samples would be poor representatives of the population with respect to attitudes toward studying. These inappropriate samples of convenience are known as **biased samples**. Data from such groups cannot be considered valid indicators of characteristics of the population. Nothing inherent in the definition of a sample specifies *how* it should be selected from a population, but methods are available to enhance the likelihood that the sample selected will be an **unbiased sample** that is representative of the population.

SAMPLING TECHNIQUES

Two sampling problems confront the researcher at the planning stage. One is concerned with *how* to select the sample from the population to maximize the chances of getting a representative sample. The second problem is *how large* a sample should be to be sufficiently representative of a defined population. Several commonly used techniques are described that address how to select the sample.

Random Sampling

A random sampling technique is commonly employed by statisticians. A key condition of **random sampling** is that each element of the population has an equal probability of being selected into the sample. In essence, a random sample is one where chance does the selecting and every element in the population has an equal chance of being sampled.

The counselor who wanted to assess attitudes toward studying could use a "lottery" model. That is, names of the entire student body could be written on individual slips of paper and put in a container. After mixing, the counselor would blindly select 35 slips of paper. The sample would be considered a random sample because each student in the school would have an *equal opportunity* of being selected. The sample results of the attitude assessments could then be generalized to the entire student body with confidence that the sample was not biased.

A more common method of random sampling consists of assigning a number to each element (student) in the population of concern. Or, in the present example, the counselor could randomly choose student ID numbers from a list to form the sample. Computer-generated lists of random numbers (called "pseudo-random" numbers) can be used for such a purpose (see Computer Focus 9 at the end of the chapter).

Stratified Random Sampling

Another commonly used technique is **stratified random sampling**. This procedure involves dividing the population into natural groupings or strata prior to selecting the sample. For example, the counselor might first get a list of students by grade level and then use a random sampling technique to select students from each of the grade levels. The reason for using this technique is to maximize the probability of selecting a representative sample, as subjects in different strata may have different perspectives. For example, seniors may have a different attitude toward studying than the recently arrived freshman or sophomore class.

Cluster Sampling

Another sampling method very similar to stratified random sampling is known as **cluster sampling**. This also is an extension of simple random sampling and involves dividing the population into "clusters" or blocks. Cluster sampling is often used with large populations such as a city where census tracts, school districts, or other "natural" clusters or boundaries exist. A random sample is taken from each of the clusters. The size of the sub-sample taken from each cluster is proportional to the number of elements in that cluster of the population.

Systematic Sampling

Systematic sampling procedures are frequently used. These are not random samples; rather the elements are selected according to a plan (systematic). The counselor could complete a systematic sample by selecting every tenth name from the school roster as a fairly representative sample of the student body.

SAMPLE SIZE

The second problem associated with sampling has to do with the question: "How many subjects are needed from the population to form a sample of sufficient size?" This is an important question, because even if an appropriate technique is used for obtaining a representative sample, the technique will be ineffective if the sample size is too small. In general, a sample size that is less than the size of the total population will produce some degree of error. A sample that includes 5% of the population will not yield inferential results that are as reliable as a sample that includes 80% of the population, assuming equivalent sampling techniques are used. On the other hand, the smaller sample may be adequate and considerably more economical to obtain in terms of time, resources, cost, etc. There are trade-offs to be made between large samples, which are more accurate but more difficult to obtain, and small samples, which are less accurate but easier to obtain.

The sample size problem occurs when one is trying to ensure that "extreme" cases from the population are included in the sample. Keep in mind that a sample should be a miniature population, which implies that population extremes are represented in the sample. Assuming that the attribute, trait, or characteristic being studied is normally distributed in the population, the chances are high that a small sample would be made up entirely of cases from the central portion of the distribution, thereby excluding extreme cases located in the tails of the curve. Figure 9.2 provides a reminder that the majority of the cases in a bell-shaped curve are in the middle portion. Common sense indicates that a sample n is needed that is large enough to provide a relatively high probability of including some

Figure 9.2—Normal Distribution of Population Characteristic

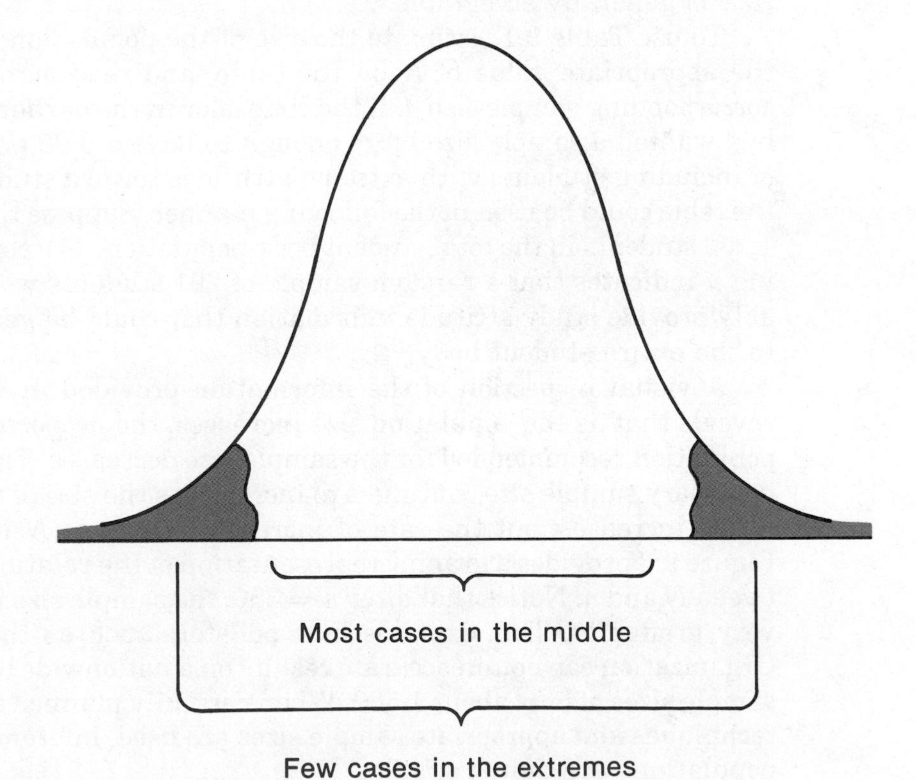

cases in the sample that are *beyond* one and two standard deviations from the mean.

The research division of the National Education Association (NEA) has simplified the process for determining sample size if the size of the population can be estimated. The formula developed by NEA can be used to generate a table that is simple to use and provides useful information about sample size (reference: "Small Sample Techniques," *The NEA Research Bulletin* 28, December 1960). This particular technique should be used as a guideline rather than a definitive answer for all situations, because there are many factors involved in the solution to the sample size problem.

Table 9.1 provides the required sample size n for a given population of size N and is to be used with a random sampling technique. The probability is 0.95 that a sample size determined by the table will contain extreme cases. That is, if 100 samples were obtained using the information in Table 9.1, 95 of them would be expected to contain the rarely occurring or extreme cases from the population. Of course, to be 100% certain that such cases were included, it would be necessary to use the entire population. A 95% confidence rate is generally acceptable.

To use Table 9.1, estimate the size of the population N, locate the appropriate value of N on the table, and read across to the corresponding sample size n. If the counselor in the earlier example had wanted a sample size large enough to have a 0.95 probability of including students with extreme attitudes toward study habits, the table could be used in the following manner. Suppose there were 1,200 students in the total student body population. The corresponding n indicates that a random sample of 291 students would probably provide study attitude information that could be generalized to the entire student body.

A visual inspection of the information provided in Table 9.1 reveals that as the population size increases, the *proportion* of the population recommended for the sample size decreases. That is, the necessary sample size continues to increase as the size of the population increases; but the rate of increase lessens as N increases. Figure 9.3 provides a graphic representation of the relationship between N and n. Notice that after $n = 380$, the sample size increases very gradually. This explains how pollsters such as the Gallup Organization can obtain accurate results on a nationwide basis with sample sizes of only about 1,500. When carefully planned sampling techniques and appropriate sample sizes are used, inferences about populations are quite reliable.

Table 9.1 Table for Determining Sample Size (*n*) from a Population Size (*N*)

N	*n*	*N*	*n*	*N*	*n*	*N*	*n*	*N*	*n*
10	10	190	127	520	221	1,450	304	8,000	367
15	14	200	132	540	225	1,500	306	9,000	368
20	19	210	136	560	228	1,550	308	10,000	370
25	24	220	140	580	231	1,600	310	11,000	371
30	28	230	144	600	234	1,650	312	20,000	377
35	32	240	148	620	237	1,700	313	30,000	379
40	36	250	152	640	240	1,750	315	50,000	381
45	40	260	155	660	243	1,800	317	100,000	383
50	44	270	159	680	246	1,850	318	1,000,000	384
55	48	280	162	700	248	1,900	320		
60	52	290	165	720	251	1,950	321		
65	56	300	169	740	253	2,000	322		
70	59	310	172	760	255	2,100	325		
75	63	320	175	780	258	2,200	327		
80	66	330	178	800	260	2,300	329		
85	70	340	181	850	265	2,400	331		
90	73	350	183	900	269	2,500	333		
95	76	360	186	950	274	2,600	335		
100	80	370	189	1,000	278	2,700	336		
110	86	380	191	1,050	281	2,800	338		
120	92	390	194	1,100	285	2,900	339		
130	97	400	196	1,150	288	3,000	341		
140	103	420	201	1,200	291	3,500	346		
150	108	440	205	1,250	294	4,000	351		
160	113	460	210	1,300	297	5,000	357		
170	118	480	214	1,350	299	6,000	361		
180	123	500	217	1,400	302	7,000	364		

Figure 9.3—Relationship Between Sample Size (n) and Population Size (N)

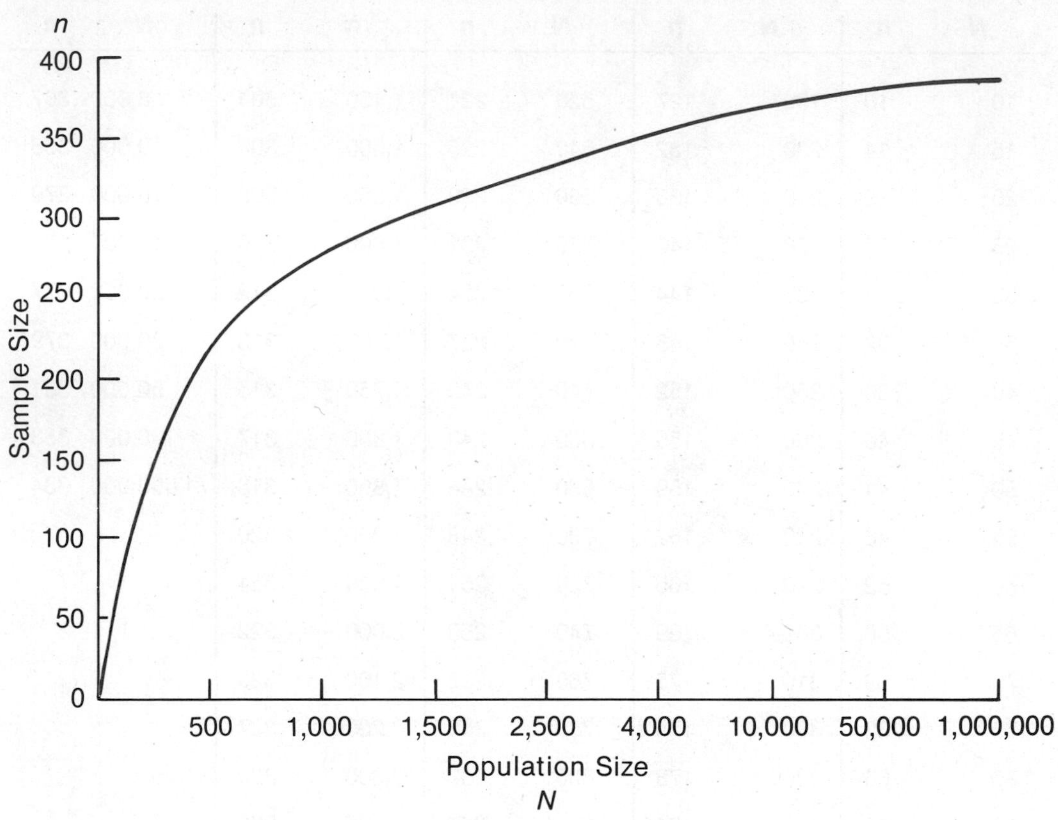

PARAMETERS AND STATISTICS

Two terms with which you should become familiar are *parameters* and *statistics*. **Parameters** are quantitative characteristics of a *population*. They are generally unknown quantities, as they cannot be directly measured using every element in the population of interest. Examples of population parameters are the percentage of all adults in America who voted during the last presidential election, the average reading difficulty level of high school textbooks, the proportion of last year's high school graduates who joined a branch of the military, the variability in temperature across the nation at a particular time, the fat content of hot dogs from a certain company, the most common (mode) eye color of past presidents of the United States, and the median height of female high school basketball players in the United States.

Statistics are quantitative characteristics of *samples*. Some examples of sample statistics are the mean age of one class section of a "representative" senior-level course, the median weight of the interior football linemen drafted by the Dallas Cowboys last year, the number of pecans in a slice of pecan pie, the percentage of students with red hair in the freshman class, and the typical mathematics aptitude score of 30 students picked at random from a student body.

This is a new definition of statistics, and it differs from earlier descriptions of the term as being synonymous with data and as a science of studying numerical data. Until now it has not been necessary to distinguish between the terms, because you have been dealing only with data you have. Now, as your statistical sophistication grows into inferential statistics, you need to be able to distinguish quantitative characteristics you *have* (data on a sample) and quantitative information you *don't have* (data on an entire population).

Return for a moment to the counselor who wanted to know about the attitudes toward study in the student body. A sample ($n = 291$) was selected from the total population ($N = 1{,}200$), and a standardized "attitudes toward studying" questionnaire was administered to the sample. The scores were then analyzed in several ways. First some summary statistics were computed, namely, the mean, median, mode, mean deviation, variance, and standard deviation. Also determined was the percentage of students scoring below the testing company's suggested minimum score.

The results of such computations are *statistics*, because they represent quantities taken from the sample. These statistics become the counselor's best estimate of the corresponding population parameters. Consequently, the mean, median, mode, mean deviation, variance, and standard deviation of the scores as well as the percentage of students below the recommended level for the entire student body (population) are *inferred* from the values computed for the sample. Any quantity determined by the analysis of data from a representative sample is called a statistic and is used to estimate (the topic of the next chapter) the value of the corresponding parameter.

As a matter of convention, symbols for statistics will make use of English letters, while Greek letters will symbolize parameters. For example, s will be used to stand for a standard deviation of a sample (statistic), whereas the lowercase Greek letter σ (sigma) will be used for the **population standard deviation** (parameter). The symbol μ represents the **population mean**, and σ^2 denotes the **population variance**. Table 9.2 provides a summary of the most commonly used symbols for statistics and parameters.

Table 9.2 Symbols for Parameters and Statistics

Unknown Population Parameter	Quantity Computed from Sample (Statistic) to Estimate Population Parameter
μ = population mean	$\overline{X}$ = sample mean
σ = population standard deviation	s = sample standard deviation
σ^2 = population variance	s^2 = sample variance

You should now understand that "representativeness" rests on sampling technique and sample size as determined by the statistical worker. To the extent that the sample is representative of the population of interest, reliable generalizations can be made from the sample data statistics to the population parameters.

COMPUTER FOCUS 9—Pseudo-random Number Generator

Purpose

At the most basic level, randomness is a construct that forms the basis for inferential statistics. Thus this construct can be and is used extensively in simulation and testing. The activity in this Computer Focus will introduce you to the role of the computer in producing random numbers. A second purpose is to introduce the INT function, which provides an efficient method of changing random-number output to whole numbers.

Problem

In the strict sense of the term, the computer generation of "random" numbers is an impossibility. In fact, it is difficult to tell exactly what a random number is. Computer generators produce numbers that have properties that characterize random numbers. Thus a more accurate term is *pseudo-random numbers*, though for brevity they will be referred to as random numbers.

Most computers have a BASIC language function that generates random numbers between 0 and 1. If values of a random variable R are generated by the statement:

```
30 LET R = RND(1)
```

then $0 < R < 1.0$ and the values of R will be uniformly distributed in the interval (excluding the end points 0 and 1).

To expand the possible range of R to the interval 0 to 2, the statement can be modified to multiply (*) the function by 2 as:

```
40 LET R = 2*RND(1)
```

and random numbers between 0 and 10 would be generated by:

```
50 LET R = 10*RND(1)
```

and so on. Statisticians, however, have found random integers (whole numbers) to have more practical utility than fractions. BASIC addresses this need with the INT function, which *truncates* (drops the fractional part of a real number); thus

```
60 LET R = INT(10*RND(1))
```

will produce random integers 0, 1, 2, 3, . . . , 9. Adding 1 to the expression on the right side of statement 60, that is,

```
INT(10*RND(1)) + 1
```

produces random integers 1, 2, 3, . . . , 10.

Problem: Assign your classmates sequential ID numbers starting with 01, 02, 03, . . . , N (where N is the number in the class). Plan, flowchart, and write a BASIC computer program that will randomly select a sample size (n) from the class by randomly generating random ID number integers.

KEY TERMS AND NOTATION

Biased sample
Cluster sampling
Inference
Parameter
Population
Population mean (μ)
Population standard deviation (σ)
Population variance (σ^2)
Random sampling
Representative
Sample
Sampling
Statistic
Stratified random sampling
Systematic sampling
Unbiased sample

APPLICATION EXERCISES

9.1 Match each item on the left with an appropriate item on the right.

a. a chance procedure for sampling
b. all cases that constitute a defined group
c. s
d. generalize to a population
e. subgroup of a population
f. parameter

(1) population
(2) population characteristic
(3) random
(4) statistic
(5) inferential
(6) sample

9.2 Complete the analogy: POPULATION is to SAMPLE as PARAMETER is to __________.

9.3 During a junior class party, 15 balloons were attached in a vertical line 10 inches apart on a wall. A student was blindfolded, given a hatpin, directed to the balloons, and asked to puncture 2 balloons. If the 15 balloons represent a population and the 2 punctured balloons represent a sample, is the sampling technique employed a random sampling method? Why?

9.4 An elementary school has 70 certified staff members. What would be the minimum size of a random sample of the staff so that you would have a 0.95 probability of including extreme cases?

9.5 A school district has about 2,000 fifth-grade students. To determine the average reading ability of the fifth graders, the school superintendent wants to know the minimum number that could be randomly sampled for testing and still yield reliable information about the reading ability of the population of fifth-grade students. What would be your advice?

9.6 Why is a random sample that is too small apt to yield misleading conclusions about the population?

9.7 A principal asked for volunteers from the faculty to work with students on a Saturday car-wash project to raise money for paving the student parking lot. Of the 40 faculty members, 16 volunteered for the project. Does this constitute a random sample of the faculty? Does it constitute a sample that would be representative of the population of faculty with respect to attitude toward extracurricular activities? Defend your answers.

9.8 Describe the procedural steps of inferential statistics.

9.9 Suppose you wanted to know how many words are in this textbook. Rather than counting the total number, describe a sampling technique (with an appropriate sample size) that would enable you to arrive at a reliable estimate of the total number of words in the text.

9.10 Describe a sampling method you could use to obtain a representative sample from each of the following populations. Include a definition of the population, give the size of the population, and specify the sample size.

a. your present statistics class
b. students in your school building who are in the same grade as you
c. all full-time students in your building
d. all students in your school district
e. honor roll students in your school
f. heads of households with children in school in your school district

10

Estimation

CHAPTER OBJECTIVES

Upon completion of the chapter, students will be able to:

1. Compute point estimates of a population mean (μ) and standard deviation (σ).
2. Distinguish between the standard deviation of a sample (**s**) and the standard error of the mean ($s_{\bar{X}}$) with regard to the elements of a distribution.
3. Describe three implications of the central limit theorem.
4. Determine the standard error of the mean from sample data.
5. Distinguish between point estimation and interval estimation.
6. Use characteristics of the normal curve to calculate confidence intervals for any reasonable level of confidence.
7. Translate information about a confidence interval around a mean into a probability statement.

Studies were conducted of two samples of college-bound seniors. One study reported that American College Testing (ACT) composite scores for the college-bound seniors had a mean of 19.4 points. However, results of the second study indicated that the mean composite of the ACT scores for college-bound seniors was 18.5 points. Both researchers were studying the same population. They were both interested in estimating the ACT average (parameter) score for all college-bound seniors (population) using data collected on samples. The two means (statistics), although different from each other, were both estimates of the same μ. How did the researchers arrive at 19.4 and 18.5 points as a point estimate of μ? Why were they different? How do statisticians deal with apparently ambiguous answers?

Recall that studying some characteristics of an entire population can be too time consuming, too expensive, or perhaps simply impossible to accomplish. Therefore a population such as college-bound seniors is studied indirectly through a representative subgroup of the population called a sample. In this chapter you will learn to compute *estimates of population parameters*. The good news is that such estimates are extensions of what you have already learned in descriptive statistics. The bad news is that, unless the sample consists of the whole population, you will have to contend with error that has been introduced by sampling.

POINT ESTIMATES OF PARAMETERS

Part 1 presented several methods of summarizing data. The most common ways were to compute the arithmetic mean to represent the "typical" number in the distribution (measure of central tendency) and to compute the standard deviation (or variance) to describe the variability of a distribution of values. These descriptive techniques are also used in the discussion of **point estimation**, that is, using a single value to estimate the value of a population parameter. Now consider the sample mean as an estimate of the population mean.

Unbiased Estimate of the Population Mean

Suppose that the number of absences in a particular school district is a topic of interest to the school board of trustees. The board directs the principal to furnish data about the number of absences in the district for the previous two years. One of the items of information requested is the average number of absences per day for the district. To avoid going through the absences for the entire district for each of the 360 days of school in the previous two years, the principal decides to select a sample ($n = 186$) of days at random. The population in this case is the 360 days of the two previous school years. Why was 186 days selected as a sample? (*Hint:* Recall the table for determining sample size from the previous chapter.) For the randomly selected sample, the principal summed the number of absences and divided that total by 186 to establish the mean ($\overline{X}$) number of absences per day for the sample.

When the random sample was drawn, obviously some individual days revealed an absenteeism rate above the population mean and some were below the average. In fact, the principal is fairly certain, because of the sample size, that the sample contains some days with extremely large numbers of absences and some with extremely low rates. Assuming that the population distribution of absences is bell shaped as shown in Figure 10.1, it also appears that most of the sample days were in the neighborhood of the population average.

Note two aspects of the sample from Figure 10.1. First, most of the sample elements are expected to come from the central portion of the population if the principal used a random selection technique,

Figure 10.1—Distribution of Absences for Two-Year Population Period

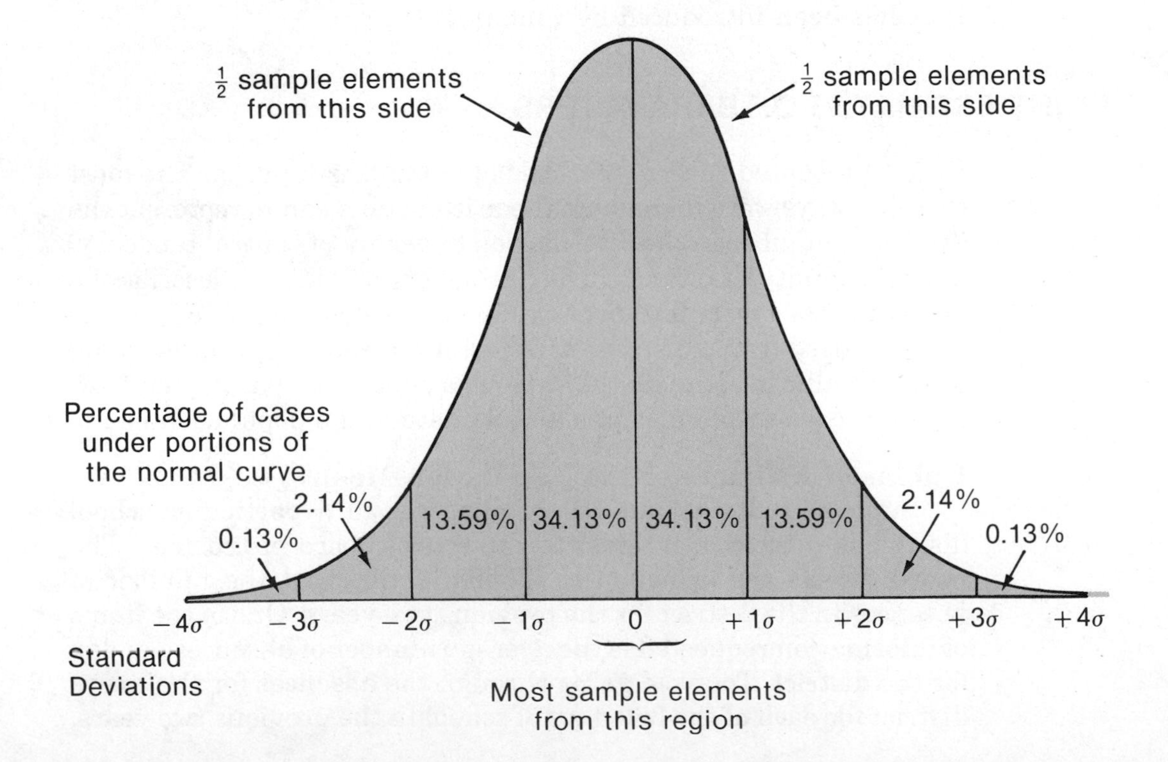

because more cases (days) are in the middle of the distribution. Second, and of more importance to the computation of a point estimate, *any element in the sample is just as likely to be above the mean as it is to be below the mean.* In other words, there is no systematic tendency for the sample elements to be either above or below the population mean. Thus the **sample mean** ($\overline{X}$) statistic is known as an **unbiased estimate** of the population mean (μ) parameter. The average (mean) number of absences per day revealed by the sample is the best estimate of the average number of absences per day in the total 360 days of school (the population).

Unbiased Point Estimates

As shown in the previous section, the point estimate of the population mean was a straightforward inference from the arithmetic mean of the sample computed just as you calculated $\overline{X}$ in Chapter 4. However, a minor adjustment must be made when estimating the population variability. By referring to Figure 10.1, you can visualize that most of the sample elements came from the central portion of the distribution. When computing the sample mean, getting a large number of sample elements from the central portion of the population was of no concern. What was at issue was the probability that an equal number of elements was obtained from above and from below the population mean.

However, when most of the sample elements come from the middle portion of the population distribution, this tends to bias the sample measure of variability. Specifically, the sample variability tends to be less than the corresponding population parameter. This is true because variability is an indication of how far the data deviate from the mean. If the data are clustered close to the mean, as is the case with sample data, the bias is toward computing a measure of variability that is smaller than the spread in the population. Further, the sample has lower variability because smaller numbers of people (days, objects, elements, etc.) do not vary as much as the large group (population) that includes all the extremes.

Degrees of Freedom. Before proceeding to the steps in calculating unbiased estimates of the variance and standard deviation, it is necessary to understand the concept of *degrees of freedom.* This concept plays an important role in inferential statistics; and, because a rigorous treatment of degrees of freedom is beyond the scope of a first course in statistics, an illustration of the concept will provide sufficient background for the inferential techniques discussed in this text.

It is important to recall two facts from earlier lessons. First, the variance and standard deviation are based on the sum of squares of the *deviations from the mean* or, more briefly stated, the sum of squares (Σx^2). Second, the sum of the deviations about the mean is zero and is expressed as $\Sigma x = \Sigma(X - \overline{X}) = 0$. An important consequence of these two facts can be shown with an example. Suppose you know that $N = 4$ for a sample and you are to guess the four *deviations* from the mean $\overline{X}$. You can guess any number for the first deviation, say

$$x_1 = 5.$$

Similarly, any values you choose can be assigned to two more deviations as:

$$x_2 = -3$$
$$x_3 = -6.$$

However, at the point where you must assign the fourth deviation, you can no longer assign any number you please as your guess for x_4. The value of x_4 *must be*:

$$\begin{aligned} x_4 &= 0 - x_1 - x_2 - x_3 \\ &= 0 - 5 + 3 + 6 \\ &= 4. \end{aligned}$$

Thus when you reached the fourth deviation, you were *no longer free* to guess. This illustrates the principle that, given the values of any $N - 1$ deviations from the mean of a set of N numbers, the value (sign and magnitude) of the Nth (last) deviation is completely determined. Therefore *there are N – 1* **degrees of freedom** *for a sample of N numbers*, which reflects the fact that only $N - 1$ of the deviations are "free to vary," but that the last number is determined. Although the N values of a variable X are independent (meaning the value of one does not depend on the value of another), the N values of x are not independent. Only $(N - 1)$ of the deviation scores (x's) are independent.

The Unbiased Estimate of the Variance. In Chapter 5 the equation for the variance was given as:

$$\sigma^2 = \frac{\Sigma x^2}{N}$$

which means that the variance equals the sum of squares of the deviation scores divided by N. This equation is correct if the data are considered a population and no inferences are going to be made.

However, the *unbiased estimate* of the population variance (σ^2) when computed from a sample is:

$$s^2 = \frac{\Sigma x^2}{(N-1)}.$$

Notice that the denominator for the sample variance is now $(N-1)$. For a given value of Σx^2, subtracting 1 from the sample size increases the value of s^2 slightly. The quantity $(N-1)$ is referred to as the *degrees of freedom* for the sample data. Because extreme values in the population will tend to be underrepresented in random samples, the $N-1$ in the denominator serves as a correction factor to produce an unbiased estimate of the variance in the population.

The Unbiased Estimate of the Standard Deviation. The previous discussion implies that the unbiased estimate of the population standard deviation (σ) can be given by an equation.

The formula for unbiased estimate of the population standard deviation is:

$$\mathbf{s} = \sqrt{\frac{\Sigma x^2}{N-1}}$$

where s is the estimate of σ, Σx^2 is the sum of squares, and N is the sample size.

The computational equations in Table 10.1 summarize the discussion to this point. From the equations it is obvious that as N tends to get very large, the value of the standard deviation will not be affected to any particular degree whether one uses the population

Table 10.1 Summary of Symbols and Formulas

For a Population (Parameter—No Inference)		For a Sample (Statistic—Estimate)
$\mu = \frac{\Sigma X}{N}$	Mean	$\overline{X} = \frac{\Sigma X}{N}$
$\sigma^2 = \frac{\Sigma x^2}{N}$	Variance	$s^2 = \frac{\Sigma x^2}{(N-1)}$
$\sigma = \sqrt{\frac{\Sigma x^2}{N}}$	Standard Deviation	$s = \sqrt{\frac{\Sigma x^2}{(N-1)}}$

where: $\overline{X}$ is an unbiased estimate of μ
s^2 is an unbiased estimate of σ^2
and s is an unbiased estimate of σ.

formula or the sample formula. However, with small samples of $N < 30$, the correction factor $(N - 1)$ should be used in the denominator when computing the standard deviation.

Virtually any statistic used to summarize sample data can be used to estimate the corresponding population parameter. Some computational equations, as with the variance and standard deviation, need to be adjusted so they will be unbiased estimators.

Unbiased Estimate of a Percentage (Proportion). In this course the primary interest is in estimating population means and standard deviations; therefore only one other estimate, the estimate of percentage (or proportion) in a population, will be briefly discussed because of its widespread use. If N is sufficiently large, the proportion of cases in the sample that belong to a particular category is the point estimate of that proportion in the population. For example, if in a random sample of students 18% were on the school's honor roll, then 18% (or 0.18) is the appropriate point estimate of the percentage of honor roll students in the population of interest.

The important concept to grasp at this time is that representative samples can yield quite accurate unbiased estimates of population parameters. Because parameters (as in the case of the average number of absences for the two-year period) are unknown, reliable estimates without systematic sources of error are invaluable for generalizing from samples to populations.

SAMPLING DISTRIBUTION OF MEANS

As previously mentioned, whenever data are being analyzed from a group that is less than the universe to which the results will be generalized, there is apt to exist a component of error. The term "error" in this case does not mean that the statistician has made a mistake in methodology. Error is a function of chance sampling events that are not controlled. There are methodologies for dealing with sampling error.

Assume that a sample ($N = 100$) is randomly drawn from a population of student rock musicians in a city. From that sample it is found that the musicians practice with a group an average of 14.5 hours per week outside of school hours. Hence 14.5 hours becomes the *point estimate* of the average number of hours all student rock musicians in the population practice per week outside of school. That is, because $\overline{X} = 14.5$ hours, the inference is that the population $\mu = 14.5$ hours. However, assume that the study needed to be replicated (repeated) to confirm or disconfirm the first results. So a second sample ($N = 100$) is selected and the mean number of practice hours is computed for the sample. If $\overline{X}_1$ is the mean of the first sample (14.5) and $\overline{X}_2$ is the mean of the second sample, do you think $\overline{X}_1 = \overline{X}_2$?

Unless the pollster happened to be very lucky, you would expect $\overline{X}_1$ and $\overline{X}_2$ to be fairly close because they are both estimates of the same population mean (μ). But it is doubtful that they would be *exactly* equal. Any difference between the two sample means would be the result of simple chance fluctuation inherent in random sampling, and the difference would be a random error component. This would be the likely cause of the difference between the two ACT mean scores discussed at the beginning of the chapter. However, in estimation the researcher is trying to capture the population mean (parameter). If the chances of two samples from the same population having the same mean are quite small, what are the chances of a point estimate exactly equaling the population mean?

Theoretically, any number of samples, say k, could be randomly selected from the population of student rock musicians. Figure 10.2 shows a hypothetical case in which many (k) samples are randomly selected from the same population. Assume that all k sample N's are equal. If the k sample means are treated as any other numerical values, you would observe variability within the set of $\overline{X}$'s. A frequency distribution of the sample means could be constructed, and the distribution could be smoothed into a curved-line graph as was done in earlier chapters. The graph would be in the shape of a bell curve with the same characteristics as the normal curve distribution.

The set of $\overline{X}$'s from the k samples is called a *sampling distribution of means*. A sampling distribution of any particular statistic from many samples could be created. The variability of a sampling distribution of statistics from a large number of samples (with the same N's) is caused by sampling error. Thus any measure of variability (spread) applied to data from a sampling distribution *describes* the sampling error component. The standard deviation is the most popular and useful measure of dispersion for sampling distributions. The sampling distribution made up of *sample means* is the most important aspect for most statistical purposes.

When used to describe the error variability within a sampling distribution of means, the standard deviation is called the **standard error of the mean**. How does one find the standard deviation (standard error) of a set of sample means (sampling distribution)? The

Figure 10.2—A Large Number of Theoretical Random Samples from a Population

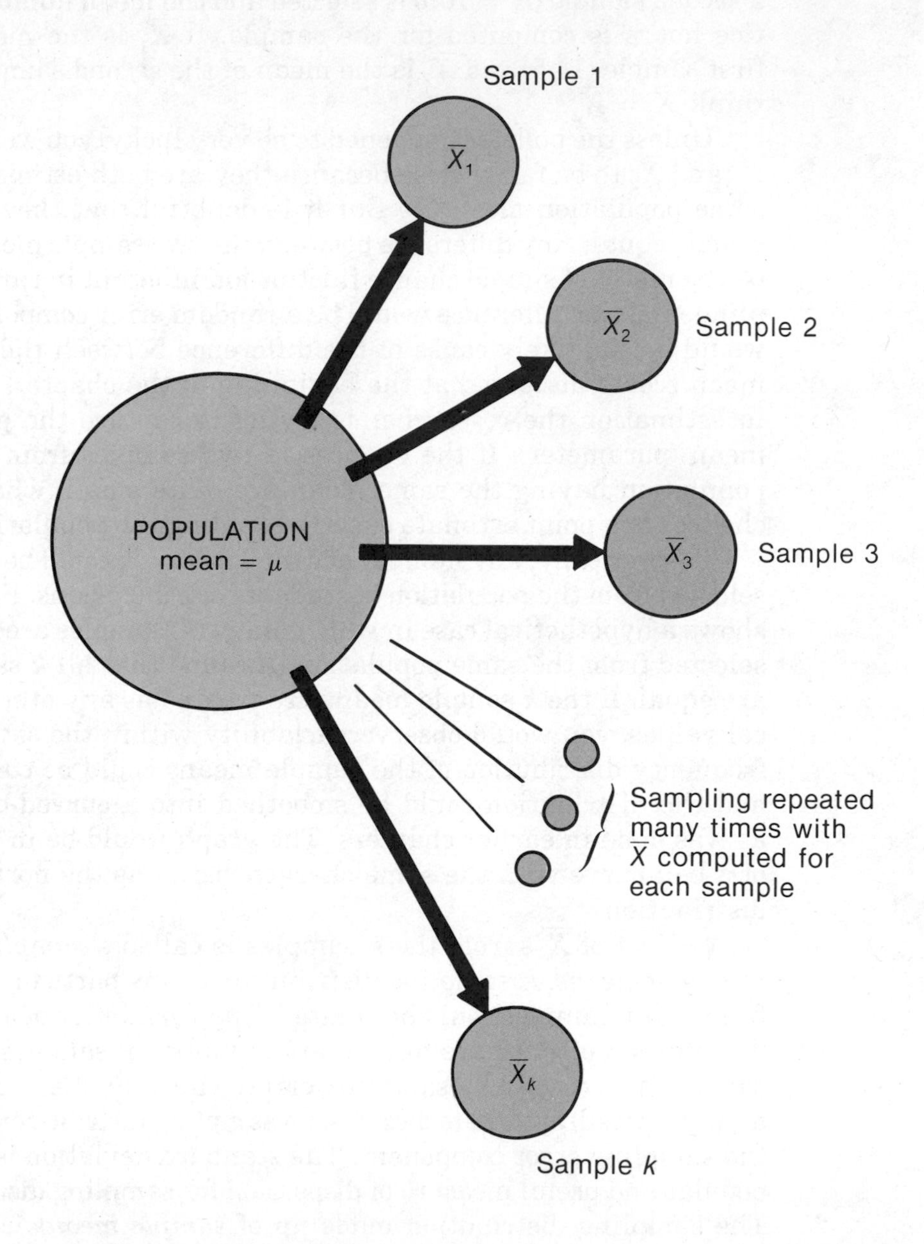

hypothetical example of student rock musicians is of no use, because who could collect data on a large number of samples (say 1,000) from a common population? This problem is addressed by the central limit theorem.

The Central Limit Theorem and the Standard Error of the Mean

A theorem is a principle or proposition that can be proved using accepted premises. One such principle in statistics is the **central limit theorem**. This theorem states: *For any population with a mean μ and standard deviation σ, the sampling distribution of means with sample sizes = N will approach a normal distribution with a mean of μ and a standard deviation of $\frac{\sigma}{\sqrt{N}}$ as N gets large.*

As an illustration, consider a small population consisting of six numbers {5, 6, 7, 8, 9, 10}. The total number of samples of size $N = 2$ is the number of combinations of the 6 numbers taken 2 at a time. Therefore there are 15 possible samples of size 2. The list of samples with their respective sample means is:

Sample	Mean ($\overline{X}$)	Sample	Mean ($\overline{X}$)
5, 6	5.5	6, 10	8.0
5, 7	6.0	7, 8	7.5
5, 8	6.5	7, 9	8.0
5, 9	7.0	7, 10	8.5
5, 10	7.5	8, 9	8.5
6, 7	6.5	8, 10	9.0
6, 8	7.0	9, 10	9.5
6, 9	7.5		

Confirm that the mean of the distribution of possible means is 7.5, the same as the mean (μ) for the population consisting of the six values. Because there are 15 possible samples (outcomes), the probability of getting a particular mean with a random sample of two is given by the distribution:

$\overline{X}$	$p(\overline{X})$
5.5	$\frac{1}{15}$
6.0	$\frac{1}{15}$
6.5	$\frac{2}{15}$
7.0	$\frac{2}{15}$
7.5	$\frac{3}{15}$
8.0	$\frac{2}{15}$
8.5	$\frac{2}{15}$
9.0	$\frac{1}{15}$
9.5	$\frac{1}{15}$

Notice that the original distribution making up the population has no distinct mode. The probability distribution for the sample means is beginning to show a probability mode and a symmetrical distribution. Intuitively, you can see from the probability distribution that samples with means close to 7.5 will have a numerical advantage over others if a large number of samples of size two is drawn from this population (of course, with replacement after each draw).

The central limit theorem implies at least three important facts. First, if random samples from a population are large enough, and if there is a large number of samples, the mean of the sample means ($\overline{X}$'s) will equal the mean of the population (μ). Second, the standard deviation of the sample means (called the standard error of the mean) will be equal to the standard deviation of the population divided by the square root of N: $\sigma_{\overline{X}} = \frac{\sigma}{\sqrt{N}}$. Third, the sampling distribution of means will be distributed as a normal curve distribution.

The standard deviation of the sampling distribution is calculated using a parameter, namely, σ. However, because population parameters are rarely known quantities and because generally there is only one sample with which to work, statistics (point estimates) are used in the equation.

The practical computational equation for the standard error of the mean is:

$$s_{\overline{X}} = \frac{s}{\sqrt{N}}$$

where $s_{\overline{X}}$ is the standard error, s is the sample standard deviation, and N is the sample size.

This equation states that the standard error of the mean is estimated by the point estimate of the standard deviation divided by the square root of N.

The example using the sample of student rock musicians to determine the amount of practice time will illustrate an application of the central limit theorem. Recall that data from a random sample ($N = 100$) of student rock musicians yielded a mean of 14.5 hours. Assume that the standard deviation of sample data was 2.0 hours. The population mean (μ) is estimated as being 14.5. However, the error in the sample mean is described by the standard deviation of the sampling distribution of means, which is the standard error of the mean. The standard error of the mean for a sample with $s = 2.0$ is found as follows:

$$
\begin{aligned}
s_{\overline{X}} &= \frac{s}{\sqrt{N}} \\
&= \frac{2.0}{\sqrt{100}} \\
&= \frac{2}{10} \\
&= 0.20.
\end{aligned}
$$

Thus the standard error of the mean is two-tenths of an hour.

Because the central limit theorem also implies that the sampling distribution of means is normally distributed, Figure 10.3 shows the theoretical distribution of sample means of practice time for the musicians.

Figure 10.3—Normal Curve Showing Standard Error of the Mean and Sampling Distribution

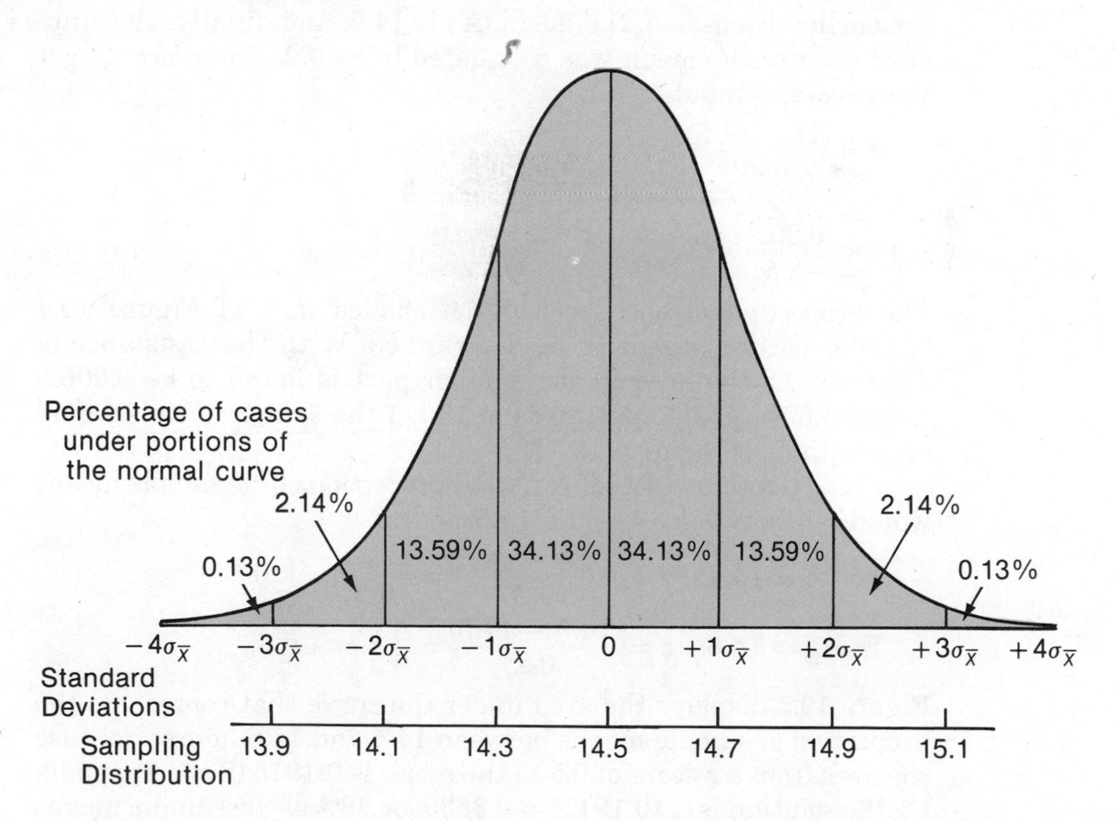

If, in a hypothetical situation, many samples of size $N = 100$ of student rock musicians were assessed as to how many hours they practiced outside school hours, Figure 10.3 shows that 68% of the sample means would be expected to fall between 14.3 and 14.7 hours. This is because about 68% of the area under the normal curve falls between the mean and ± one standard deviation (standard error of the mean for a sampling distribution).

Because μ and σ can be estimated from $\overline{X}$ and s, respectively, and because the resulting distribution of the sample means is normally distributed, the same operations can be performed with a sampling distribution as with normal curves from any other sources of data. For example, what proportion of randomly drawn samples would be expected to practice an average of 14 hours or less? A z-score is computed in the customary fashion:

$$z = \frac{(X - \overline{X})}{s_{\overline{X}}}$$

because the standard deviation of a sampling distribution is the standard error of the mean. In the example, X (the score) is 14; from the earlier discussion, the mean ($\overline{X}$) is 14.5; and, finally, the standard error of the mean was calculated to be 0.2. Therefore, to get the z-score, compute:

$$\begin{aligned} z &= \frac{(14 - 14.5)}{0.20} \\ &= \frac{-0.5}{0.20} \\ &= -2.5. \end{aligned}$$

The proportion of the area in the shaded part of Figure 10.4 (smaller portion) needs to be determined. With the assistance of Appendix C, the area in the smaller part is found to be 0.0062. Hence 0.62% (about two-thirds of 1%) of the sample means would be less than 14 hours.

Using the same procedures, what proportion of the sample means would fall between 14.4 and 14.6 hours?

$$\text{For } X = 14.4\text{:} \quad z = \frac{(14.4 - 14.5)}{0.2} = \frac{-0.1}{0.2} = -0.5$$

$$\text{For } X = 14.6\text{:} \quad z = \frac{(14.6 - 14.5)}{0.2} = \frac{0.1}{0.2} = +0.5.$$

Figure 10.5 displays the area under the curve that represents the proportion of sample means between 14.4 and 14.6 hours. Because the area from a z-score of 0.5 to the mean is 0.1915 (from Appendix C), the solution is $(2)(0.1915) = 0.3830$, or 38% of the sample means would fall between 14.4 and 14.6 hours.

Figure 10.4—Sampling Distribution: Area Below 14.0 Hours

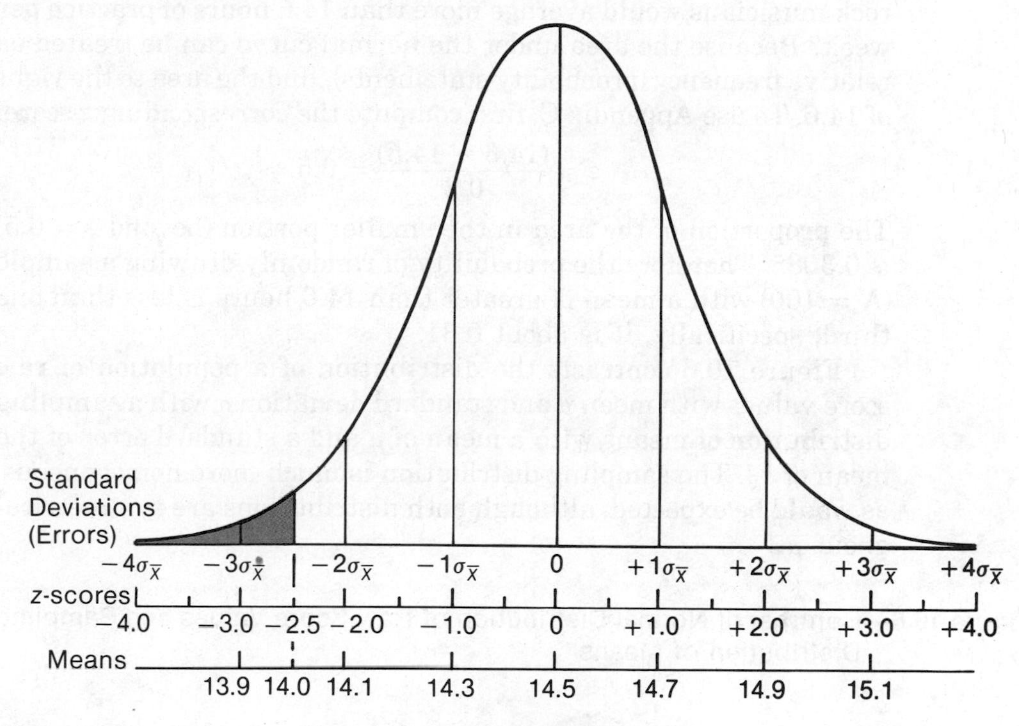

Figure 10.5—Sampling Distribution: Area Between 14.4 and 14.6 Hours

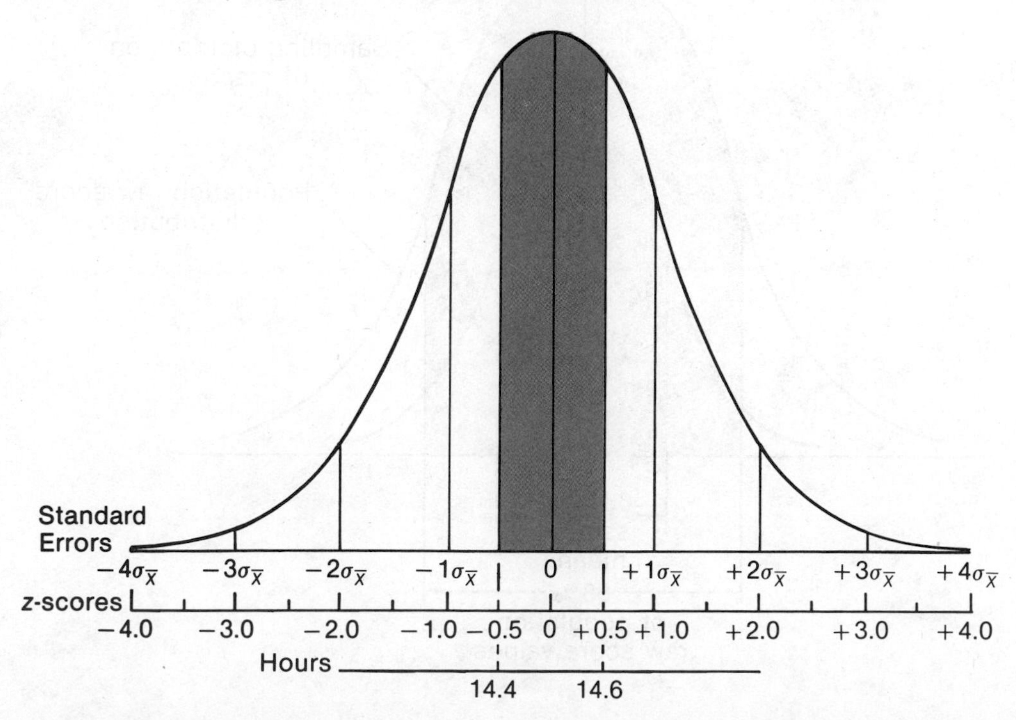

What is the probability that a random sample of 100 student rock musicians would average more than 14.6 hours of practice per week? Because the area under the normal curve can be treated as relative frequency (probability statements), find the area to the right of 14.6. To use Appendix C, first compute the corresponding *z*-score:

$$z = \frac{(14.6 - 14.5)}{0.2} = 0.5.$$

The proportion of the area in the smaller portion (beyond $z = 0.5$) is 0.3085. Therefore the probability of randomly drawing a sample ($N = 100$) with a mean of greater than 14.6 hours is less than one third; specifically, it is about 0.31.

Figure 10.6 contrasts the distribution of a population of raw score values with mean μ and standard deviation σ with a sampling distribution of means with a mean of μ and a standard error of the mean of $s_{\overline{X}}$. The sampling distribution is much more homogeneous, as would be expected, although both distributions are symmetrical about μ.

Figure 10.6—Contrast of Normal Distribution of Raw Score Values and Sampling Distribution of Means

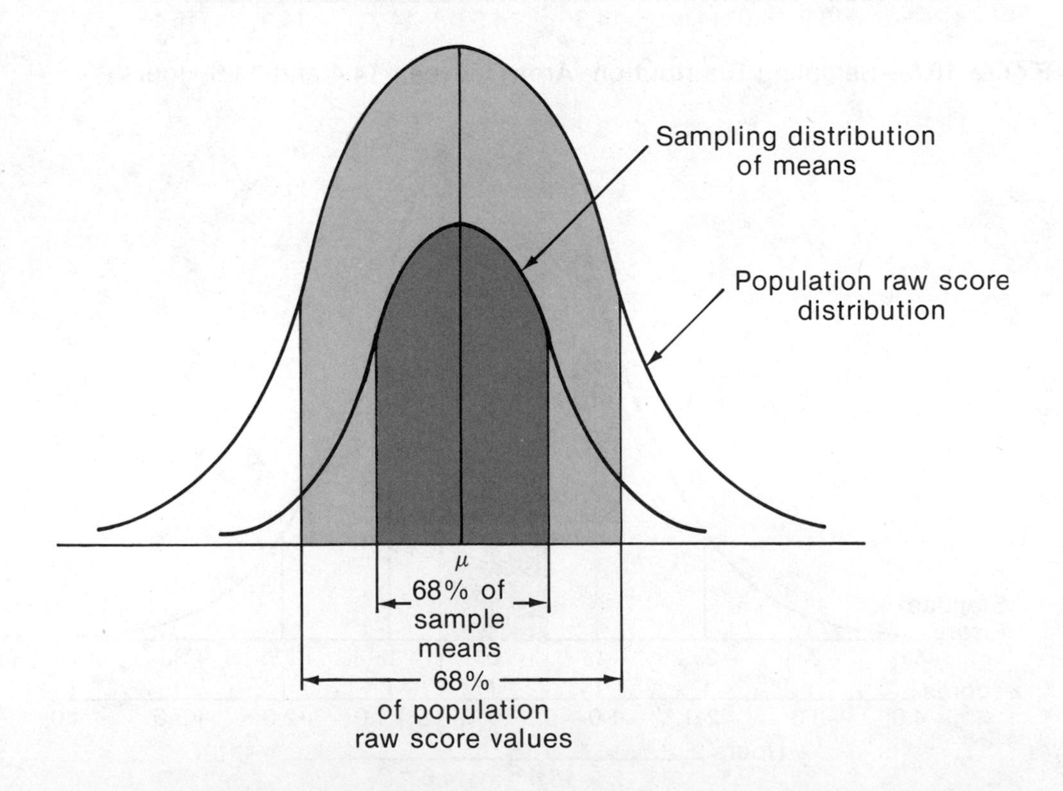

Notice that $\mu \pm \sigma$ on the population distribution contains about 68% of the individual population elements. The interval $\mu \pm \sigma_{\overline{X}}$ contains approximately 68% of the hypothetical sample means. Because of this phenomenon, such an interval ($\mu \pm 1\sigma_{\overline{X}}$) is referred to as a 68% confidence interval in the discussion that follows.

INTERVAL ESTIMATES OF THE MEAN: CONFIDENCE INTERVALS

As indicated earlier, the point estimate of the musicians' practice time was 14.5 hours per week. Further, Figure 10.5 illustrated that about 68% of the hypothetical sample means would fall within an interval from 14.4 to 14.6 hours. How confident would you be about the validity of the point estimate of 14.5 as compared to an interval estimate of 14.4 – 14.6? That is, which would you have more confidence in for capturing the population mean?

Statisticians are often concerned with estimating population parameters using sample statistics in such a way that a probability statement can be made about the amount of error in the estimate. One kind of estimate has been described as a point estimate: a single-valued statistic that is an unbiased estimate of the corresponding parameter. While the existence of sampling error has been pointed out, specific ways of making probability statements about the size of potential sampling error have not been explained.

The most common method used to arrive at a statement about the amount of possible sampling error is **interval estimation**. As you have seen, sampling error for a distribution of means is normally distributed. This enables the statistician to establish a range of values on each side of the point estimate of the mean and to determine the probability that the parameter μ lies within this range. Such a probability is expressed as a percentage and is referred to as the **level of confidence**. The interval between the upper and lower limits of the range of values on either side of the point estimate of the mean is called the **confidence interval** for the mean. Because the sample mean is an unbiased estimate of μ (does not have a systematic tendency to be larger or smaller than the parameter), $\overline{X}$ is located in the middle of the interval. The higher the specified level of confidence, the wider the confidence interval.

At the end of the discussion in the previous section, a 68% confidence interval was described as an interval ranging from one standard error below the mean to one standard error above the mean. Of course the standard error of the mean is a standard deviation.

With the aid of Appendix C, you can determine that 68% of the area under the normal curve lies within ± one standard deviation from the mean. In the present instance, the range specified by a 68% confidence interval implies that 68% of the means of a large number of randomly selected samples from a population would fall within that interval. Because the relative frequency of the sample statistics ($\overline{X}$'s) is 0.68 within one standard error above and below the mean, the probability of getting a sample mean in that interval is also 0.68. Conceptually, such an interval will have a 68% chance of including μ.

Actually, 90%, 95%, or 99% confidence intervals are more widely used than 68% confidence intervals. A 95% confidence interval would consist of a lower limit and an upper limit to form a range within which 95% of means of samples of size N from a population would fall. Similarly, a 90% confidence interval would be an interval around the mean in which 90% of the hypothetical sample means would fall. When constructing a 99% confidence interval, the researcher is attempting to define an interval that has a 0.99 probability of containing the population parameter. Naturally, after the interval is constructed, μ either is or is not in that interval; therefore the probability is either 1.00 or 0.00 that μ is in the interval after it is determined.

Many possible levels of confidence can be used to establish confidence intervals with the aid of Appendix C. As an illustration of this procedure, suppose a 95% confidence interval for a mean is desired. By referring to Appendix C, you will determine that the 95% confidence interval (which extends 0.475 of the area on each side of the mean) corresponds to 1.96 standard errors (z-scores) from the mean. In general, the limits of a confidence interval of the mean are:

$$\text{upper limit} = \overline{X} + (z)(s_{\overline{X}})$$
$$\text{lower limit} = \overline{X} - (z)(s_{\overline{X}})$$

where $z = 1.65$ for the 90% confidence interval,
$z = 1.96$ for the 95% confidence interval, and
$z = 2.58$ for the 99% confidence interval.

What value of z would be required for an 80% confidence interval? By using Appendix C, you find that the z-score closest to including 40% on either side of the mean is 1.28; therefore the limits of an 80% confidence interval are:

$$\overline{X} \pm (1.28)(s_{\overline{X}}).$$

The following example will help synthesize the important statistical procedures of estimation. Assume the values in Table 10.2 are the results of a random sampling from a population.

Table 10.2 Hypothetical Sample Elements (*X*-values)

20	22	25	26	26	23
25	30	29	23	25	26
22	24	23	24	28	21
28	27	25	28	35	20
22	26	25	25	24	27
24	25	24	27	20	26

The following computations are performed to two-decimal-place accuracy, which is usually sufficient for integer data.

Because

$$\overline{X} = \frac{\Sigma X}{N} = \frac{900}{36} = 25$$

the point estimate of μ is 25. The sample variance statistic (s^2) is the unbiased estimate of σ^2; thus:

$$s^2 = \frac{\Sigma x^2}{(N-1)} = \frac{320}{35} = 9.14.$$

Taking the square root of the variance provides the unbiased point estimate of σ:

$$s = \sqrt{9.14} = 3.02.$$

Therefore the unbiased point estimates of the population mean, variance, and standard deviation are 25, 9.14, and 3.02 respectively.

The standard error of estimate is computed as follows:

$$s_{\overline{X}} = \frac{s}{\sqrt{N}} = \frac{3.02}{\sqrt{36}} = \frac{3.02}{6} = 0.50.$$

The 95% confidence interval for the mean is:

$$\overline{X} \pm (1.96)(s_{\overline{X}}) = 25.00 \pm (1.96)(0.5) \text{ or } \{24.02 \text{ to } 25.98\}.$$

The 95% confidence interval for μ is an interval from 24.02 to 25.98.

The 99% confidence interval computation and calculator keystroke sequence are shown in Figure 10.7. Notice that as the confidence level increases, so does the width of the interval.

Figure 10.7—Calculator Keystrokes

PROBLEM:

Compute the upper and lower limits of a 99% confidence interval when $\overline{X} = 25$, $s = 3.02$, and $N = 36$; that is, solve $\overline{X} \pm (2.58)\left(\frac{s}{\sqrt{N}}\right)$

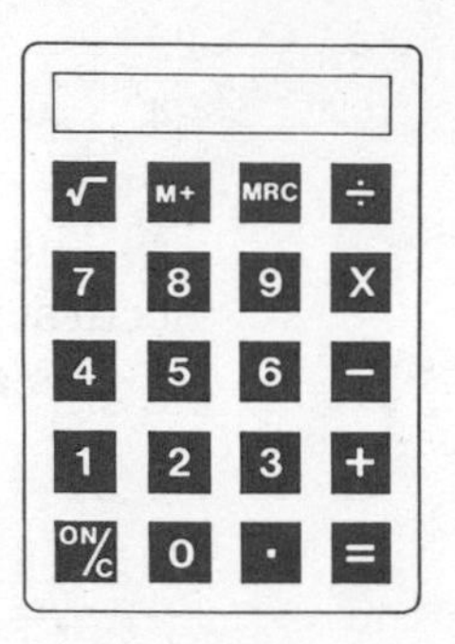

Procedure	Arithmetic Operation	Calculator Keystrokes	Display
1. Divide s by $\sqrt{N}$	**1.** $\frac{3.02}{\sqrt{36}} = \frac{3.02}{6}$ $= 0.5033333$ $= 0.50$	**1.** 3.02 [÷] 36 [√]	3.02 6
2. Multiply quotient in 1. by 2.58	**2.** $0.50 \times 2.58 = 1.29$	**2.** [×] 2.58 [=]	0.5033333 1.2985999
3. Add product in 2. to the mean	**3.** $1.29 + 25 = 26.29$ (upper limit)	**3.** [M+] [+] 25 [=]	1.2985999 1.2985999 26.298599 (Upper)
4. Subtract product in 2. from mean	**4.** $25 - 1.29 = 23.71$ (lower limit)	**4.** 25 [−] [MRC] [=]	25 1.2985999 23.701401 (Lower)

Note: Carrying several decimal places throughout the computation causes the final results to differ somewhat as a result of rounding error.

The information just presented not only provides single-valued estimates for the important population parameters, but also provides an interval with upper and lower bounds within which μ lies for a given level of probability. In inferential statistics, techniques are available for establishing confidence intervals for other parameters such as variance, median, and proportions, but such techniques are beyond the scope of this text. However, a conceptual understanding of confidence intervals will help you be a wiser consumer and user of statistical reports.

COMPUTER FOCUS 10—Confidence Intervals

Purpose

If a practical need exists for a program, modification of existing programs is often a more efficient means of implementing the program than starting over from the beginning. While the task of writing a program that would output the confidence intervals of the mean of a distribution is not an imposing one, the purpose of this Computer Focus is to provide an opportunity to modify a program to change the output to the desired results.

Problem

In Computer Focus 5 you completed a BASIC program for computing the standard deviation (SD) and the mean (M) for a set of *N* scores. These values are primary ingredients for computing a standard error of the mean (SE) and confidence intervals.

Therefore your task is to modify the program in Computer Focus 5 to accomplish the computations of the standard error of the mean and the upper and lower limits for 95% and 99% confidence intervals.

Procedures

1. Remove statement 55 from the program in Computer Focus 5.
2. Modify statement 47 to correctly compute the sample variance:

```
V = SS/(N - 1)
```

3. Use the values from the program in Computer Focus 5 to compute the standard error of the mean (SE):

```
SE = SD/SQR(N)
```

4. Use the mean (M) from the program and SE from step 3 to find confidence interval limits:

```
L1 = M - (1.96)*SE
U1 = M + (1.96)*SE
```

`L2` = You supply the expression for finding the lower limit of the 99% confidence interval.

`U2` = You supply the expression for finding the upper limit of the 99% confidence interval.

5. Print out the standard error and the two confidence intervals:

```
PRINT "95% CONFIDENCE INTERVAL: ";L1;" TO ";U1
PRINT "99% CONFIDENCE INTERVAL: ";L2;" TO ";U2
```

6. Reattach the END statement as the last line.
7. Test the program using data from a problem with known results.

KEY TERMS AND NOTATION

Central limit theorem
Confidence interval
Degrees of freedom
Interval estimation
Level of confidence
Point estimation
Population mean (μ)
Population standard deviation (σ)
Population variance (σ^2)
Sample mean ($\overline{X}$)
Sample standard deviation (s)
Sample variance (s^2)
Standard error of the mean ($s_{\overline{X}}$)
Unbiased estimate

APPLICATION EXERCISES

10.1 A sample size N is to be drawn from a population with a mean of 250 and a variance of 100. Complete the following table by calculating the standard error of the mean for the various sample sizes:

	N	$s_{\overline{X}}$
a.	4	______
b.	16	______
c.	25	______
d.	36	______
e.	64	______
f.	100	______
g.	500	______
h.	1,000	______

10.2 Referring to Application Exercise 10.1, what would you conclude about the relationship between sample size and the amount of potential sampling error as measured by the standard error of the mean?

10.3 Seventy-five students were randomly selected from a senior class to participate in an experiment being conducted by the psychology club. Each member of the sample was given the task of assembling a jigsaw puzzle as quickly as possible. The sample data for the time variable were: $\overline{X} = 50$ and $s = 8$. What are the point estimates for the entire senior class for the parameters:

a. μ?
b. σ?
c. σ^2?

10.4 A random sample of 64 students were asked to rate the school cafeteria food on a scale from 1 to 30. The sample mean was 22, and the sum of squares (Σx^2) was 1,008.

a. Calculate s, the standard deviation statistic.
b. Compute the standard error of the mean.
c. Determine the limits of a(n):
   (1) 68% confidence interval.
   (2) 80% confidence interval.
   (3) 95% confidence interval.

10.5 The standard error of the mean is $\frac{s}{\sqrt{N}}$, but the standard error of the median is 1.25 times $\frac{s}{\sqrt{N}}$. Suppose that for a particular problem, $\overline{X} = 100$, median = 100, $s^2 = 25$, and $N = 25$.

a. Compare the magnitudes of the standard errors of the mean and median. Which is larger? By what factor?
b. Based on your answer to a., which sample statistic (mean or median) would generally provide the more accurate estimate of the population's central tendency? Why?

10.6 Assume that a sample ($N = 100$) is drawn from a city's PTA population. They are administered a set of 100 questions about the public schools that require factual answers. The sample mean ($\overline{X}$) for the group was 80, and the sample standard deviation (s) was 15 points.

a. What is the probability that a *single member* drawn from the population of PTA members would score higher than 95 on the survey exam?
b. What is the probability that a randomly selected *sample* of 100 cases would have a *mean* score higher than 81.5?

10.7 A high school principal wishes to estimate the average number of hours spent watching television by the sophomore class. A random sample of 100 sophomores are asked to record their number of TV viewing hours daily for a period of two weeks. The mean number of hours reported for the two-week period by the students was 20.6, and the standard deviation was 6.0.

a. Assuming a normal distribution, about what percentage of the sophomore students watched TV from 8.84 to 32.36 hours during the two-week period?
b. Suppose that academic grades begin to drop if more than 26 hours per two-week period are spent watching TV. What percentage (or proportion) of the sophomore population of students would be running the risk of lowering their grades because of an excess of TV viewing?
c. Construct the 95% and 99% confidence intervals that the principal could use in estimating the *mean* number of TV viewing hours for the sophomore population during a two-week period.

10.8 If the standard deviation of a sample was 9.0 and the standard error of the mean was 1.5, what was the sample size N?

10.9 Suppose that a school bus, on the average, arrives at a particular bus stop at 8:30 a.m. However, as a statistics class project, actual times of arrival were recorded for a period of one semester. The students found that the bus arrival time has a standard deviation of 5 minutes. Given these data:

a. About what are the chances that the bus will arrive at 8:32 a.m. or later on any particular day?
b. Find the approximate probability that the *average* arrival time for a two-month period in which the bus comes 25 times is between 8:29 and 8:31 a.m.

10.10 As a fund-raising project, students in the arts and crafts club decided to sell boxes of candy. A sample of 100 boxes of candy were randomly drawn from a large shipment of candy. The average weight of the boxes in the sample was 0.98 pound. The standard deviation of the weights was 0.1 pound.

a. Construct a 95% confidence interval for the average weight of the candy.
b. How confident would you be in telling your customers that the *average* weight of the boxes in the shipment was *at least* 1 pound?
c. How confident would you be in telling a customer that one randomly selected box would weigh at least 1 pound?

11

Hypothesis Testing

CHAPTER OBJECTIVES

Upon completion of the chapter, students will be able to:

1. Use the sampling distribution characteristics to test hypotheses about differences between sample and population means.
2. Distinguish between a null hypothesis and the alternate (alternative) hypothesis.
3. Define the significance level (α) of a statistical test.
4. Distinguish between one- and two-tailed statistical tests and know when each is used.
5. Define and contrast type I and type II statistical errors.
6. Perform the steps of a z-test of significance at a specified α level.

The faculty lounge and student commons area at Jesse Mitchell High School were buzzing with news about the school-wide achievement test results that had been received on Monday morning. Tests on various subject matter areas had been commercially scored, and press-on labels were being prepared for distribution of the test results to parents, students, and counselors.

The composite score shown on the report of the test results was a total of the scores on all tests that had been administered during the early fall term. On a nationwide basis, the composite test score was scaled on a standard *T*-score system with a mean of 50 and a standard deviation of 10. However, the 900-member student body at Jesse Mitchell High had a mean of 51.1 with a standard deviation of 10. Furthermore, the 25 students taking statistics had an average composite achievement score of 54.2. The principal speculated that the Jesse Mitchell student body was, on the average, scoring higher than average for the national population of students who had taken the test. Was the principal correct? By using some of the techniques from the previous chapter, it is possible to make a probability statement about the principal's conjecture.

In this case, $\mu = 50$ and $\sigma = 10$ for the entire population of students. For the Jesse Mitchell students, $\overline{X} = 51.1$, $s = 10$, and $N = 900$. The question of central concern is whether or not the difference between the population mean (μ) of 50.0 and the sample mean for the school of 51.1 is the result of simple sampling error and/or chance fluctuations. The standard error of the mean is the measure of variability for the sampling distribution of means and can be calculated by:

$$\begin{aligned} s_{\overline{X}} &= \frac{\sigma}{\sqrt{N}} \\ &= \frac{10}{\sqrt{900}} \\ &= \frac{10}{30} \\ &= 0.33 \quad \text{(rounded to two decimal places).} \end{aligned}$$

Using the sample mean for the school as an element from the sampling distribution of means, Figure 11.1 displays the comparison of μ and $\overline{X}$. What is the likelihood that a sample mean would deviate that far from the population mean just by chance?

Figure 11.1—Sampling Distribution of Means Showing National Mean Achievement (μ) and Jesse Mitchell Student Body Achievement ($\overline{X}$)

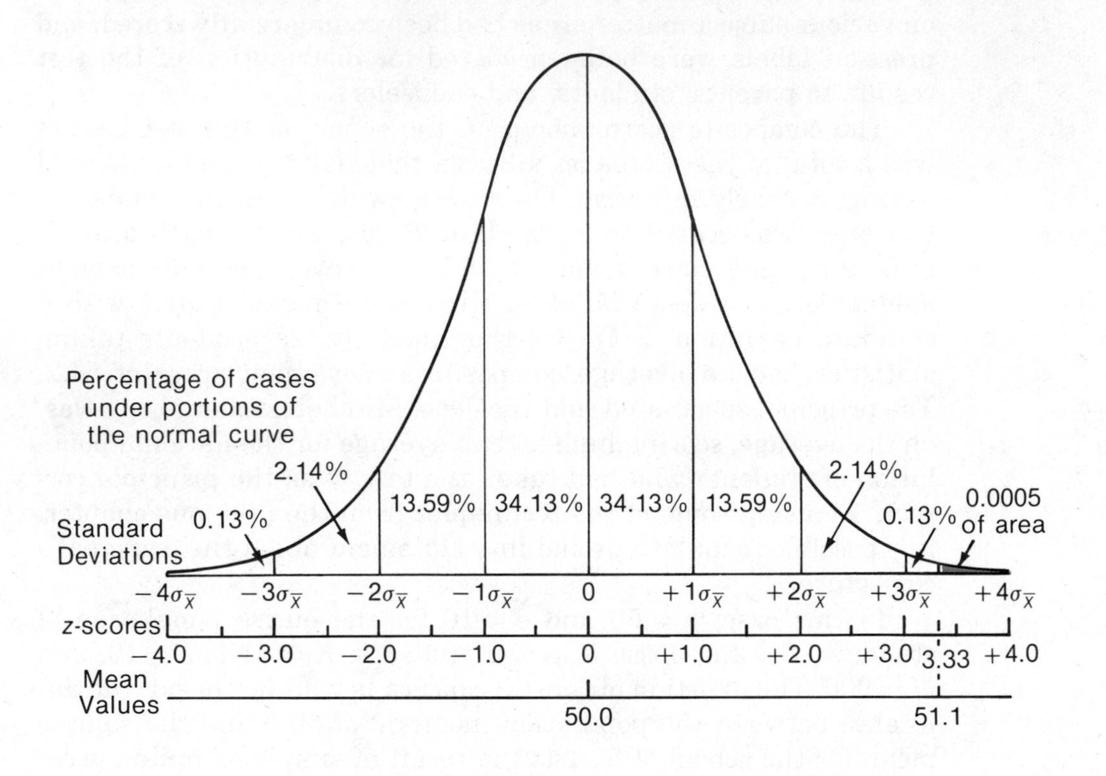

A slight modification of the *z*-score formula used in Chapter 6 with the appropriate symbols for representing a *sampling* distribution is:

$$z = \frac{(\overline{X} - \mu)}{s_{\overline{X}}}$$

where $\overline{X}$ is the sample mean, μ is the population mean, and $s_{\overline{X}}$ is the standard error of the mean.

The corresponding *z*-score for the typical sample value of 51.1 is shown in Figure 11.1 and computed as:

$$z = \frac{(51.1 - 50.0)}{0.33}$$
$$= +3.33.$$

The sample mean is 3.33 standard deviations above the population mean. By consulting Appendix C, you can determine that the probability of getting a *z*-score of 3.33 or higher by chance on the normal distribution is less than 0.0005. This implies that the probability of getting a difference between the means as large as 1.1 (51.1 – 50.0) is less than 0.0005, which is sometimes shown symbolically as ($p < 0.0005$). Because the probability that such a difference would have occurred by chance is so small, the conclusion would be that the principal was correct in believing that Jesse Mitchell students are achieving at a higher level than the national population on the average. If you were able to follow the above discussion, congratulations! You have just completed a statistical test of a hypothesis.

Another problem is presented for illustration purposes before the rationale for and details of statistical hypothesis testing are explained. There was also speculation that the statistics students were achieving at a higher level than the Jesse Mitchell student body as a whole. To test this hypothesis, the population is now defined as the high school student body at Jesse Mitchell High School, and the sample is defined as the students enrolled in the statistics class. Using these conditions, the mean $\mu = 51.1$ and the standard deviation $\sigma = 10$, while the statistics for the problem are $\overline{X} = 54.2$ and $N = 25$. Using the formula for finding the standard error of the mean yields:

$$\begin{aligned} s_{\overline{X}} &= \frac{\sigma}{\sqrt{N}} \\ &= \frac{10}{\sqrt{25}} \\ &= \frac{10}{5} \\ &= 2.0. \end{aligned}$$

In this case the question whether or not 54.2 and 51.1 are different because of error is answered with the same technique as the earlier example. Given that $\mu = 51.1$, $\sigma = 10$, $\overline{X} = 54.2$, and $N = 25$, the corresponding standard *z*-score can be computed by the sequence of keystrokes shown in Figure 11.2.

Thus the *z*-score equivalent to 54.2 on the sampling distribution curve shown in Figure 11.3 is:

$$\begin{aligned} z\text{-score} &= \frac{(\overline{X} - \mu)}{s_{\overline{X}}} \\ &= \frac{(54.2 - 51.1)}{2} \\ &= 1.55. \end{aligned}$$

This is the calculated value of *z*.

The information in Appendix C indicates that the probability of getting such a difference (3.1 or more) by chance is the same as the probability of randomly getting a *z*-score of 1.55 or more, which is about 0.06 (that is, $p = 0.06$). Hence there is only about a 6% chance that the difference in achievement of the statistics students and the entire student body occurred by chance. Thus you would be inclined to think that the statistics group is outperforming the student body as a whole. However, the confidence level is not as strong in making such a statement as it would be with the probability statement made in the first example with the comparison of the Jesse Mitchell High School mean and the national population mean.

Figure 11.2—Calculator Keystrokes

PROBLEM:

Compute *z*-score equivalent to a given sample mean:

$$z = \frac{(\overline{X} - \mu)}{\frac{\sigma}{\sqrt{N}}} = \frac{(\overline{X} - \mu)}{s_{\overline{X}}}$$

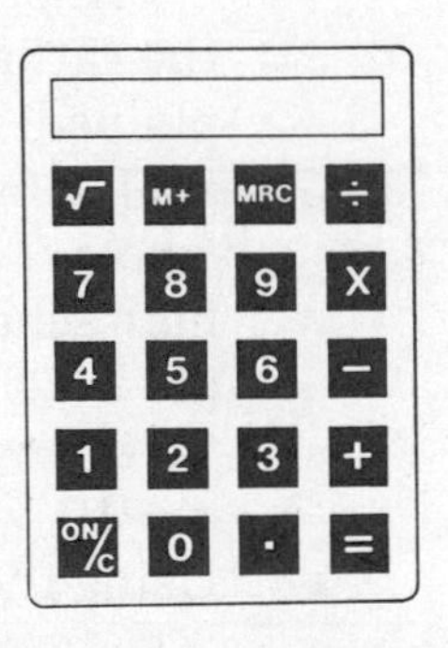

Procedure	Arithmetic Operation	Calculator Keystrokes	Display
1. Compute the value of $s_{\overline{X}} = \frac{s}{\sqrt{N}}$	**1.** $\frac{10}{\sqrt{25}} = \frac{10}{5} = 2$	**1.** 25 [√]	5
		[M+]	5
		10 [÷]	10
		[MRC]	5
		[=]	2
		[CLM]	2
		[M+]	2
2. Find $(\overline{X} - \mu)$	**2.** 54.2 − 51.1 = 3.1	**2.** 54.2 [−]	54.2
		51.1 [=]	3.1
3. Divide numerator by denominator: $z = \frac{\overline{X} - \mu}{s_{\overline{X}}}$	**3.** $\frac{3.1}{2} = 1.55$	**3.** [÷] [MRC]	2
		[=]	1.55

Figure 11.3—Sampling Distribution of Means Contrasting Student Body Mean Score (μ) and Statistics Class Mean ($\overline{X}$) Achievement Score

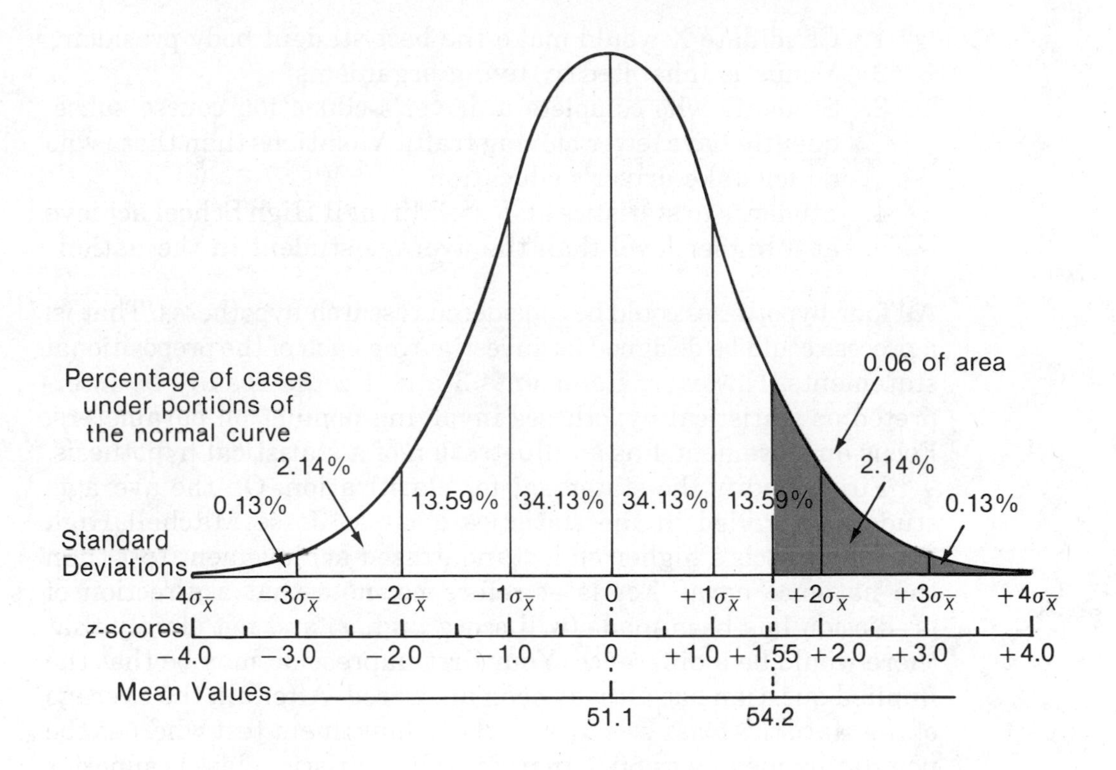

HYPOTHESES

In general, a **hypothesis** is a conjecture or a supposition tentatively accepted to explain certain phenomena. It is an unproven proposition that provides a foundation for investigation or argument. Throughout recorded history, stated hypotheses have stimulated scientific inquiry and thought that have expanded the frontiers of knowledge in every aspect of life. With the twentieth-century development of inferential statistical techniques, drawing conclusions about hypotheses for *populations* from samples is now commonplace.

Research Hypotheses and Statistical Hypotheses

A **research hypothesis** is a conjectural statement made about an unknown or unproven phenomenon. A special kind of research hypothesis is called a *statistical hypothesis*. A **statistical hypothesis** is a supposition about some population parameter. Consider the four hypotheses that follow:

1. Candidate X would make the best student body president.
2. Venus is inhabited by living organisms.
3. Students who complete a driver's education course subsequently have fewer moving traffic violations than those who do not take driver's education.
4. Students in statistics at Jesse Mitchell High School achieve at a higher level than the average student in the nation.

All four hypotheses could be considered research hypotheses. That is, a process could be designed for investigating each of the propositional statements. However, statements 3 and 4 could readily be interpreted as statistical hypotheses involving population parameters. Focus on statement 4 as an illustration of a statistical hypothesis.

First, modify the statement for clarification: On the average, students enrolled in the statistics class at Jesse Mitchell High School will score higher on a standardized achievement test than the national norm. For later reference, note that a direction of prediction has been made (will score higher) and not simply that there would be a difference. Your first impression may be that the implied question has already been answered. After all, the average of the statistics class was 54.2 on the achievement test whereas the population mean was 50.0; therefore the statistics class is superior to the national norm group with respect to academic achievement. But wait a moment before jumping to what seems to be an obvious conclusion. That difference may have a high probability of being a function of error. If so, the conclusion might be that the 4.2-point difference between $\overline{X}$ and μ is not a real difference; it might be the result of sampling error. In such a case there would be no evidence of higher achievement on the part of the statistics class. Follow this example further.

The Null Hypothesis

Unfortunately, a statistical hypothesis cannot be directly tested with statistics. Such a hypothesis is generally stated in "positive" terms according to the way in which the investigator perceives the phenomenon may occur. Statistical tests do not directly address such "positively" phrased statements. Specifically, statistical techniques provide tests only for statements called *null hypotheses*. A **null hypothesis** (expressed $\mathbf{H_0}$) is a statement of the expectation of how a phenomenon would occur if only random events were operating.

As such, it is usually a statement of *no* relationship between variables, or of *no* difference between two or more groups on a particular measurement. For example, if an "honest" coin is flipped 100 times, the expected outcome, if the events are random, would be close to an equal number of heads and tails. The null hypothesis would be an expression of that expectation: There will be no difference between the number of heads and tails. Or it could be phrased as: Number of heads = number of tails = 50.

In the present example of the statistics class achievement, *if* achievement is a random phenomenon without regard to enrollment in a statistics class, then the expectation would be that there is no difference between the national mean (μ) and the class mean ($\overline{X}$). The null hypothesis would state: There is no difference between the statistics class mean and the national population mean score on the achievement test. Symbolically, the "testable" null hypothesis is H_0: $\mu = \overline{X}$, where H_0 is the symbol for the null hypothesis, μ is the mean achievement of the national population, and $\overline{X}$ is the mean achievement of the statistics class at Jesse Mitchell High School.

The statistical test of a null hypothesis will eventually lead to either the *rejection* of H_0 or the *failure to reject* H_0. The wording may seem a bit clumsy, but notice that if the null hypothesis is *rejected*, this implies that μ *is not equal* to $\overline{X}$, or, in this case, the statistics class mean is higher. If H_0 is rejected, the alternative decision is to accept the original statistical hypothesis (sometimes called the **alternate hypothesis** or *alternative hypothesis*). Of course, such decisions and conclusions will have to be expressed in statements of probability, because inferential statistics are risk-defining procedures that are not designed to definitely prove a hypothesis. How is the test of H_0 accomplished? How is a null hypothesis rejected? The concept of statistical significance will help clarify the answers to these questions.

STATISTICAL SIGNIFICANCE

The term **statistical significance** means that what is observed is departing too much from what is expected to happen under random circumstances. Statistical significance means that the null hypothesis has been rejected because H_0 is a statement of expected outcomes under random conditions. How much of a departure from chance outcomes is "too much"? If the probability that the difference between what is observed (that is, the statistics class mean and what would be expected under random circumstances ($\overline{X} = \mu$) is small, H_0 is rejected and there is a statistically significant difference. In the illustration, if the probability that the resulting difference between 54.2 ($\overline{X}$) and 50.0 (μ) is small under random conditions, then we will claim that $\overline{X} > \mu$, which confirms the higher achievement of the statistics class.

Significance Probabilities

Before the statistician performs a statistical test of a null hypothesis, what is meant by a "small" probability must be defined. Traditionally, either 0.05 (5%) or 0.01 (1%) probabilities have been used, although they are arbitrary. Before computers became universally accessible, statisticians relied on published probability tables for testing null hypotheses. Most of these tables report significance levels of 0.05 and 0.01; consequently, tradition and convenience have made these two values very popular with statistical workers.

Assume that 0.01 was chosen as the significance level in the example. This would imply that if the difference between the national population mean and the statistics class mean was so large that the probability of such a difference occurring by chance was 1% or less, the null hypothesis (H_0) would be rejected and the result would be statistically significant.

On the other hand, if the probability was greater than 0.01 ($p > 0.01$) that the observed difference would occur by chance, the data would *fail to reject* the null hypothesis. In this case H_0 would remain tenable and the statistical evidence would not support the contention stated in the alternate (statistical) hypothesis.

Be aware that hypothesis testing will require an explicit statement of the significance level *before* the experiment is conducted. This statement defines what is meant by a "small" probability or the probability of the occurrence of a "rare" event.

Decisions

Now it is possible to go ahead with the task of testing H_0: There is *no* difference between the statistics class mean (54.2) and the national population mean (50.0). The significance level has arbitrarily been established as 0.01 (1%). If it can be shown that a difference of *at least* 4.2 points [54.2 − 50.0] would occur less than 1% of the time by chance, the difference will be classed as statistically significant and H_0 will be rejected.

***z*-test.** One technique for accomplishing the test was demonstrated at the beginning of this chapter; for convenience, call it a ***z*-test.** Following the procedures set forth earlier, the *z*-test can be computed as follows:

$$z = \frac{(\overline{X} - \mu)}{s_{\overline{X}}} \quad \text{where } s_{\overline{X}} = \frac{s}{\sqrt{N}}.$$

1. Compute $s_{\overline{X}}$: $s_{\overline{X}} = \frac{10}{\sqrt{25}} = \frac{10}{5} = 2.0$
2. Compute *z*-test value: $z = \frac{(54.2 - 50.0)}{2} = \frac{4.2}{2} = 2.1$
3. Notice in Figure 11.4 that the probability of obtaining a *z*-score of 2.1 or more is the proportion of the area in the smaller part of the normal distribution.
4. Consult Appendix C to determine *p*, the probability of obtaining a *z*-score of 2.1 or greater ($p = 0.0179$).

Figure 11.4—Comparison of Normal Curve Percentiles, Standard Deviations, z-scores, and Means

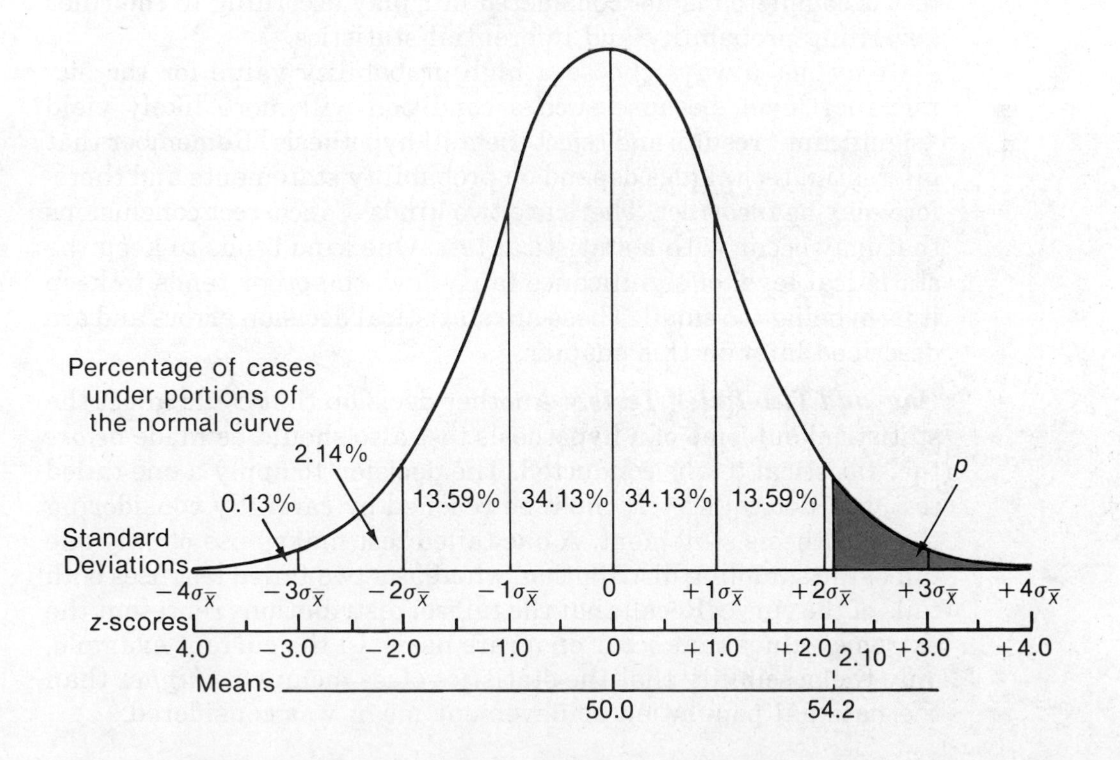

5. Because $p > 0.01$, the chance of getting a z-value of 2.1 or more is greater than the preselected level of significance. Therefore the null hypothesis *is not* rejected at the 0.01 level of significance. This is because a difference of 4.2 (or more) points between the parameter μ and the statistics class mean ($\overline{X}$) would occur by chance more than 1% of the time. Consequently, the data do not provide a statistically significant difference at the 0.01 level of significance. The bottom-line conclusion is that the data do not provide sufficient evidence that the statistics class outperformed the national population on the achievement test.

Using the previously cited values of the z-test, *what if* a 0.05 (5%) level of significance had been selected for the hypothesis test before the analysis was performed? Obviously, because $p < 0.05$, the statements of the results and conclusions would have been drastically different. Think about it for a moment: does this illustration give a clue as to why the significance level is determined *before* the hypothesis test? Do you think all statistical researchers abide by the ethics of preselecting the significance level and then live with the consequences? To change levels of significance after the hypothesis test is completed is not considered fair play according to the rules governing probability and inferential statistics.

Why not always choose a high probability value for the significance level, because such a condition will more likely yield "significant" results and reject the null hypothesis? Remember that inferential techniques depend on probability statements and therefore may be incorrect. There are two kinds of incorrect conclusions that may occur with a statistical test. One kind tends to keep the statistical level of significance fairly low; the other tends to keep it from being too small. These are statistical decision errors and are described later in this chapter.

One- and Two-tailed Tests. Another decision that could affect the statistical outcome of a hypothesis test also should be made *before* the statistical test is conducted. The decision to apply a one-tailed test or a two-tailed test must be reached by carefully considering the hypothesis statement. A one-tailed test makes use of only one tail of the sampling distribution, whereas a two-tailed test uses both tails of the curve. Recall that the tails of distributions represent the extreme values that occur on a rare basis. In the current example, only the possibility that the statistics class mean was *higher* than the national population achievement mean was considered.

In a realistic research situation, the investigator would probably not know if the statistics class would score higher than, lower than, or the same as the national population on the achievement test. Therefore provisions would have to be made for potential outcomes in which the statistics class mean was lower as well as higher than μ. In this case a two-tailed test would be appropriate. This implies that the significance level area would be divided into both tails of the normal distribution instead of only one. Figures 11.5 and 11.6 show the **area(s) of rejection** for a 5% significance level test.

In Figure 11.5, the area of rejection at the 0.05 level is beyond a z-test value of 1.645 (because 5% of the area of a normal curve lies beyond $z = 1.645$). In Figure 11.6, 2.5% of each tail of the normal curve denotes the total 5% area of rejection. This means that if $|z| > 1.645$, the one-tailed results would be statistically significant at the 0.05 level. Or, if $z < -1.96$ or $z > 1.96$ (that is, $|z| > 1.96$), a two-tailed test would be significant at the 5% level. The notation $\boldsymbol{p < 0.05}$ is used in statistics to denote statistical significance at the 0.05 level and indicates that *the probability that the results are due to random chance is less than 5%.*

The z-scores in Appendix C can be used in a similar way to determine the area of rejection for various levels of significance. The

Figure 11.5—One-Tailed Statistical Test at 0.05 Level of Significance

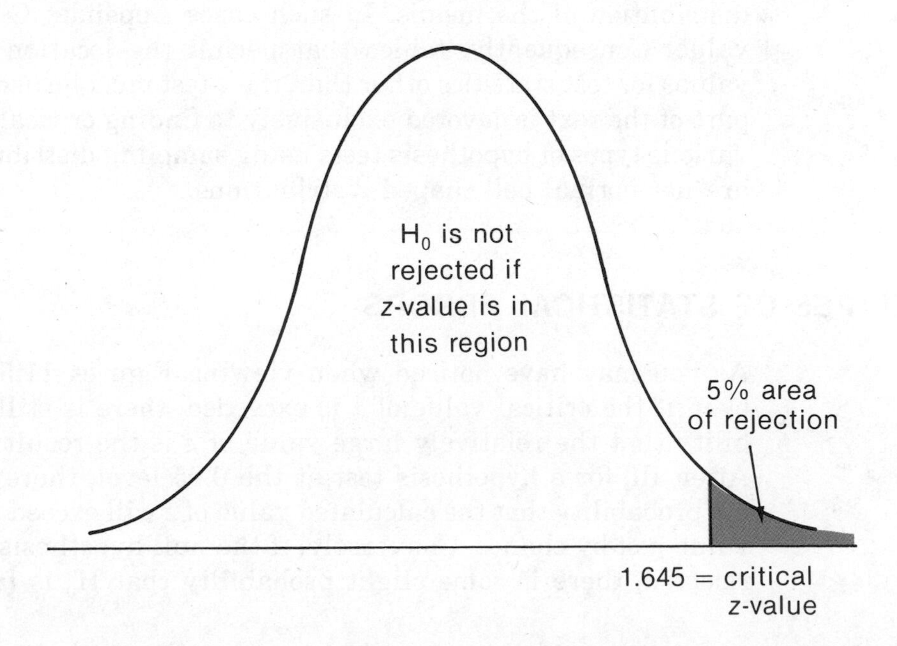

Figure 11.6—Two-Tailed Statistical Test at 0.05 Level of Significance

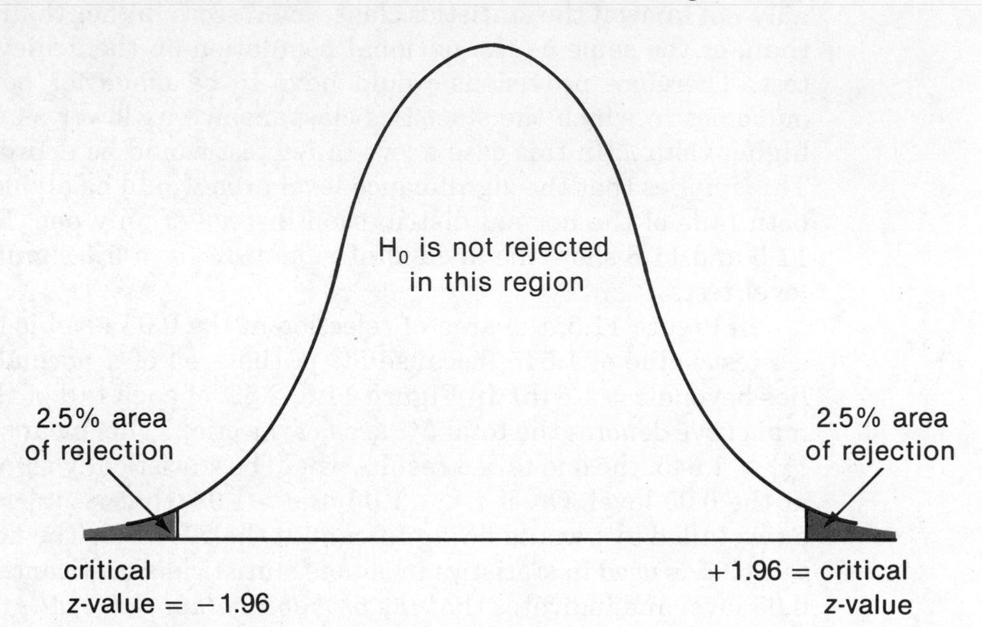

values that divide the middle part of the curve from the area(s) of rejection in the tails are called **critical values**. Similar interpretations of hypothesis testing using techniques other than *z*-tests are employed. Not all data are normally distributed as is the sampling distribution of the means. In such cases Appendix C is of little value. Consequently, tables that permit the location of critical values for test statistics other than the *z*-test must be used. The last part of the text is devoted exclusively to finding critical values for various types of hypothesis tests using sampling distributions that are not normal bell-shaped distributions.

TYPES OF STATISTICAL ERRORS

As you may have noticed when viewing Figures 11.5 and 11.6, even if the critical value of *z* is exceeded, there is still the possibility that the relatively large value of *z* is the result of chance. After all, for a hypothesis test at the 0.05 level, there remains a 5% probability that the calculated value of *z* will exceed the critical value just by chance. Conversely, if the null hypothesis fails to be rejected, there is some slight probability that H_0 is false in the population and the wrong conclusion has been reached. These wrong conclusions are possible errors in hypothesis testing and are known as type I and type II errors, respectively.

Type I Errors

A **type I error** occurs when the data reject a null hypothesis that should not have been rejected. These are inferential statistics and as such are making probability statements about the nature of the population from a smaller sample. "The luck of the draw" will sometimes provide incorrect information about the population.

The predetermined significance level (denoted by the Greek letter alpha—α) for a problem is the probability of committing a type I error if H_0 is rejected. Thus if $\alpha = 0.05$ and the null hypothesis is rejected, there is a 5% risk of having committed a type I error; if $\alpha = 0.01$ and H_0 is rejected, there is a 1% risk of a type I error. This means that the statistician controls the risk of a type I error with the selection of α. Obviously, the higher the value of α, the greater the risk of a type I error.

In the earlier example it was concluded that the Jesse Mitchell student body had a higher achievement mean than did the national population ($p < 0.0005$). If 0.0005 ($\frac{5}{100}$ of 1%) had been preselected as the α-level, there would be a 0.0005 probability that the difference was in fact the result of chance and that there is really no difference between the sample and the population. In other words, there is a 0.0005 probability of the occurrence of a $z > 3.33$ just by chance. If in this instance the calculated z of 3.33 just happened to be one of those 5 out of 10,000 randomly distributed outcomes, then a type I error would have been committed.

Because statisticians generally have only one sample from a population with which to work, whether or not a type I error has been committed remains unknown. What is known, and what the statistician accepts, is α—the probability that the rejection of H_0 is the result of a type I error. Knowing this risk tends to keep α relatively small—in the neighborhood of 0.10, 0.05, 0.01, 0.001, and very rarely more than 0.10. Why not "tighten up" on this risk and choose very small α levels such as 0.0001 or less? The response to that question has to do with another risk called a type II error.

Type II Errors

A **type II error** occurs when a false null hypothesis is *not rejected.* In other words, the observed phenomenon is not a function of random chance, but the calculated z-value does not exceed the critical value. The probability of committing a type II error is symbolized by the Greek letter beta (β). The value of β can be determined, but the methodology is beyond the scope of this text. In the hypothesis test to determine if the Jesse Mitchell statistics class had a higher mean achievement score than the national population, the null hypothesis was not rejected at the 0.01 level ($p > 0.01$). The fact remains that the statistics class *might be higher* achievers than the national population. If this is the case, the results have led to a type II error.

One dilemma for the statistician then is this: If the predetermined level of α (significance level) is set too high, statistical significance becomes easy to achieve even with randomly occurring events. This increases the probability of a type I error to an unacceptable level. However, if the statistician establishes an α-level that is too low, such restrictions make it very difficult to achieve statistical significance even in cases where H_0 should be rejected. This circumstance increases the chances for committing a type II error to unacceptable levels.

In reality the null hypothesis is either true or false. Here reality refers to the unknown status of H_0 in the population. So the statistician collects data from a sample and attempts to find out about "reality" with a hypothesis test. The null hypothesis either is rejected or is not rejected. If **H_0 is rejected**, the evidence supports the alternate hypothesis (statistical or research hypothesis). If the null hypothesis is not rejected, it remains tenable and the researcher concludes that the evidence is not sufficient to support the alternate hypothesis.

CONSEQUENCES OF DECISIONS

Suppose that a school principal wants to increase the ability of students in mathematics. The principal believes that a program called "Mathemate" has sufficient potential for increasing mathematics reasoning and computational skills to justify its use in an experiment. Students ($N = 152$) are randomly selected to participate in a one-year program using Mathemate. Suppose that the average score for mathematics students in the school on a year-end mathematics achievement test for the past five years is 80. A design is implemented to investigate the hypothesis about μ, the population mean achievement test score for students in the Mathemate program.

The *research hypothesis* states: "On the average, students in the Mathemate program will perform better on a standardized mathematics achievement test than students in the regular program." The symbolic inequality is: $\mu > 80$. The *null hypothesis* (H_0) states: "There is no difference between the mean score for the Mathemate group and the mean of students in the regular program." Symbolically, H_0: $\mu = 80$.

Reasoning for the experimental implementation follows this logic. *If* sufficient evidence can be found in support of the research hypothesis, the Mathemate program will be implemented throughout the secondary school level in the district. In this case the additional $4.65 per student cost will be considered justified and the necessary funds will be budgeted for the project. On the other hand, the Mathemate project may not be shown to be any better than the traditional program, in which case the null hypothesis will remain tenable. Obviously the additional cost could not be justified if H_0 was not rejected, and the current program would continue.

In either of the previous cases, the decisions and consequences are straightforward. If the experimental program (Mathemate) is more effective than the traditional curriculum, it should be implemented even though there is an additional cost. In this case mathematics achievement is expected to increase as a result of the program. If the program does not work any better than the present curriculum, it should not be implemented and the school district's money should be saved.

However, as a third alternative, what if the null hypothesis was rejected just because of a chance fluctuation in the data or because of sampling error? That is, the Mathemate program is not effective, but because of random error the statistical test indicated that $\mu > 80$. The principal would have committed a type I error. The consequences would include *the expenditure of additional money for a program that is no more effective than the traditional one*. Thus, if 3,000 students per year are enrolled in mathematics courses, nearly $14,000 of the district's funds would essentially be wasted on an annual basis. Therefore, to guard against a type I error and the accompanying consequences, the principal would want to preselect a low alpha level, maybe 0.01 or 0.001.

A fourth alternative would be to fail to reject the null hypothesis when it should be rejected. In such a case the evidence would not be sufficient to justify implementing the program when in fact it does enhance mathematics achievement. Although the students in the district would miss the opportunity to be involved in Mathemate, they would be *no worse off* than they were originally, because the status quo would be maintained. Because the "truth" about the effectiveness of the experimental program would be unknown, the principal probably would not risk the additional expenditure for the program and the probable increases in achievement would not be realized as a consequence of the type II error.

Table 11.1 provides a summary of the possible outcomes of a hypothesis test and indicates two conditions where no statistical error occurs. However, two cells show possible errors. Ideally, α and β, the probabilities of committing a type I and type II error, respectively, are small compared to the probability of not committing an error.

Table 11.1 Summary of Hypothesis Test Results

Results of Null Hypothesis Test (Sample)		"Reality" in the Population	
	H_0 is:	True	False
	Rejected	Type I error (α)	No error ☺
	Not Rejected	No error ☺	Type II error (β)

The statistician rarely if ever knows if either error has occurred. That is why the results must be expressed in probability statements and not absolutes. Null hypothesis testing can confirm or support an alternate hypothesis; or the null hypothesis may remain tenable and nothing is "proved" with certainty.

COMPUTER FOCUS 11—Central Limit Theorem

Purpose

Because of the speed at which computers operate internally, they can be used to simulate certain sampling tasks that would be extremely time consuming if done manually. This means that to illustrate certain theoretical concepts requiring large samples or large numbers of samples, computers have been indispensable. The purpose of this Computer Focus is to illustrate such a use of the computer for estimatng μ and the standard error of the mean with theoretical concepts from the central limit theorem.

Problem: Generating a Random Sample

Use the computer to simulate the selection of k random samples of size N from a box containing 10 balls numbered from 0 to 9. The process would involve selecting a single ball, recording the number, accumulating X, and then replacing the ball before the next draw. A random integer X such that $0 \leq X \leq 9$ can be generated by:

```
LET X = INT(10*RND(1))
```

The process would be repeated N times to complete the sample. The sample mean will be computed and stored for subsequent use.

Estimating Mu (μ) and the Standard Error of the Mean

The experiment would involve finding the mean of k such samples. Then the mean of the k sample $\overline{X}$'s would be a good estimate of μ; the standard deviation of the $\overline{X}$'s would be a good estimate of the standard error of the mean (SE).

In your simulation, try sample sizes (N) of 5, 10, 15, 20, and so on. Obtain summary information from $k = 100$ means of each sample size. Have the computer program compute the mean and standard deviation of the k means. Explain how the central limit theorem is illustrated.

KEY TERMS AND NOTATION

Alpha (α)
Alternate hypothesis
Area(s) of rejection
Beta (β)
Critical value(s)
Hypothesis
Null hypothesis (H_0)
$p < 0.05$
Reject H_0
Research hypothesis
Statistical hypothesis
Statistical significance
Type I error
Type II error

APPLICATION EXERCISES

11.1 Using Appendix C, find the critical values of z for testing a null hypothesis at each of the following levels of significance:

alpha	Critical Values One-tailed Test	Critical Values Two-tailed Test
a. 0.05	______	______
b. 0.01	______	______
c. 0.001	______	______
d. 0.10	______	______

11.2 Suppose a researcher tested several null hypotheses at the 0.05 alpha level. For the following *calculated* z-test values, what should be the researcher's decision about H_0—reject or not reject? (Assume a one-tailed test of significance.)

a. $z = 1.14$
b. $z = -2.03$
c. $z = 0.00$
d. $z = 2.54$
e. $z = -1.90$
f. $z = 4.59$

11.3 Suppose a researcher rejected a null hypothesis at the predetermined alpha level of 0.01. What is the probability that the researcher committed a:

a. type I error?
b. type II error?

11.4 In hypothesis testing, translate the following symbolic notation to verbal statements:

a. $p < 0.05$
b. $p < 0.001$
c. $p > 0.10$
d. $p < 0.01$

11.5 For a normally distributed population with $\mu = 75$ and $\sigma = 5$, what is the probability that, in a random sample of 100, the sample mean ($\overline{X}$) will be

a. greater than 77?

b. What would be your decision about a null hypothesis test of H_0: $\mu = 77$ at the 0.05 level on a two-tailed test?

11.6 Suppose your world history teacher has given a particular exam for several years and has determined that the scores are normally distributed and that the population mean (μ) score on the exam is 84 and $\sigma = 6$. Her present class of 36 students obtains an $\overline{X} = 86$. Should she retain the hypothesis that the class is representative of the population as defined by previous classes? Test at the 0.05 level with a two-tailed test to see if the null hypothesis is tenable.

11.7 A car manufacturer claims that its cars use an average of 5.20 gallons of gasoline for each 100 miles driven. A local dealer for this company tests 49 cars for gas mileage and finds that the mean gas consumption for 100 miles is 5.35 with a standard deviation of 0.35 gallon. Assuming that gas mileage data are normally distributed, do these results cast doubt on the manufacturer's claim (at the 0.01 significance level)? Explain.

11.8 A vending machine distributor knows that standard machines in schools sell an average of 650 drinks per week. Twenty-five new machines are tested that have the school mascot flashing in neon lights. Results with the new machines show a normal distribution with $\overline{X} = 656$ and $s = 20$.

a. Testing at the 0.05 significance level with a one-tailed test, should you conclude that the flashing mascot machine improved sales?

b. Does the experimenter run the risk of a type I or a type II error?

11.9 Describe a hypothetical example of a:

a. type I error.

b. type II error.

11.10 Identify two decisions that should be made before data are analyzed in a hypothesis test. How could an unethical researcher misuse these decisions to reinforce his or her bias?

11.11 Recall from the text that the Jesse Mitchell High School students with an average score of 51.1 outperformed the national norm, while the statistics class with a score of 54.2 did not necessarily outperform the national norm.

a. Mathematically, how does the sample size (N) influence the size of $s_{\overline{X}}$ and eventually the size of the corresponding z-score (assuming other factors are held constant)?

b. How does the size of the z-score relate to whether or not a "significant" difference is found?

PART FOUR
Inferential Statistical Methodology

12

Correlation

CHAPTER OBJECTIVES

Upon completion of the chapter, students will be able to:

1. Distinguish between univariate and bivariate distributions.
2. Explain the meaning of positive and negative relationships.
3. Construct a scattergram for a bivariate distribution.
4. Compute a product-moment correlation coefficient (r) for a bivariate distribution.
5. Determine the degrees of freedom and test a correlational null hypothesis.
6. Define and compute the coefficient of determination.
7. Demonstrate an awareness that correlation does not imply cause and effect.
8. Determine the appropriate sampling distribution statistics and compute a confidence interval for the correlation coefficient.
9. (Optional) Compute and interpret a rank order correlation coefficient.

During his many years of algebra teaching experience, Mr. Lawson had observed a general tendency that students who earned high marks in algebra seemed to like the subject more than those who did not achieve as well. Although there were isolated exceptions to the general rule, the parallel between attitude toward algebra and achievement in algebra was a consistently occurring phenomenon in Mr. Lawson's experience. Do you think these two variables "go together"; or, in statistical terms, are they related? Another way of asking the same question is: "Are attitude and achievement *correlated*?"

The study of **relationships between variables**, or **correlation**, is an important topic because of its widespread applicability. In the list that follows, do you think that any of the pairs of variables might follow some pattern of relationship:

1. number of cigarettes smoked and incidence of lung cancer?
2. number of hours of rock music listening and amount of hearing loss?
3. amount of drug use and loss of memory (retention) capability?
4. school grades and number of absences?
5. age and number of deaths caused by driving while intoxicated?
6. exposure to direct sunshine and risk of melanoma?

A methodology for addressing such questions is presented in this chapter.

Up to this point in your study of statistics, most of the discussion has focused on populations and samples consisting of measurements on a single variable (univariate distributions). However, in the study of correlation, the need arises for at least two measures per case. If for every measurement of a variable X, a corresponding value of a second variable Y is known, the resulting set of pairs of variables is called a **bivariate distribution**. For example, if we have a measurement of height and a measurement of weight for a set of individuals, the resulting distribution is a bivariate distribution.

The study of correlation is concerned with how two variables of a bivariate distribution relate to each other. In the example just cited, correlation techniques would permit a study of the tendencies of the variation in weight as heights varied. Thus the material in this chapter could be accurately described as the study of covariation of bivariate distributions. Except for the optional section titled "Rank Order Correlation", this chapter will consider the technique known as **product-moment correlation, Pearson *r*** (named after the mathematician Karl Pearson), or simply ***r***. Other techniques are available for studying bivariate distributions, but the Pearson r is by far the most important.

THE RELATIONSHIP BETWEEN TWO VARIABLES

Correlation addresses two primary questions: (1) *Does a relationship exist* between the two variables of a bivariate distribution? and (2) if a relationship does exist, *what is the direction* of the relationship? The first question depends on the magnitude of the relationship as expressed by a *correlation coefficient*, r. The second question is answered by the algebraic sign of r. The following three definitions are basic to the understanding of correlation:

1. Two variables are said to **correlate positively** when as one variable increases in size, the other shows some systematic tendency to increase correspondingly in a uniform way.
2. Two variables are said to **correlate negatively** when as one variable increases in size, the other shows some systematic tendency to decrease correspondingly in a uniform way.
3. Two variables are said *not to correlate* when as one of the variables increases in size, the other shows no overall tendency to increase systematically or decrease systematically in a uniform way.

The value of a correlation coefficient (r) can range from -1.00 to $+1.00$. The more r departs from zero, the stronger the relationship. If r has a positive algebraic sign (+), the relationship between the two variables is positive. Conversely, r's with negative algebraic signs ($-r$'s) indicate negative relationships, and r's close to zero basically show no relationship.

GRAPHING THE RELATIONSHIP BETWEEN TWO VARIABLES

Chapter 3 introduced a *bivariate plot* or **scattergram** as a pictorial means of describing a bivariate distribution. These graphs are valuable for providing a visual overview of the relationship between two variables. Using the methodology for constructing a scattergram that was presented in Chapter 3, consider the bivariate distribution that follows.

Suppose that Mr. Lawson, the algebra teacher introduced at the beginning of the chapter, administered a questionnaire to 25 randomly selected mathematics students. This questionnaire assessed the students' attitude toward mathematics, and the scores were designated as variable X. Further, Mr. Lawson collected standardized mathematics achievement test scores from the school's records on the same 25 students and designated those numbers as variable Y (achievement in mathematics). The hypothetical data are presented in Table 12.1.

Table 12.1 Hypothetical Bivariate Distribution

Attitude Toward Math (X)	Achievement in Math (Y)
66	75
82	74
48	55
55	60
76	85
56	74
65	59
40	56
88	85
58	53
65	68
95	97
71	64
50	59
57	51
87	74
40	54
48	50
58	58
53	64
80	65
92	85
69	55
64	52
72	79

The question of primary concern to Mr. Lawson was whether attitude toward mathematics was related to math achievement; or, is X related to Y? Mr. Lawson first constructed a bivariate scattergram of the data so he could *see* how the two variables were related. The bivariate plot (scattergram) is shown in Figure 12.1.

As you can see in Figure 12.1, there does seem to be a tendency toward a positive relationship. As scores on the attitude scale (X) increase, there is a tendency for scores on the mathematics achievement variable (Y) also to increase. Although there are a few exceptions, the general pattern of the scattergram points is from lower left (low scores on both X and Y) to upper right (high scores on both variables). Such is the nature of a positive relationship, or a positive correlation.

Figure 12.2 provides illustrations of positive and negative relationships as well as a case in which there is virtually no relationship.

Figure 12.1— Scattergram of Mathematics Attitude (X) Scores and Mathematics Achievement (Y) Scores

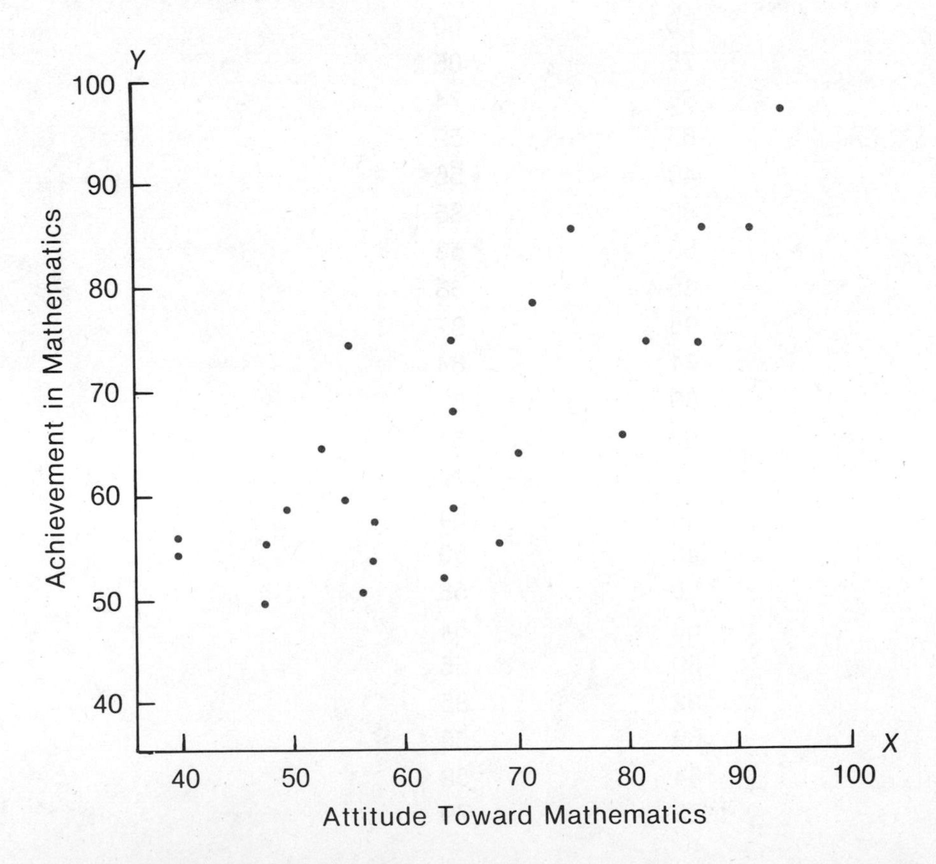

Figure 12.2—Scattergrams Illustrating Bivariate Relationships

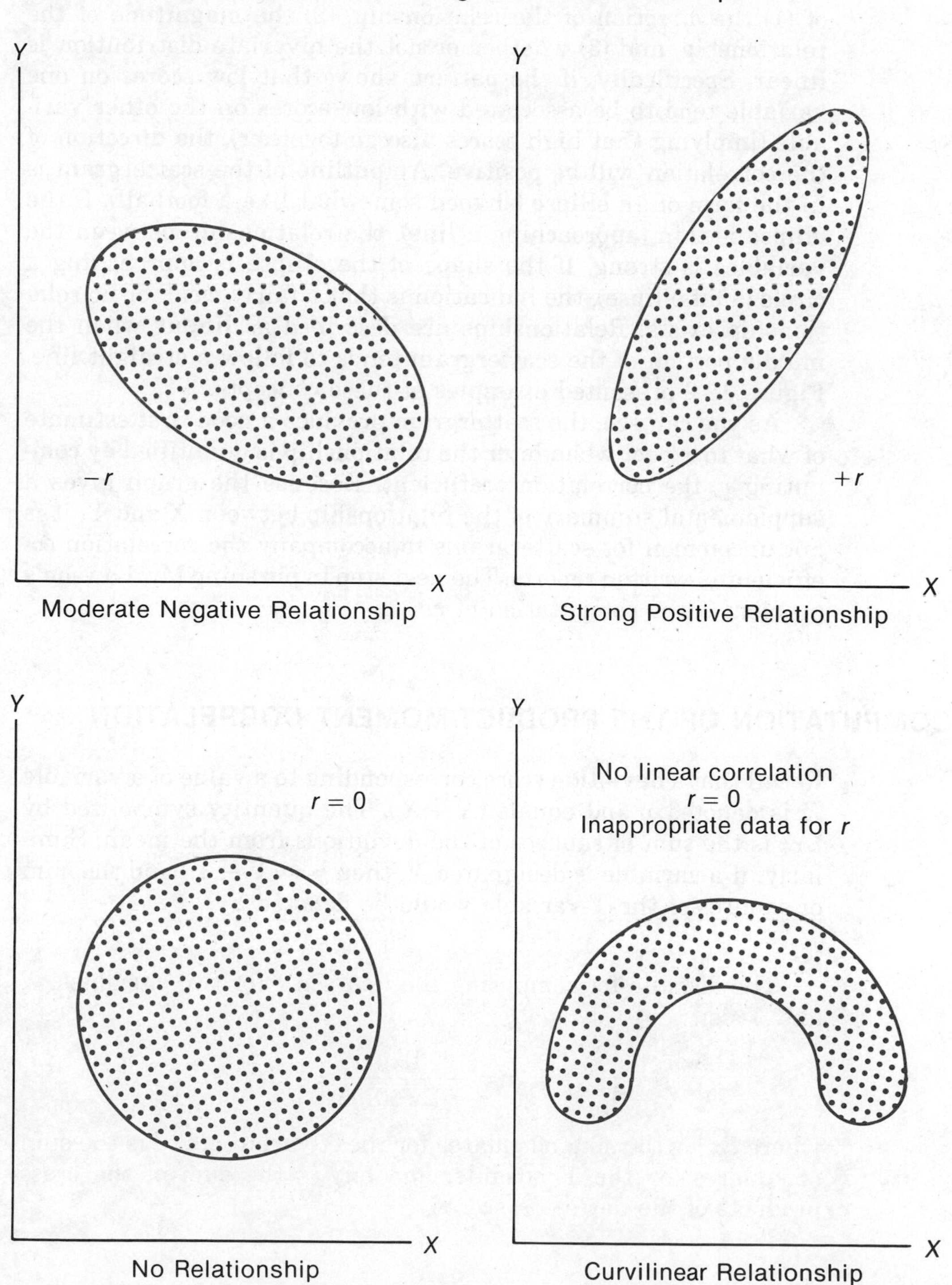

Examination of a scattergram also can provide an indication of (1) the direction of the relationship, (2) the magnitude of the relationship, and (3) whether or not the bivariate distribution is linear. Specifically, if the pattern shows that low scores on one variable tend to be associated with low scores on the other variable (implying that high scores also go together), the direction of the correlation will be positive. An outline of the scattergram is in the form of an ellipse (shaped somewhat like a football). If the ellipse is thin (approaching a line), the relationship between the variables is strong. If the shape of the ellipse is approaching a circle (a fat ellipse), the indication is that a fairly weak or no relationship exists. Relationships are described as **linear** when the middle portion of the scattergram tends to follow a straight line. Figure 12.2 presented examples of these concepts.

As you can see, the scattergram provides a good first estimate of what to expect when later the relationship is quantified by computing r, the correlation coefficient. Because the graph gives a supplemental summary of the relationship between X and Y, it is not uncommon for scattergrams to accompany the correlation coefficient in written reports. The next step in pursuing Mr. Lawson's problem is the computation of r.

COMPUTATION OF THE PRODUCT-MOMENT CORRELATION

Recall that a deviation score corresponding to a value of a variable X is denoted x and equals $(X - \overline{X})$. The quantity symbolized by Σx^2 is the sum of squares of the deviations from the mean. Similarly, if a variable is designated Y, then $y = (Y - \overline{Y})$ and the sum of squares of the Y-variable would be denoted Σy^2.

The formula for computing the Pearson r for two variables, X and Y, is:

$$r = \frac{\Sigma xy}{\sqrt{(\Sigma x^2)(\Sigma y^2)}}$$

where Σx^2 is the sum of squares for the X-variable, Σy^2 is the sum of squares for the Y-variable, and Σxy is the sum of the cross products of the deviation scores.

Or, in equation form:

$$\Sigma x^2 = \Sigma X^2 - \frac{(\Sigma X)^2}{N}$$

$$\Sigma y^2 = \Sigma Y^2 - \frac{(\Sigma Y)^2}{N} \text{ and}$$

$$\Sigma xy = \Sigma XY - \frac{(\Sigma X)(\Sigma Y)}{N}.$$

The first two identities are familiar, although this is the first introduction to the sum of the cross products of the deviation scores ($\Sigma[X - \overline{X}][Y - \overline{Y}]$). Using Mr. Lawson's data from Table 12.1, a three-step computational procedure can be illustrated.

Step 1–Find the sum of the raw scores [ΣX and ΣY], the sum of the raw scores squared [ΣX^2 and ΣY^2], and the sum of the cross products of the raw scores [ΣXY] as shown in Table 12.2.

The first two columns in Table 12.2 are the bivariate raw score values repeated from Table 12.1. The third and fourth columns are the squares of X and Y, respectively. The fifth column consists of cross products of the corresponding X- and Y-values. To complete step 1, the five columns are summed. The resulting column totals provide the necessary input values for determining the *deviation* sums of squares and cross products in step 2.

Step 2–Using the deviation score formulas presented previously, compute:

$$\begin{aligned}\Sigma x^2 &= 112{,}845 - \frac{(1{,}635)^2}{25} \\ &= 112{,}845 - 106{,}929 \\ &= 5{,}916.00\end{aligned}$$

$$\begin{aligned}\Sigma y^2 &= 113{,}061 - \frac{(1{,}651)^2}{25} \\ &= 113{,}061 - 109{,}032.04 \\ &= 4{,}028.96\end{aligned}$$

$$\begin{aligned}\Sigma xy &= 111{,}812 - \frac{(1{,}635)(1{,}651)}{25} \\ &= 111{,}812 - 107{,}975.4 \\ &= 3{,}836.6\end{aligned}$$

Table 12.2 Hypothetical Bivariate Distribution with Squares and Cross Products

Attitude Toward Math (X)	Achievement in Math (Y)	X^2	Y^2	XY
66	75	4,356	5,625	4,950
82	74	6,724	5,476	6,068
48	55	2,304	3,025	2,640
55	60	3,025	3,600	3,300
76	85	5,776	7,225	6,460
56	74	3,136	5,476	4,144
65	59	4,225	3,481	3,835
40	56	1,600	3,136	2,240
88	85	7,744	7,225	7,480
58	53	3,364	2,809	3,074
65	68	4,225	4,624	4,420
95	97	9,025	9,409	9,215
71	64	5,041	4,096	4,544
50	59	2,500	3,481	2,950
57	51	3,249	2,601	2,907
87	74	7,569	5,476	6,438
40	54	1,600	2,916	2,160
48	50	2,304	2,500	2,400
58	58	3,364	3,364	3,364
53	64	2,809	4,096	3,392
80	65	6,400	4,225	5,200
92	85	8,464	7,225	7,820
69	55	4,761	3,025	3,795
64	52	4,096	2,704	3,328
72	79	5,184	6,241	5,688
$\Sigma X = 1{,}635$	$\Sigma Y = 1{,}651$	$\Sigma X^2 = 112{,}845$	$\Sigma Y^2 = 113{,}061$	$\Sigma XY = 111{,}812$

Step 3–Compute r using the Pearson r formula and the appropriate substitutions from step 2.

$$r = \frac{\Sigma xy}{\sqrt{(\Sigma x^2)(\Sigma y^2)}}$$

$$= \frac{3{,}836.6}{\sqrt{(5{,}916)(4{,}028.96)}}$$

$$= \frac{3{,}836.6}{4{,}882.1}$$

$$= 0.79$$

To use a calculator to compute the Pearson r using the sums of squares and sum of the cross products of the deviation scores, follow the sequence shown in Figure 12.3.

Figure 12.3—Calculator Keystrokes

PROBLEM:

Calculate r given Σx^2, Σy^2, and Σxy:

$\Sigma x^2 = 5{,}916$
$\Sigma y^2 = 4{,}028.96$
$\Sigma xy = 3{,}836.6$

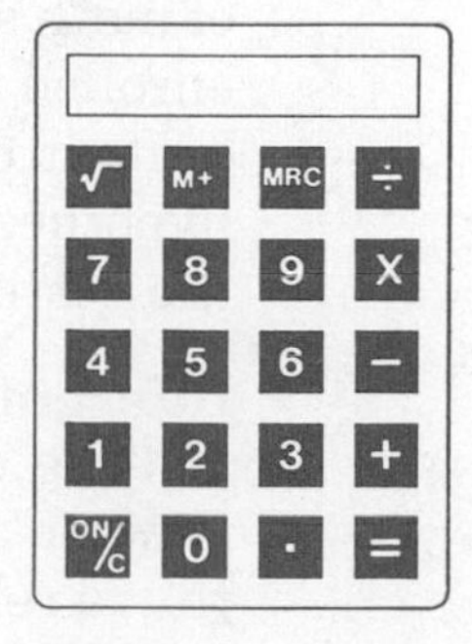

Procedure	Arithmetic Operation	Calculator Keystrokes	Display
1. Multiply $\Sigma x^2 \cdot \Sigma y^2$	1. 5,916.00 × 4,028.96 = 2,385,327	1. 5,916 [×]	5,916
		4,028.96 [=]	2,385,327
2. Find square root of product: $\sqrt{(\Sigma x^2)(\Sigma y^2)}$	2. $\sqrt{2{,}385{,}327}$ = 4,882.1436	2. [√]	4,882.1436
		[M+]	4,882.1436
3. Divide numerator (Σxy) by denominator $\sqrt{(\Sigma x^2)(\Sigma y^2)}$	3. $\frac{3{,}836.6}{4{,}882.1436} = 0.79$	3. 3,836.6 [÷]	3,836.6
		[MRC]	4,882.1436
		[=]	0.7858433

Presto! You have the correlation coefficient. As you will learn later in this chapter, an $r = +0.79$ is indicative of a moderately high positive relationship between mathematics achievement and attitude toward mathematics. Consequently, in Mr. Lawson's sample there is a tendency for achievement and attitude to covary in a positive direction. This means that as attitude scores get higher, there is a corresponding *tendency* for achievement scores to get higher.

INTERPRETATION OF *r*

The outcome of the correlation computation ($r = 0.79$) is actually a quantitative description of how X and Y relate to each other in a *sample*. Hence $r = 0.79$ is a *statistic*. From an inferential perspective, Mr. Lawson wants to generalize the findings to the population of math students (recall that the sample was selected from students enrolled in mathematics classes). To address the problem within the context of inferential statistics, Mr. Lawson can (1) test the null hypothesis and/or (2) establish a confidence interval estimate for the correlation coefficient.

Inferential Analysis

The correlation coefficient r is known for the sample in Mr. Lawson's experiment. Can he say with a high degree of certainty that a relationship between achievement and attitude exists in the population? If a sample could be hand picked, the resulting r could be made positive and large by selectively choosing subjects that would fulfill the necessary requirements. If such a sample could be obtained intentionally, *it is possible* that a random selection would also result in the same sample just by chance. A test of the null hypothesis takes such a circumstance into consideration and directly addresses the question.

Hypothesis Testing. *If* achievement and attitude have a random relationship in the population, it would be expected that the correlation coefficient *parameter* would be near zero. If that is the case, the sample r of 0.79 was due to random sampling error. Let the Greek letter **rho** (ρ) symbolize the correlation parameter; then the null hypothesis is:

$$H_0: \rho = 0.$$

The task now is to provide evidence that H_0 should either be rejected or not be rejected.

Unfortunately, the *sampling distribution* of r's is not normally distributed; therefore the z-test technique would not be an appropriate procedure to use. Values in Appendix D have been computed and provide the necessary information. These values are the *minimum* r's that are statistically significant for the respective alpha levels (levels of significance) and degrees of freedom. *Degrees of freedom* were previously discussed for a univariate distribution and were interpreted to mean "freedom to vary." In the case of a bivariate distribution, the degrees of freedom are $N - 2$, because two variables are involved. For a particular test at the 0.05 level of significance, the probability is 5% that a value of r (or greater) would occur just by chance if the null hypothesis was true.

To use Appendix D, you need to determine the degrees of freedom (df) for the problem, the significance (alpha) level, and the value of r. The degrees of freedom for all product-moment correlation tests are:

$$\text{df} = N - 2$$

the number of subjects minus two (because there are two variables in the problem). To test the null hypothesis for Mr. Lawson:

$r = 0.79$ (calculated value from sample data)
$\text{df} = 25 - 2 = 23$ ($\text{df} = N - 2$) and
$\alpha = 0.05$ (predetermined significance level).

By examining Appendix D, you find that the *critical value* for a two-tailed test at the 0.05 level of significance with 23 df is 0.40. This implies that the probability of the absolute value of r being equal to or greater than 0.40 just by chance is 5%. An r as large as the *calculated* value (0.79) *would occur less than 5%* of the time by chance ($p < 0.05$). The results are *statistically significant* at the 0.05 level, because the probability is small ($p < 0.05$) that such a result would have occurred by chance if $\rho = 0$. Thus H_0 is rejected in favor of the alternate hypothesis that mathematics achievement and attitude toward mathematics are positively correlated in the *population*. So Mr. Lawson is relatively secure in his belief that the two variables are related. If they are not correlated in the population and $r = 0.79$ happened to be one of the rare 5 out of 100 cases that would occur by chance, a type I error has been committed, a risk Mr. Lawson is willing to take by virtue of the predetermined $\alpha = 0.05$.

Point and Interval Estimation. Imagine a large number of samples drawn from a population as was the case in Chapter 10. This time the sample statistics will be correlation coefficients. Figure 12.4 depicts the hypothetical process of obtaining a large number of correlation coefficients between an X- and a Y-variable.

If many samples were obtained, the sample r's would show variation, thus forming a sampling distribution of r's. As previously mentioned, this particular sampling distribution is not a normal distribution. More specifically, if the population correlation coefficient is near zero, the sampling distribution of correlation coefficients will closely approximate a normal distribution. If the population parameter is positive (greater than zero), the sampling distribution tends to be very skewed in the negative direction. Conversely, when $\rho < 0$, the sampling distribution of r's becomes positively skewed.

Figure 12.4—Sample Statistic r's

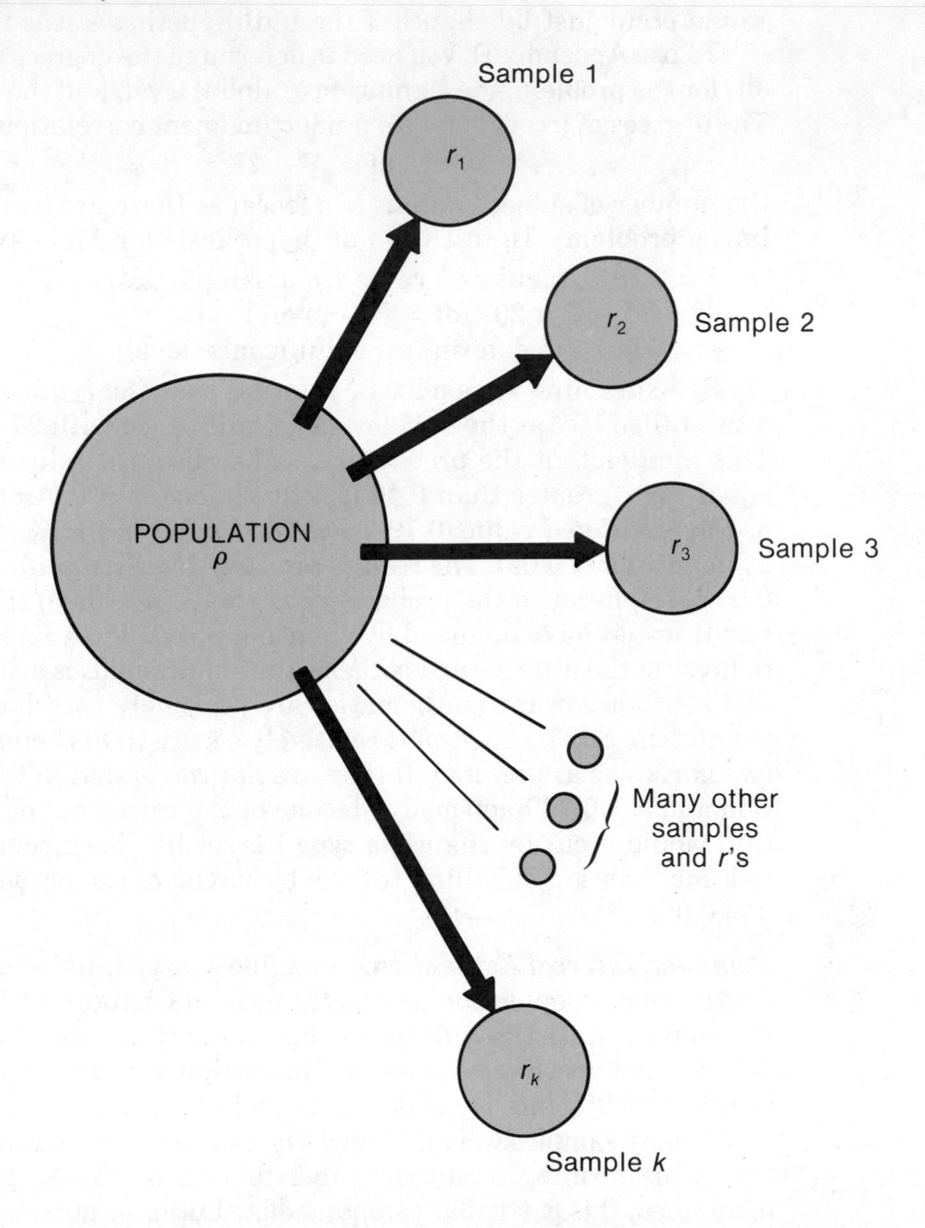

The fact that the particular form of the sampling distribution changes depending on the size and algebraic sign of the population parameter rho makes it impossible to construct a confidence interval using the table of normal curves used in determining the confidence interval of the mean. Although this situation does not create a significant problem with the point estimation of ρ, it does create a little inconvenience for interval estimation tasks. R. A. Fisher, an English statistician, developed a logarithmic transformation for the distribution of r that is approximately normally distributed and can be used to solve the confidence interval problem, as will be illustrated. Appendix H is an easily used conversion table that eliminates the need for actually using logarithms to make such transformations.

The correlation coefficient computed between attitude toward and achievement in mathematics is used as the **point estimate** of the parameter ρ. Because Mr. Lawson has only one sample, the product-moment r is an unbiased estimate of ρ. However, in recognition of the fact that some random error is present, confidence intervals will be determined.

The steps in constructing the *confidence intervals around the correlation coefficient* are as follows:

Step 1–Compute r (sample statistic = point estimate).

Step 2–Use Appendix H to transform r to a normalized z_r-score. This transformation now allows working with normally distributed z-scores as has been done in the previous chapters. Ignore the algebraic sign of r for the time being. (Note that if $r < 0.25$, $r = z_r$ to two decimal places.)

Step 3–Compute the standard error of the normalized z_r sampling distribution with: standard error $= \dfrac{1}{\sqrt{N-3}}$. (Note that the standard error of the normalized sampling distribution is a function of the sample size.)

Step 4–Multiply the standard error by the appropriate constant from the table of normal z-scores (Appendix C) to establish the desired confidence interval for a normal curve, such as 1.96 for a 95% confidence interval, 2.58 for a 99% confidence interval, and so on.

Step 5–*Add and subtract* the value obtained in step 4 to/from the transformed point estimate in step 2—this establishes the confidence interval for z_r.

Step 6–Don't forget to use Appendix H (in reverse) to convert the z_r-values found in step 5 back to correlation coefficients. Assign the appropriate algebraic signs to the r-values.

Using these six steps and the data collected by Mr. Lawson, a 95% confidence interval is found as follows:

Step 1–$r = 0.79$

Step 2–$z_r = 1.07$ (Appendix H)

Step 3–Standard error of $z_r = \frac{1}{\sqrt{25 - 3}} = \frac{1}{\sqrt{22}} = 0.21$

Step 4–For a 95% confidence interval: $(1.96)(0.21) = 0.41$

Step 5–$1.07 \pm 0.41 = [0.66 \text{ to } 1.48]$

Step 6–Convert z_r's to r's using Appendix H:

z_r		r	
0.66	converts to	0.58	(lower limit)
1.48	converts to	0.90	(upper limit)

The procedures shown in Figure 12.5 may be used when solving the problem with a calculator.

Thus Mr. Lawson can be 95% confident that ρ is in the range 0.58 to 0.90. Not only has he demonstrated that there is a relationship between X and Y in the population with the hypothesis test, he also has determined an interval that has a high probability ($p = 0.95$) of including the value of the parameter rho (ρ).

Coefficient of Determination

In a bivariate distribution, each variable (X and Y) has a measurable degree of variability. This variation may be quantified with the mean deviation, variance, range, and standard deviation; and, because N is the same for both variables, Σx^2 and Σy^2 may be used to describe the dispersion of the distributions. As will be addressed shortly, a popular method of "explaining" variability (sum of squares) in one variable with variation in another variable is the quantity r^2. The quantity $\mathbf{r^2}$ is called the **coefficient of determination** and is expressed as either a percentage or a proportion. The coefficient of determination is the percentage of variability in one variable that is "explained" by variability in another related variable. Because the coefficient of determination is entirely a function of r, the range of r^2 is from 0 to 1.00 (or 0 to 100%). If X and Y do not correlate (that is, if $r = 0$), $r^2 = 0$ means that they have no common or shared variance. If X and Y are perfectly correlated ($r = 1.00$), then one variable accounts for 100% of the variability of the other.

In the example, $r = 0.79$ implies that 62% (0.79^2) of the variability in achievement can be explained by variation in attitude. Conversely, 62% of the variation in attitude is in common with achievement variance. As will be discussed in the next section, the term "explains" does not imply that one variable *causes* the other. To "explain" in the context of r simply means that common variability can be accounted for mathematically and may have nothing to do with explaining actual cause and effect.

Figure 12.5—Calculator Keystrokes

PROBLEM:
Determine upper and lower bounds for a 95% confidence interval of z_r for a correlation coefficient parameter when $r = 0.79$

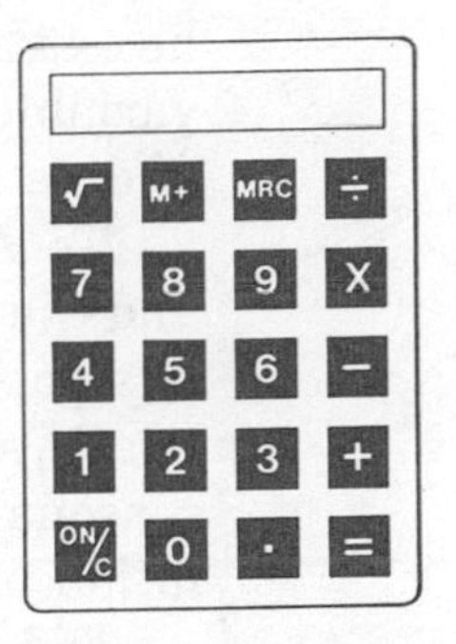

Procedure	Arithmetic Operation	Calculator Keystrokes	Display
1. Compute the standard error of z_r: $\frac{1}{\sqrt{N-3}}$	**1.** $\frac{1}{\sqrt{22}} = 0.2132007$	**1.** 1 [÷]	1
		22 [√]	4.6904157
		[=]	0.2132007
2. Multiply by constant from Appendix C for 95% confidence interval: 1.96 × 0.2132007	**2.** 0.2132007 × 1.96 = 0.4178733	**2.** [×] 1.96	1.96
		[=]	0.4178733
		[M+]	0.4178733
3. Find upper bound for z_r	**3.** 1.07 + 0.4178733 = 1.4878733 (upper bound)	**3.** [+] 1.07	1.07
		[=]	1.4878733
4. Find lower bound for z_r	**4.** 1.07 − 0.4178733 = 0.6521267 (lower bound)	**4.** 1.07 [−]	1.07
		[MRC]	0.4178733
		[=]	0.6521267

(Use Appendix H to find corresponding r's.)

CAUTIONS AND LIMITATIONS

Alan, a park ranger at the county recreation area, was enrolled in an evening statistics course offered at the local university. As a homework assignment, he had to collect data and perform a correlation analysis. From a population of 365 days in the calendar year, Alan selected a random sample of 100 days. For each day in the sample, he checked the records of the first-aid station on the lake shore to get the number of waterskiing accidents, which he labeled variable X. From the concession stand records for the same days, he extracted the number of ice cream cones sold, which he labeled variable Y. For the 100 days, he then had a bivariate distribution of data for his correlation assignment.

To Alan's surprise, the number of waterskiing accidents (X) and the number of ice cream cones sold (Y) were highly correlated in a positive direction for his sample. In fact, the result was statistically significant at the 0.01 level, which means that he could be reasonably certain that a positive correlation existed for the population of an entire year.

Did X cause Y? Did Y cause X? Although Alan computed a high positive correlation between X and Y, most thinking people would not be inclined to believe that eating ice cream would lead to a waterskiing accident. Similarly, it is doubtful that when someone is injured while waterskiing, he or she develops a craving for ice cream. Obviously, a third factor (temperature) would be a likely suspect. When weather is warm or hot, more people are involved in waterskiing, and it stands to reason that more accidents will occur. Also, more ice cream is sold in hot weather than during cold temperatures.

The point of this illustration is that correlation does not necessarily imply causation. While it is tempting to attribute cause and effect to correlation, to do so is inappropriate reasoning. A better notion is that a high correlation *does not rule out* causation. For example, the occurrence of lightning and of thunder over a period of time has a high positive correlation that clearly is indicative of cause and effect. On the other hand, the absence of a relationship between two variables rules out cause and effect (within the probability limits of having committed a type II error).

Another misconception of many beginning students is that negative relationships are not as important as positive correlations. The direction of a relationship is important, but only for interpretation purposes. The size of r (or r^2) is the primary concern. For example, if $r = 0.75$ between homework average and scores on the final exam for a class, it can be said that better performance on the homework tended to be associated with better performance on the exam. If homework average was correlated with the *number of items missed* on the final, the relationship would be -0.75. However, when the scaling of the variables is considered, the interpretation would be exactly the same; that is, good homework performance tends to be associated with *fewer* wrong anwers on the final exam.

Correlation is a technique for quantifying relationships. The concept is nothing new to you, as related variables are everywhere in your daily life. If you have a record of poor achievement in mathematics, how do you think you would do in higher mathematics classes? If you speed through a downtown street in a car, what is apt to happen sooner or later? Why? With a little thought, you can undoubtedly create a lengthy list of related events that you observe every day. The product-moment r is simply an unambiguous method for expressing such relationships between measured variables.

RANK ORDER CORRELATION—OPTIONAL

Suppose six students have been ranked by their classmates on two traits: their school spirit, which is designated as variable X, and popularity, denoted as variable Y. The results of the ranking are as follows:

Student	Rank on X	Rank on Y
A	1	1
B	2	2.5
C	3	5
D	4	2.5
E	5	4
F	6	6

Is there a relationship between school spirit and popularity demonstrated in the data? This problem calls for correlating rank order (ordinal) bivariate data, which is a frequently encountered circumstance in statistics. Because the concept of sums of squares used in the computation of a product-moment r is not meaningful when used with rank order data, a technique known as Spearman's rank order correlation is an appropriate alternative. The Spearman's correlation coefficient is simply an estimate of what r would have been if a product-moment method could have been used. To compute the Spearman rank order correlation coefficient, two columns of additional information are required and can be derived from the ranked data as follows:

Student	Rank on X	Rank on Y	$\lvert$Rank on X – Rank on $Y\rvert$ $\lvert d \rvert$	d^2
A	1	1	0.0	0.00
B	2	2.5	0.5	0.25
C	3	5	2.0	4.00
D	4	2.5	1.5	2.25
E	5	4	1.0	1.00
F	6	6	0.0	0.00
				$\Sigma d^2 = 7.50$

The d column is created by taking the absolute difference between the corresponding ranks. The algebraic sign is unimportant, because Σd^2 is used in the computational equation.

The formula for Spearman's correlation is:

$$1 - \frac{(6)(\Sigma d^2)}{N(N^2 - 1)}$$

where Σd^2 is the sum of the d^2 column and N is the number of bivariate cases.

For the present case, the Spearman rank order correlation is computed as follows:

$$\text{rank order } r = 1 - \frac{(6)(7.50)}{(6)(36 - 1)}$$
$$= 1 - 0.21$$
$$= 0.79$$

The Spearman correlation is designed to correlate rank order data. Computation involves the following steps:

1. Rank all individuals (objects) on the X-variable and on the Y-variable.
2. Form a column of the difference (d) between each pair of corresponding ranks—ignore +'s and –'s.
3. Square each value of d; then sum d^2.
4. Substitute the values of Σd^2 and N in the formula and complete the computations.

The difference between the Spearman rank order correlation and the Pearson r correlation will generally not be dramatic. They will be fairly similar if there is not an excessive number of "tied" ranks as with student B and student D on variable Y in the numerical example presented earlier.

COMPUTER FOCUS 12—Product-Moment Correlation

Purpose

The purpose of this Computer Focus is to provide a program for finding the product-moment correlation coefficient (r) from a bivariate distribution. You will have a chance to study the algorithms involved in the calculation and to improve the program by modifying it to be more "user friendly" under certain circumstances. The task also gives you the opportunity to test and debug the program.

Problem

The accompanying BASIC computer program accepts a bivariate distribution as input and yields a Pearson product-moment r as output.

1. Study the processes in the program and for each of the following values symbolized, indicate which line (by number) of the BASIC program performs that calculation (the first one, ΣX, is already completed as an example).

Function	BASIC Line Number
ΣX	55
ΣY	______
ΣXY	______
Σx^2	______
Σy^2	______
Σxy	______
r	______

2. What happens if either X or Y is a constant?
3. Modify the program to eliminate the error message under the condition in 2.
4. **Given:** a. r (symbolized R) can be transformed to z_r by:

```
Z = 0.5*LOG((1 + R)/(1 - R))
```

b. z_r can be converted back to r by:

```
R = (EXP(2*Z) - 1)/(1 + EXP(2*Z))
```

and

c. $Z_R = R$ IF $R \leq 0.25$

Task: Modify the program to compute the 95% and 99% confidence intervals for the correlation coefficient.

BASIC Program

```
5   REM  START OF CORRELATION PROGRAM
10  N = 0
15  SX = 0
20  SY = 0
25  XSQ = 0
30  YSQ = 0
35  XY = 0
40  INPUT "ENTER A BIVARIATE PAIR SEPARATED BY A
    COMMA: ";X,Y
50  N = N + 1
55  SX = SX + X
60  SY = SY + Y
65  XSQ = XSQ + X*X
70  YXQ = YSQ + Y*Y
75  XY = XY + X*Y
80  INPUT "IS THIS THE LAST DATA ENTRY? (YES OR NO): ";D$
85  IF (D$ = "YES") THEN 90 ELSE 40
90  XY = XY - SX*SX/N
100 XSQ = XSQ - SX*SX/N
105 YSQ = YSQ - SY*SY/N
110 R = XY/(SQR(XSQ*YSQ))
115 PRINT "CORRELATION COEFFICIENT = ";R
200 END
```

KEY TERMS AND NOTATION

Bivariate distribution
Coefficient of determination (r^2)
Correlation
Cross products of deviation scores (Σxy)
Linear relationship
Negative correlation
Pearson r
Point estimate
Positive correlation
Product-moment correlation (r)
Relationship between variables
Rho (ρ)
Scattergram
z_r

APPLICATION EXERCISES

12.1 Read each statement that follows and code your belief regarding the relationship between the two implied variables in the space provided according to this scheme:

+ = positive relationship
0 = little or no relationship
– = negative relationship

______ a. social status and the proportion of income spent on food for a group of families

______ b. ability to run the 100-yard dash and ability to add two-digit numbers

______ c. height and chronological age of boys ages 6 to 14

______ d. height and age of adult males

______ e. city population and number of elementary schools

______ f. coaches poll and sports writers poll of "top 20" football rankings

______ g. fatigue and motor coordination (manual dexterity)

______ h. rankings of two Olympic diving judges for 20 contestants

______ i. "common sense" and number of factory accidents suffered

______ j. weight and interest in cowboy novels of sixth graders

12.2 Can the Pearson r be used to analyze a univariate distribution? Why or why not?

12.3 Given the following values of X and Y:

X	Y
7	16
11	8
9	12
12	6
10	10
11	8
10	10
13	4
9	12
8	14

a. Calculate the Pearson product-moment r.

b. Plot the bivariate points in a scattergram. How does the scattergram confirm the value of r?

12.4 An investigator was examining the relationship between social maturity and assertiveness among a population of students. A correlation coefficient of 0.50 was computed from a sample of 27 pairs of numbers. Can the population parameter be regarded as different from zero on the basis of a 5% level of significance?

12.5 For a sample size of $n = 32$, how large must r be to be statistically significant at the 0.05 level on a two-tailed test?

12.6 Determine the coefficient of determination for each value of r that follows.

a. 0.90

b. 0.15

c. -0.63

d. -0.85

12.7 If $\Sigma x^2 = 25$, $\Sigma y^2 = 16$, and $\Sigma xy = 15$, find r.

12.8 If $n = 103$ and $r = 0.60$ between neatness and employability, find:

a. the 95% confidence interval for the correlation coefficient

b. the 99% confidence interval

12.9 In a sample of 52 bivariate pairs, $r = 0.45$ between self-esteem and grade-point average. Find the 95% confidence interval for rho.

12.10 One teacher found an $r = -0.84$ between the number of instances of tardiness and course grades. The teacher concluded that material missed because of tardiness to class was the cause of lower grades. Comment on this conclusion.

12.11 Given the following data (X = number of push-ups; Y = number of sit-ups):

X	Y
8	18
10	25
14	38
9	21
10	26
9	20
11	28
12	32
9	20
10	23
13	28
9	19

a. What is the correlation between the number of sit-ups and the number of push-ups?
b. Test H_0: $\rho = 0$ at the 0.05 level of significance on a one-tailed test.
c. What percentage of the variation in sit-ups can be explained by variation in push-ups?
d. Construct
   (1) an 80% confidence interval for ρ.
   (2) a 95% confidence interval for ρ.

12.12 Psychological studies of college males have shown a positive correlation between orderliness (a desirable trait) and stinginess (an undesirable trait). Discuss this finding in light of the traits involved; that is, provide an explanation of the $+r$ between traits with apparent opposing social valences.

12.13 *Optional* Ten projects from the woodshop class were entered in an exhibit. Two judges ranked the projects, with a score of 1 being best, 2 being second best, and so on through a rank of 10, which was the last-place finisher. Here are the two judges' rankings of the projects:

Project	Rank of Judge 1	Rank of Judge 2
A	4	3
B	9	10
C	2	4
D	10	8
E	3	2
F	7	9
G	1	1
H	8	7
I	5	6
J	6	5

a. Compute Spearman's rank order correlation coefficient.

b. Were the judges reasonably consistent (reliable)?

13

Regression Analysis

CHAPTER OBJECTIVES

Upon completion of the chapter, students will be able to:

1. Compare and contrast correlation and regression.
2. Describe the least-squares criterion for regression.
3. Recognize the slope-intercept form of a linear equation.
4. Define the slope and intercept of a line of regression.
5. Compute the necessary constants for a regression equation from a bivariate distribution.
6. Compute a predicted value of the dependent variable from a regression equation.
7. Compute the standard error of estimate from a set of bivariate data.
8. Test the null hypothesis that $\beta = 0$ with two different methods.
9. Given a regression line and standard error of estimate for the regression, construct confidence bands using characteristics of the normal curve.
10. Explain how the size of a correlation coefficient is inversely related to the size of the standard error of estimate for a given size sample.
11. (Optional) Develop a linear prediction formula using the median fit technique.

Jan, after receiving a favorable report of her achievement test score from her homeroom teacher, scheduled an appointment with the high school counselor. The purpose of the conference was to discuss the entrance exam at an academically elite university. The counselor explained that admissions data from recent years revealed that a relationship existed between performance on the achievement test and subsequent performance on the admissions test administered by the university. The counselor further explained that because high scorers on the achievement test *tended* to be higher scorers on the admissions exam, Jan's chances of passing the admissions exam would be quite high. In fact, students who applied to the university and had scores similar to Jan's had about an 80% rate of passing the admissions exam. Jan was encouraged by these results and was cautiously optimistic about her chances of passing the admissions exam.

In statistical terms, the correlation (relationship) between performance on the achievement test and subsequent performance on the admissions exam enabled the counselor to make an informed *prediction* about Jan's performance on the latter assessment. *Regression*, a statistical technique for formally addressing the topic of prediction, is the subject treated in this chapter.

Sir Francis Galton (1822–1911) became involved in a series of studies in which hypotheses about heredity posed by his cousin, Charles Darwin, were tested. During the studies, Galton found that the heights of offspring tended to regress toward the mean of the general population. That is, he found a general trend that tall parents had children who were taller than average but were not as tall as their parents. And he found that the children of short parents were generally taller than their parents but still below average in height. This phenomenon of moving toward the general population mean was often referred to as the *law of filial regression*.

The term regression came to be used whenever the prediction of one variable from another variable was discussed. For example, when the value of a variable X is two standard deviations above the mean and when this value is used to predict the value of Y using regression techniques, the resulting **predicted value of Y** may be only one standard deviation above the mean. Such a tendency to move closer to the mean when predicting Y-values from related X-values is known as regression toward the mean, and it is characteristic of any bivariate relationship for which the correlation (r) is less than perfect (± 1.00).

If the topics presented in Part 4 of this text could be considered a family of techniques, correlation and regression would surely be considered siblings, and probably twins. Bivariate correlation and regression are more alike than they are different in that both deal with linear relationships between two variables. Consider the primary purpose of correlation to be one of determining the size and direction of a bivariate linear relationship. Then, if the size of a correlation coefficient between two variables is sufficiently large (that is, different from zero), knowledge of a measurement on one variable would permit a fairly accurate prediction of a value of the other variable. In that context, regression will be considered as the technique for making predictions.

THE CONCEPT OF PREDICTION

The more quality study time devoted to statistics before the next exam, the higher the predicted exam performance. The more practice a musician logs, the better the predicted performance. If you make a good grade in Algebra I, it would be predicted that you would do well in Algebra II. As fatigue goes up, eye-hand coordination can be predicted to decrease. You make predictions daily. Predictions about what will happen in a given situation are usually based on some related circumstance. Regardless of whether cause-and-effect conditions exist, if the relationship is strong enough, the prediction can be quite accurate. The point here is that the concept of prediction is not new to you. The technique of regression will simply formalize and quantify operations that you perform on a less objective basis every day.

Linear Predictions

One of the major objectives of any science is to be able to predict. That is, the scientist gathers information that is believed to be related to the phenomenon to be predicted. If the data are related, then variation in one variable can be explained by variation in another variable with the coefficient of determination (r^2). Consequently, given a particular value for one variable, one should be able to predict a value for another variable with greater accuracy than if knowledge of the original variable's value were not available. Correlation is the technique for determining the degree to which two variables are related in a *linear* fashion. With regression, it is possible to take advantage of any such linear relationship and develop an equation with one variable as a linear function of the other.

Linear Equations

In algebra, the general form of a **linear equation** is:

$$AX + BY = C$$

where A, B, and C are constants; X and Y are variables. With some algebraic manipulation, one can derive a more convenient form of the equation known as the **slope-intercept form of a linear equation**: $Y = a + bX$. In this form, Y is called the *dependent* variable and X is the *independent* variable. In the chapter on correlation, independent and dependent variables were not explicitly designated. However, in regression, it is convenient to designate the variable standing alone on the left side of the equation as the **dependent variable**, whereas the variable multiplied by a constant (b) on the right side is called the **independent variable**. As the names imply, the values of the dependent variable are determined, or are dependent upon, the values of the independent variable. The independent variable takes on values throughout its range without regard to the values of other variables. For example, the time required for a dip of ice cream to melt is *dependent on* the temperature, the amount of a traffic citation is *dependent on* excessiveness of the speed above the speed limit, and how well you would perform in Auto Mechanics II is *dependent on* the quality of your performance in Auto Mechanics I. As is frequently the case with bivariate data, the independent variable does not have to cause the dependent variable; however, variation in the dependent variable can be statistically "explained" by variation in the independent variable (recall r^2).

Working with slope-intercept forms of linear equations is not a new activity. Recall from Chapter 6 that z-scores were transformed to T-scores by: $T = 10z + 50$; or, by virtue of the commutative law of addition, an equivalent form is $T = 50 + 10z$. In essence, z-scores served as the independent variable and T-scores were dependent on the values of z.

Another familiar linear equation is the conversion from Celsius temperature to Fahrenheit temperature:

$$F = 32 + 1.8C.$$

In this equation, F is the dependent variable; C is the independent variable. Note that any numerical value of C can be substituted in the formula and when the arithmetic is performed, a value of F is determined. In other words, the value of F *depends* on the value of C substituted into the equation. Table 13.1 shows the corresponding values of F when six arbitrarily chosen values of C are 0, 10, 20, 25, 40, and 100.

Table 13.1 Six Celsius Temperatures and Corresponding Fahrenheit Temperatures

C°	[formula]	F°
0	[32 + 1.8(0)]	32
10	[32 + 1.8(10)]	50
20	[32 + 1.8(20)]	68
25	[32 + 1.8(25)]	77
40	[32 + 1.8(40)]	104
100	[32 + 1.8(100)]	212

GRAPHIC REPRESENTATION OF REGRESSION

Figure 13.1 shows the bivariate plot of the six pairs of temperature values displayed in Table 13.1. Note that the points appear to determine a straight line, which illustrates the meaning of the phrase **linear** equation. When the line is inserted through the points, the line now represents all of the values in the range C: $0 \leq C \leq 100$ with the corresponding F-values. What does this line have to do with prediction? Because the line can be considered an infinite number of bivariate points, each point would represent both C and a corresponding F-value. This implies that an F-value can be determined for any C-value in the appropriate range of the graph. Just choose a particular value of C somewhere along the horizontal axis (C-axis); move vertically until the line is reached. Then move to the left to determine the value of F on the vertical axis. You have, in the context of regression, *predicted* an F-value from a given value of C.

Slope of the Regression Line

Examine some important characteristics of the graph in Figure 13.1. First, notice that the line is not parallel to the horizontal axis; the line has slope. If you begin from any point on the line and move 1 unit to the right horizontally, you must move 1.8 units up to reach the line again. Let dX represent an amount of change in the X-variable and let dY represent an increment in the Y-variable. Notice that, in general, if you move dX units to the right, dY vertical units would be required to reach the line. Some texts call this operation "rise-over-run," and it is simply the ratio of the change in Y to the change in X. The quotient $\frac{dY}{dX}$ is the **slope** of the line; and in the case of C to F conversion, this ratio is 1.8 because for every 10 degrees change on the Celsius scale, the corresponding change

Figure 13.1—Graph of $F = 32 + 1.8C$

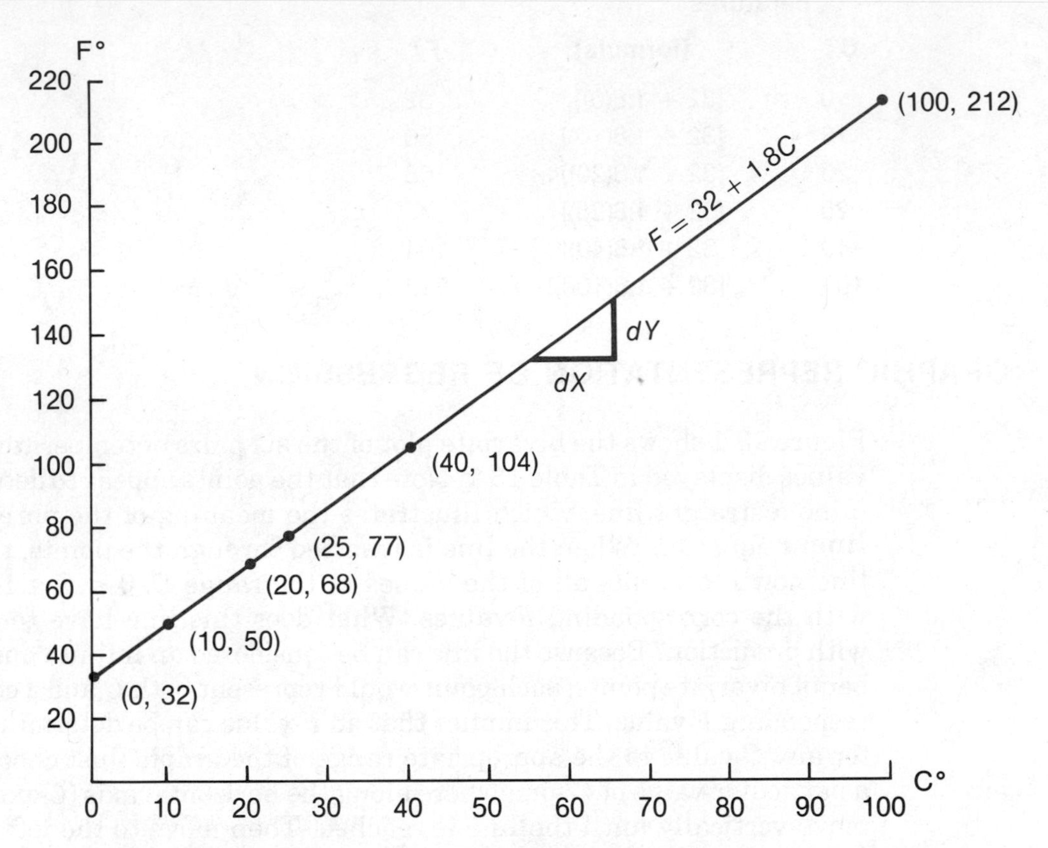

on the Fahrenheit scale is 18 degrees. In Figure 13.1, $dX = 10$ and $dY = 18$; therefore $\frac{dY}{dX} = 1.8$, which is the slope of the line.

Is it coincidental that 1.8 is the slope of the line and is also the constant in the slope-intercept form, that is, the coefficient of *C*, the independent variable ($F = 32 + 1.8C$)? No, this was no accident; in the slope-intercept form, the coefficient of the independent variable *is* the slope of the resulting graph. There exists an infinite number of lines with slope of 1.8; therefore, to properly define the particular line shown in Figure 13.1, some additional information is needed, namely, a point on the line. One point has particular significance and is called the *intercept*.

Intercept

With reference to Figure 13.1, notice that the line crosses the vertical axis at 32° Fahrenheit. The value of the dependent variable at the point where the line intersects the axis that represents the dependent variable is called the **intercept**. In Figure 13.1 the intercept is 32. Another look at the slope-intercept form of the conversion equation reveals that the intercept is a constant: $F = 32 + 1.8C$.

The major points illustrated by this example are:

1. An equation of the form $Y = a + bX$ will produce a straight-line graph.
2. In the slope-intercept form of a linear equation, $Y = a + bX$, b is the slope of the resulting graph.
3. A graph of $Y = a + bX$ will intersect the axis of the dependent variable at the value a, the intercept.

The correlation between Celsius and Fahrenheit temperatures is +1.00, so the example has shown a special relationship. The scattergram of F and C forms a line called a *line of regression*. However, in most statistical situations the correlation is less than one. Still, the task is to determine a and b in the equation $Y = a + bX$.

THE LEAST-SQUARES CRITERION

If X and Y do not correlate perfectly (+1.00 or −1.00), the points in a scattergram do not fall on a straight line. When values of Y are predicted from values of X, the predicted values of Y (denoted Y' and called "Y prime") will form a line; however, the actual measured values of Y may not fall on this line. In other words, the predicted values of Y (Y') will not generally be equal to the actual values of Y. The difference between any particular Y and the corresponding Y' is symbolized as e_i (error) $= Y_i - Y_i'$.

Figure 13.2 illustrates a prediction line, which is called the **line of best fit**. The line of best fit can be used to estimate values of Y from known values of X. Actual Y-scores shown as the plotted points in Figure 13.2 do not all fall on the line. The vertical distances between the line (predicted Y's) and the actual plotted points are the errors (e_i) associated with predicting Y from X. The aim of regression is to find a prediction line of best fit and an equation that will make the errors of prediction as small as possible.

A good first guess might be to determine the line so that the algebraic sum of the errors would be zero, or $\Sigma e_i = 0$. In such a case the sum of the "positive" errors would be equal to the sum of the "negative" errors. However, such a criterion would not provide a unique solution, because there is an infinite number of lines that would satisfy that condition. The solution to the problem is similar

Figure 13.2—Regression Line with Errors in Prediction

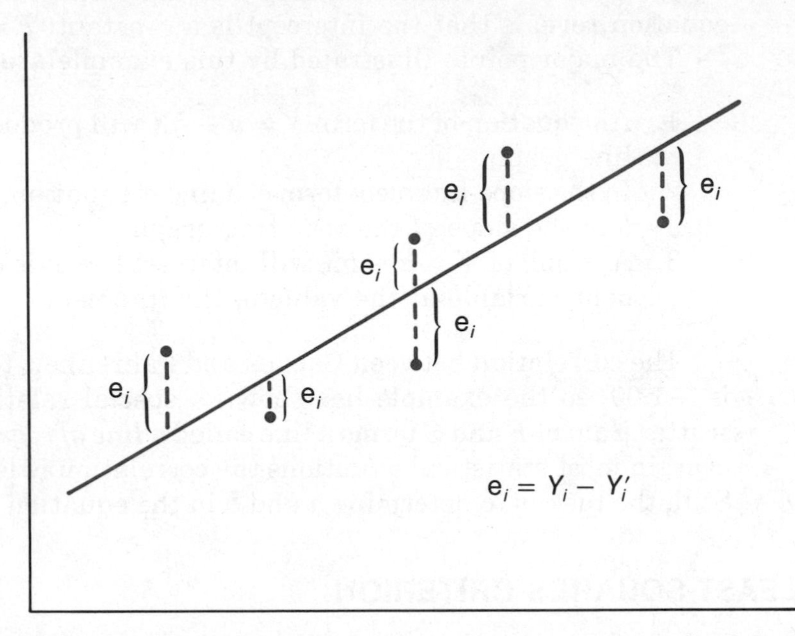

to the approach used for computing variance and standard deviation; specifically, to use the squares of distances called sum of squares. In the case of regression, all values of e_i (+ and −) are squared and a prediction line is drawn for which the *sum of the squared errors* (Σe_i^2) *is a minimum.* This uniquely defines the line of best fit according to the **least-squares criterion** and is called a *regression line.* The method for showing that this line does minimize the squared errors is a topic for calculus and is therefore beyond the scope of this text. Incidentally, the least-squares criterion is also one of the infinite number of cases in which $\Sigma e_i = 0$.

COMPUTATION OF REGRESSION EQUATIONS

The formula for the slope-intercept form of the least-squares regression line is:

$$Y' = a + bX$$

where Y' = the *predicted* value of Y (dependent variable),
X = the known value of X (independent variable),
a = the Y-intercept (constant), and
b = the slope of the line (constant).

The quantity designated as b is the slope of the regression line and is called the **regression coefficient**. From a computational perspective, the purpose of regression is to determine the values of a and b in the prediction equation.

Computational Procedures

The process of developing the regression equation requires a sample bivariate distribution as a starting point. From this sample, the Y-intercept (a) and the slope (b) can be calculated for use in the equation for other individuals generally considered to be from the same statistical population. Consequently, prediction with regression is inferential, although the equation used for prediction is computed with a representative sample bivariate distribution. Assume the data for a bivariate distribution are available. The steps for computing the regression coefficient (b) and the Y-intercept (a) are similar to the steps for calculating r described in the previous chapter.

Step 1–For the raw data, compute ΣX, ΣY, ΣXY, ΣX^2, ΣY^2, $\overline{X}$, and $\overline{Y}$. These values serve as inputs for the next step.

Step 2–Compute the sums of the deviation squares and cross products—Σx^2, Σy^2, and Σxy.

Step 3–From the values obtained in step 2, compute b:

$$b = \frac{\Sigma xy}{\Sigma x^2}.$$

Step 4–Using results from step 1 and step 3, calculate a:

$$a = \overline{Y} - b\overline{X} \quad \text{(mean of } Y \text{ minus } b \text{ times mean of } X\text{)}$$

Numerical Illustration

As an illustration of the computation procedures, suppose you wanted to develop an equation that would predict achievement in mathematics from scores of attitude toward mathematics. For convenience, the bivariate data used in Chapter 12 along with summary calculations are provided in Table 13.2.

Recall that the computations for Mr. Lawson's sample r were identical to those shown in Table 13.2. Steps 1 and 2 are shown in the table.

Table 13.2 Hypothetical Bivariate Distribution with Squares, Cross Products, and Summary

Attitude Toward Math (X)	Achievement in Math (Y)	X^2	Y^2	XY
66	75	4,356	5,625	4,950
82	74	6,724	5,476	6,068
48	55	2,304	3,025	2,640
55	60	3,025	3,600	3,300
76	85	5,776	7,225	6,460
56	74	3,136	5,476	4,144
65	59	4,225	3,481	3,835
40	56	1,600	3,136	2,240
88	85	7,744	7,225	7,480
58	53	3,364	2,809	3,074
65	68	4,225	4,624	4,420
95	97	9,025	9,409	9,215
71	64	5,041	4,096	4,544
50	59	2,500	3,481	2,950
57	51	3,249	2,601	2,907
87	74	7,569	5,476	6,438
40	54	1,600	2,916	2,160
48	50	2,304	2,500	2,400
58	58	3,364	3,364	3,364
53	64	2,809	4,096	3,392
80	65	6,400	4,225	5,200
92	85	8,464	7,225	7,820
69	55	4,761	3,025	3,795
64	52	4,096	2,704	3,328
72	79	5,184	6,241	5,688

Step 1 (same as in Chapter 12)

$\Sigma X = 1{,}635$ $\Sigma Y = 1{,}651$ $\Sigma X^2 = 112{,}845$ $\Sigma Y^2 = 113{,}061$ $\Sigma XY = 111{,}812$

$\overline{X} = 65.40$ $\overline{Y} = 66.04$

Step 2 (same as in Chapter 12)

$\Sigma x^2 = 5{,}916$ $\Sigma y^2 = 4{,}028.96$ $\Sigma xy = 3{,}836.6$

The summary computations in Table 13.2 can be used to compute a and b for the regression equation.

Step 3–$b = \frac{\Sigma xy}{\Sigma x^2} = \frac{3{,}836.6}{5{,}916} = 0.648$ or 0.65 to two-decimal place accuracy.

Because the calculator keystroke sequences for the sum of cross products and sum of squares of the deviation scores have already been shown, the sequence for computing b is simplified to the steps shown in Figure 13.3.

Figure 13.3—Calculator Keystrokes

PROBLEM:

Calculator procedures for computing b, the regression slope, with Σxy and Σx^2 given (see Figure 5.2):

$\Sigma xy = 3{,}836.6$
$\Sigma x^2 = 5{,}916$

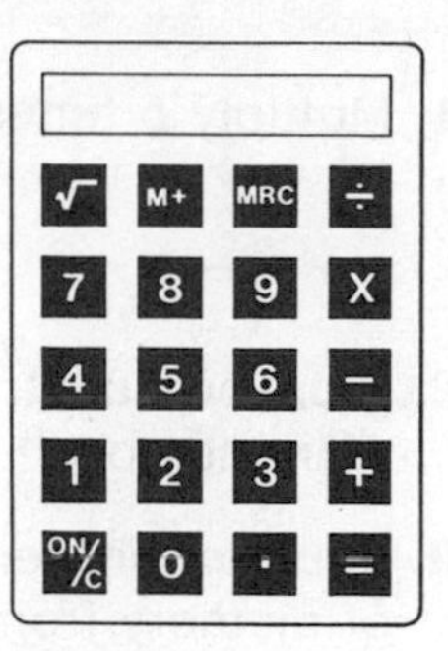

Procedure	Arithmetic Operation	Calculator Keystrokes	Display
1. $b = \frac{\Sigma xy}{\Sigma x^2} = \frac{3{,}836.6}{5{,}916}$	1. 3,836.6 ÷ 5,916	1. 3,836.6	3,836.6
2. Divide Σxy by Σx^2 $b = 0.65$	2. $b = 0.65$	2. [÷]	3,836.6
		5,916	5,916
		[=]	0.6485125 or 0.65

Step 4–$a = \overline{Y} - b\overline{X}$
$= 66.04 - (0.65)(65.40)$
$= 23.53.$

The calculator keystroke sequence for finding a is provided in Figure 13.4.

Figure 13.4—Calculator Keystrokes

PROBLEM:

Calculator sequence for computing *a*, the intercept, with $\overline{X}$, $\overline{Y}$, and *b* known:

$\overline{X} = 65.40$
$\overline{Y} = 66.04$
$b = 0.65$

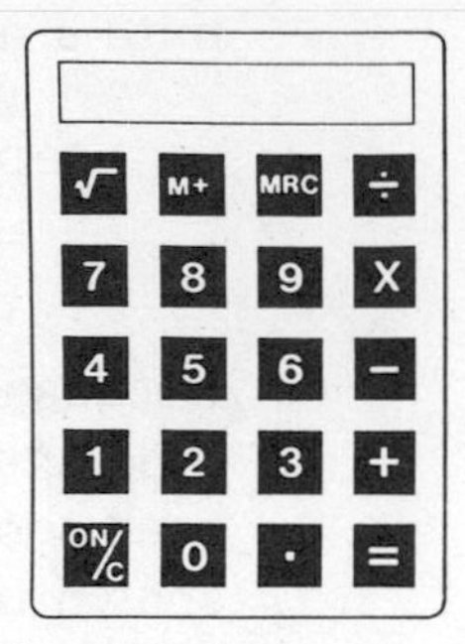

Procedure	Arithmetic Operation	Calculator Keystrokes	Display
1. Multiply *b* times $\overline{X}$	1. 65.40 × 0.65 = 42.51	1. 65.40 [×]	65.40
		0.65	0.65
		[=]	42.51
2. Store product from 1. in memory	2. Store 42.51	2. [M+]	42.51
3. Subtract contents of memory from $\overline{Y}$	3. 66.04 − 42.51 = 23.53	3. 66.04 [−]	66.04
		[MRC]	42.51
		[=]	23.53

Therefore appropriate substitutions in the slope-intercept form of the prediction equation yield the least-squares equation:

$Y' = a + bX$
$Y' = 23.53 + 0.65X.$

INTERPRETATION OF REGRESSION

Predicting *Y* from *X*

A graph of the regression equation just developed is shown in Figure 13.5. The line can be determined by locating the *Y*-intercept (23.53) and one other point. The other point can be determined by substituting an arbitrary value of *X* in the regression equation and solving for *Y'*. A line drawn through the two points determines the regression line. The graph may be used to estimate the predicted value of *Y* (*Y'*) for any given value of *X*. For example, suppose you were interested in predicting achievement (*Y*) for an individual who had an attitude score (*X*) of 45. The dotted line in Figure 13.5 shows

Figure 13.5—Regression Line for Predicting Achievement from Attitude

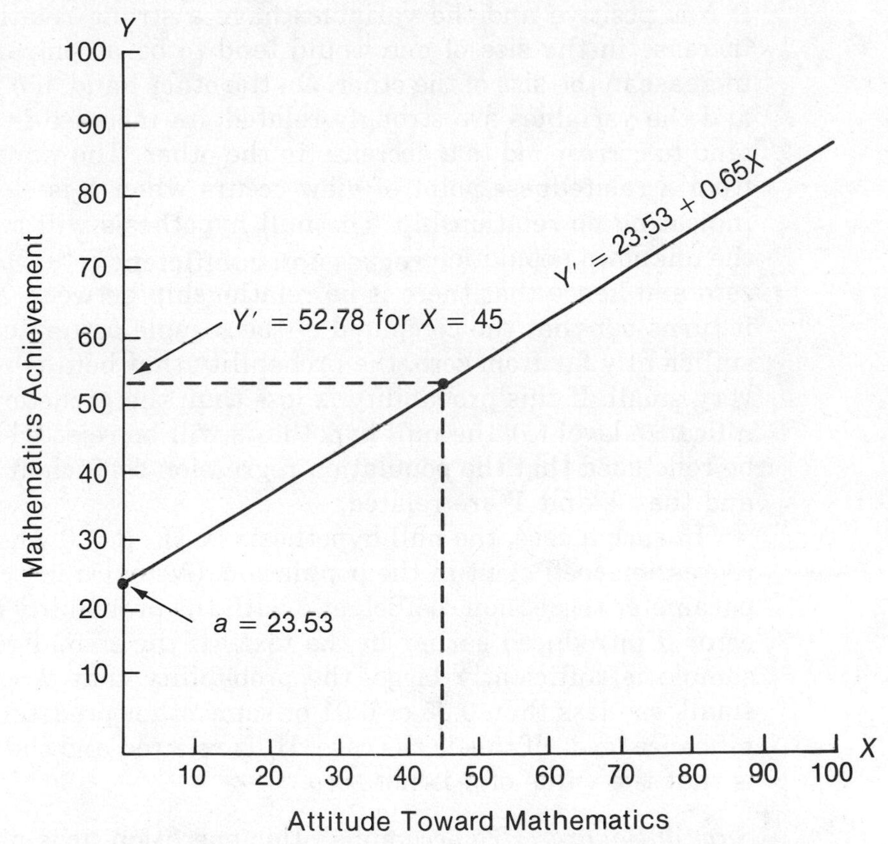

that the point that is directly above $X = 45$ has a corresponding Y-value of about 53. To verify this, or to obtain more precision than is possible on the graph, one could substitute 45 for X in the equation $Y' = a + bX$.

$$
\begin{aligned}
Y' &= 23.53 + 0.65X && \text{(regression equation)} \\
&= 23.53 + (0.65)(45) && \text{(substitute 45 for } X\text{)} \\
&= 23.53 + 29.25 && \text{(multiplication)} \\
&= 52.78 && \text{(addition)}
\end{aligned}
$$

Similarly, any appropriate value of X could be used to predict its related value of Y' either by estimating Y' from the graph or by direct substitution in the regression equation.

Inferential Analysis

Hypothesis Test. Because regression equations are derived from sample data, it is often useful to test a null hypothesis for regression. Remember that b, the slope of the regression line, is an indication of the degree of relationship between the two variables X and Y. If b is positive and the variables have a strong relationship, an increase in the size of one would tend to be accompanied by an increase in the size of the other. On the other hand, if b is negative and the variables are strongly related, an increase in one would tend to correspond to a decrease in the other. The worst situation from a relatedness point of view occurs when b is close to zero, indicating no relationship. The null hypothesis will propose that the unknown population **regression coefficient** beta (β), is actually zero and hence that there is no relationship between X and Y. If it turns out that the computed b (the sample estimate of beta) is sufficiently far from zero, the probability that beta is zero will be very small. If this probability is less than the predetermined significance level (α), the null hypothesis will be rejected and it will be concluded that the population regression coefficient is not zero and that X and Y are related.

In such a case, the null hypothesis is: H_0: $\beta = 0$ where β is the regression coefficient in the population. (*Note:* Do not confuse the parameter regression coefficient β with the probability of a type II error β introduced earlier in the text.) If the computed b for the sample is sufficiently large, the probability that $\beta = 0$ becomes small, say less than 0.05 or 0.01 or some other predetermined significance level. If this is the case, H_0 is rejected and the conclusion is that the value of β is not zero.

Sum of Squares. To accomplish this operation, it is necessary to introduce the two components of variance in the dependent variable. The total amount of variation in the dependent variable can be expressed as the sum of squares—$\Sigma y^2 = \Sigma(Y - \overline{Y})^2$. This variability can be separated into two components: **sum of squares due to regression (SS_r)** and **sum of squares due to error (SS_e)**. If the **total sum of squares (Σy^2)** is expressed as SS_t, the following relationship exists:

$$SS_t = SS_r + SS_e.$$

The SS_r is the amount of variability in Y that can be explained by variability in X. You may remember from Chapter 12 that a similar statement was made about the coefficient of determination (r^2). Indeed, the equation defining the coefficient of determination is:

$$r^2 = \frac{SS_r}{SS_t}.$$

Degrees of Freedom. Degrees of freedom (df) are associated with each source of variation in the regression model. The **total number of degrees of freedom (df_t)** for the Y-variable is $N - 1$ (see Chapter 11) where N is the total number of cases. These $N - 1$ degrees of freedom can be partitioned into two sources: (1) variation in Y due to the influence of X (df_r); and (2) error variation in Y (df_e) that is not explained by variation in the X-variable. The former is referred to as **degrees of freedom for regression (df_r)** and is equal to the number of variables in the regression model minus one: $2 - 1 = 1$. The **degrees of freedom** remaining (the difference between df_t and df_r) are **attributed to error**:

$$\begin{aligned} \mathbf{df_e} &= df_t - df_r \\ &= (N - 1) - 1 \\ &= N - 2. \end{aligned}$$

The following formulas show the sources of variation and the degrees of freedom associated with each:

$$SS_t = \Sigma y^2 = \Sigma(Y - \overline{Y})^2 = \Sigma Y^2 - \frac{(\Sigma Y)^2}{N}; \text{ with df} = N - 1.$$

$$SS_r = \Sigma(Y' - Y)^2 = b(\Sigma xy); \text{ with df} = 1.$$

$$SS_e = \Sigma e_i^2 = \Sigma(Y - Y')^2 = SS_t - SS_r; \text{ with df} = N - 2.$$

In the data used previously in this chapter (attitude toward and achievement in mathematics),

$$SS_t = \Sigma y^2 = 4{,}028.96 \quad (\text{df} = 24),$$

$$SS_r = 0.65(3{,}836.6) = 2{,}493.79 \quad (\text{df} = 1), \text{ and}$$

$$SS_e = 4{,}028.96 - 2{,}493.79 = 1{,}535.17 \quad (\text{df} = 23).$$

Notice that $r^2 = \frac{SS_r}{SS_t} = \frac{2{,}493.79}{4{,}028.96} = 62\%$, which is the value obtained by squaring r in Chapter 12.

Mean Square. With that introduction, you are now ready to test H_0: $\beta = 0$. The variance for regression $\mathbf{MS_r}$ (called **mean square regression**), is the SS_r divided by the associated degrees of freedom:

$$MS_r = \frac{SS_r}{df_r}.$$

Similarly, the $\mathbf{MS_e}$ **(mean square error)** is:

$$MS_e = \frac{SS_e}{df_e}.$$

The F-*ratio Test.* The F-ratio is designed to test the null hypothesis that two variances are equal. The F-ratio statistic plays a role in testing for differences between variances (which are called *mean squares*) in both Chapter 14 (t-test) and Chapter 15 (Analysis of Variance). Therefore, in the present problem, an F-ratio is computed to determine if MS_r is significantly larger than the error (MS_e):

$$F = \frac{MS_r}{MS_e}.$$

Substituting the values from the current problem:

$$MS_r = \frac{2{,}493.79}{1} = 2{,}493.79 \text{ and}$$

$$MS_e = \frac{1{,}535.17}{23} = 66.75.$$

Then

$$F = \frac{MS_r}{MS_e} = \frac{2{,}493.79}{66.75} = 37.36.$$

A summary table is generally provided that concisely displays this information. Table 13.3 presents such a summary.

Table 13.3 Regression Summary Table for Null Hypothesis Test

Source of Variation:	Degrees of Freedom (df)	Sum of Squares (SS)	Mean Squares (MS)	F
Regression	1	2,493.79	2,493.79	37.36
Error	23	1,535.17	66.75	
Total	24	4,028.96		

If the *calculated value of* F (37.36 in this case) *is greater than the corresponding critical* (tabled) *value of* F shown in Appendix F, *then* H_0 *will be rejected* and it would be concluded that the relationship between the independent and dependent variables is statistically significant at the predetermined alpha level.

Notice that the number of degrees of freedom associated with regression (the numerator of the F-ratio) is 1. The df associated with error (the denominator of the F-ratio) is 23. To find the **critical value of F** in Appendix F, find 1 on the horizontal key at the top of the table; this row is labeled df for numerator. Then find 23 in the left column (df for denominator) of the table. The two values shown at the intersection of the row and column just located are 4.28 and 7.88. The smaller value (4.28) is the critical value of F for $\alpha = 0.05$; the larger value (7.88) is the critical value of F for $\alpha = 0.01$.

Clearly the computed value of F (37.36) is larger than the value required for significance at both the 0.01 and 0.05 levels. Consequently, the null hypothesis ($\beta = 0$) is rejected in favor of the alternate hypothesis that a relationship does exist between the independent and dependent variables. This is exactly the same conclusion reached in Chapter 12 when you tested the null hypothesis $\rho = 0$. In fact, a test of the correlation null hypothesis $\rho = 0$ is identical to testing the regression null hypothesis $\beta = 0$. So if r has been tested for statistical significance, a significance test for b in regression is redundant, because the two procedures will yield identical results.

Standard Error of Estimate. As has been previously noted, rarely will the actual Y-scores be identical to the predicted Y-scores. There is error in the predictions, and the extent of this error is measured by a statistic known as the *standard error of estimate.* When the size of the correlation coefficient r is large, the magnitude of the error is small; and conversely, when r is low, the size of the standard error of estimate is large. As shown with the Fahrenheit to Celsius temperature conversion, a perfect relationship between X and Y will yield the same Y and Y' values. On the other hand, if there is no relationship between the two values, the maximum amount of error is present in the predictions. In fact, within the context of these two extreme cases, it can be shown that the minimum standard error of estimate is zero (if $r = \pm 1.00$) and the maximum is the standard deviation of the dependent variable (if $r = 0$).

Consider a scattergram with a large number of bivariate points, and further consider a few representative values of the independent variable (X). For any particular value of X, there will be many values of actual Y. Four such values of X with the accompanying Y-values are shown in Figure 13.6. You can see that the Y-values tend to cluster close to the regression line and become less dense as they depart from the line. The mean value of the Y-distributions for the respective values of X falls on the regression line if the assumption of linearity has been satisfied. This means that for any value of X, the predicted value of Y (Y') is simply the mean value of the Y-distribution corresponding to that X-value.

As you might guess, because the distribution of Y for some particular value of X has a mean and variation, one could construct a frequency distribution, a histogram, and a smooth-line graph of the frequency of Y-values. The result would be a normal curve as shown in Figure 13.7. The standard deviation of the bell-shaped curve shown in Figure 13.7 is called the **standard error of estimate** (symbolized as $\mathbf{S_e}$). This is tantamount to saying that the standard

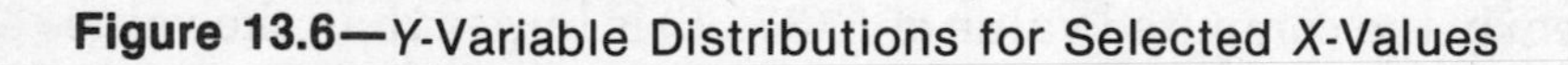

Figure 13.6—Y-Variable Distributions for Selected X-Values

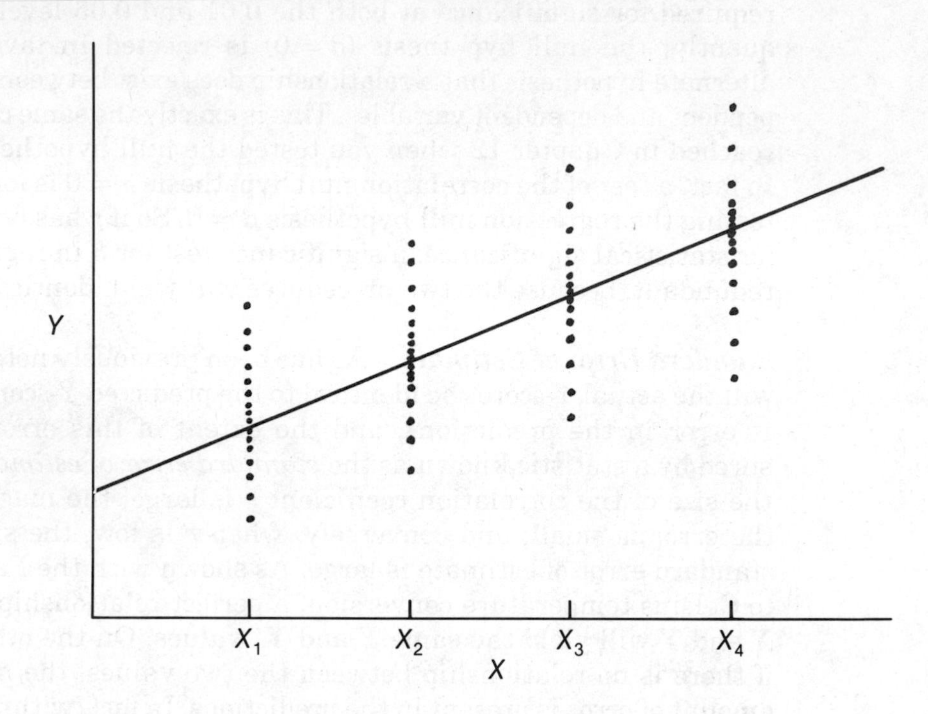

Figure 13.7—Bell-Shaped Frequency Curve of Error Around Regression Line for a Given Value of X

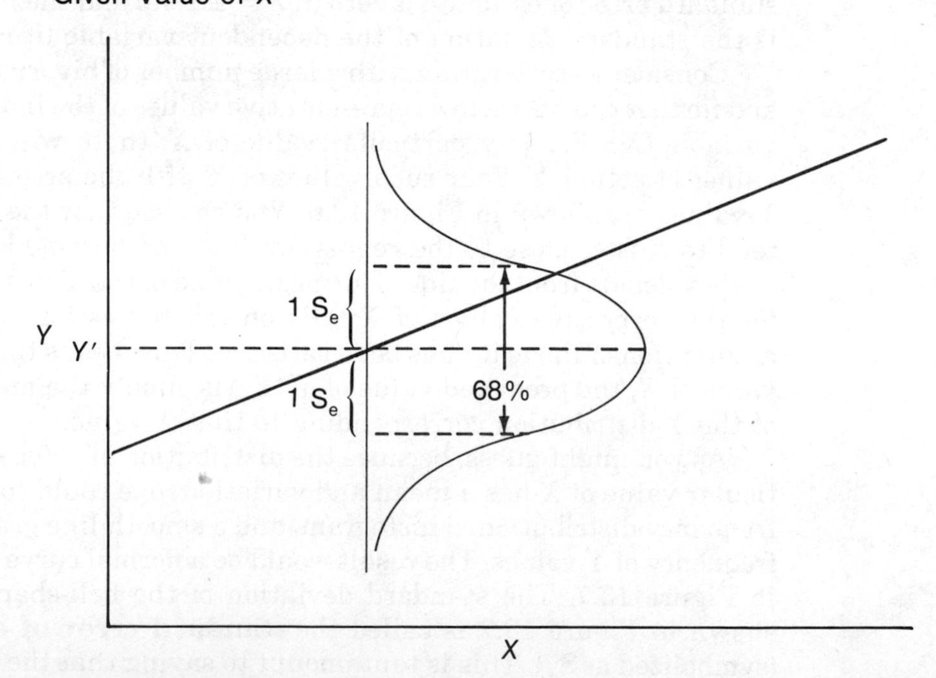

error of estimate is the standard deviation of the errors (e_i) around the regression line. By moving vertically one standard error of estimate above and below the regression line, one has included 68% of the *Y*-distribution for that particular *X*-value.

Another assumption for regression (in addition to linearity of means of the *Y*-distributions) is that the variability of error is the same over the appropriate range of the regression line. This is sometimes called **homoschedasticity** (meaning "equal variance") and is illustrated in Figure 13.8. As shown in Figure 13.8, a line that is one standard error of estimate vertically above and below the regression line now forms a 68% **confidence band**. This means that about 68% of the points on the scattergram would be within the band. Because errors are normally distributed around the regression line, we can use the characteristics of the normal curve once again to establish various levels of confidence bands. Figure 13.8 also shows that 1.96 standard errors of estimate above and below the regression line would include about 95% of the points on the scattergram.

Once the standard error of estimate (S_e) has been determined, *z*-scores vertically along the *Y*-distribution of scores at a particular *X* can be determined. Recall that a *z*-score is a deviation score divided

Figure 13.8—Regression Line and Confidence Bands

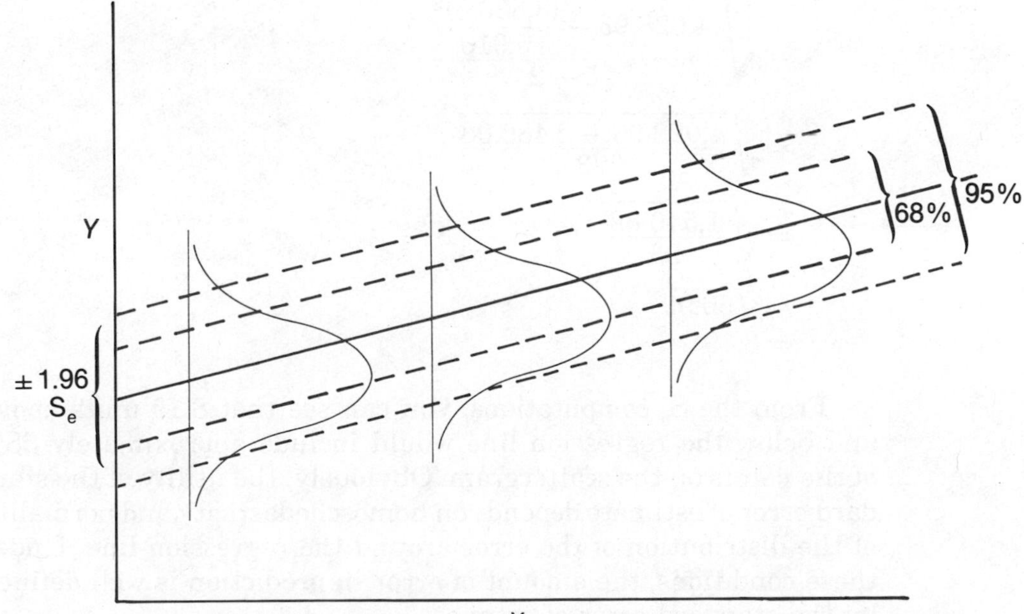

by the standard deviation; thus the z-score formula for dealing with regression error simplifies to:

$$z = \frac{(Y - Y')}{S_e}.$$

Computation of the standard error of estimate appears fairly imposing; however, if the sum of squares of the deviation scores and cross products of the deviation scores are known from the computation procedures for r and regression, the arithmetic is fairly straightforward. The equation for the standard error of estimate is:

$$S_e = \sqrt{\frac{\Sigma y^2 - \frac{(\Sigma xy)^2}{\Sigma x^2}}{N - 2}}$$

You now can compute the standard error of estimate for the mathematics attitude and achievement data used earlier in this chapter.

$$\begin{aligned}
S_e &= \sqrt{\frac{\Sigma y^2 - \frac{(\Sigma xy)^2}{\Sigma x^2}}{N - 2}} \\
&= \sqrt{\frac{4{,}028.96 - \frac{(3{,}836.6)^2}{5{,}916}}{25 - 2}} \\
&= \sqrt{\frac{4{,}028.96 - 2{,}488.08}{23}} \\
&= \sqrt{\frac{1{,}540.88}{23}} \\
&= \sqrt{66.99} \\
&= 8.19
\end{aligned}$$

From the S_e computations, you can see that 8.19 units above and below the regression line would include approximately 68% of the points on the scattergram. Obviously, the utility of the standard error of estimate depends on homoschedasticity and normality of the distribution of the error around the regression line. Under these conditions, the amount of error in prediction is well defined by the standard error of estimate.

THE MEDIAN FIT—OPTIONAL

An alternate procedure for determining a regression line uses medians to define the line. Using the attitude and achievement bivariate data from Table 13.2, the process of finding the median fit is as follows.

Construct a bivariate graph (scattergram) of the data. Then divide the bivariate graph into three approximately equal sections or regions. As shown in Figure 13.9, an *odd* number of points (seven in this case) are located in the first (bottom) and third (top) groups of points. Because you are going to be finding the medians, the odd number of points will make the median of the data easier to find (see Chapter 4 if you need a review of the median).

Examine section I of the scattergram and note that the seven X-values are 40, 40, 48, 48, 50, 53, and 55. The median of these seven X-values is 48. Further note that the median of the Y-values in section I is 56, because the seven values are 50, 54, 55, 56, 59, 60, and 64. Therefore *the median data point for section I is* $X = 48$, $Y = 56$ *and is designated with a ⊛ symbol on the graph.*

In similar fashion, the median data point for section III of the scattergram can be determined ($X = 87$ and $Y = 85$) and is noted on the graph as ⊛. The regression line is formed by connecting the two median data points as shown in Figure 13.9.

To find the equation for the line, first calculate the slope:

$$\begin{aligned} \text{slope} = b &= \frac{\text{(change in vertical scale between medians)}}{\text{(change in horizontal scale between medians)}} \\ &= \frac{(Y\text{-median for section III}) - (Y\text{-median for section I})}{(X\text{-median for section III}) - (X\text{-median for section I})} \\ &= \frac{(85 - 56)}{(87 - 48)} = \frac{29}{39} \\ &= 0.74. \end{aligned}$$

Because $Y = a + bX$ is a form of the linear equation and because one point on the line is known to be $X = 48$, $Y = 56$ (from section I), the intercept may be determined by solving:

$$\begin{aligned} Y' &= a + bX \\ 56 &= a + 0.74(48) \\ a &= 56 - 0.74(48) \text{ or} \\ a &= 20.48. \end{aligned}$$

Hence the equation for the median fit regression line is:

$$Y' = 20.48 + 0.74X$$

(compared to $Y' = 23.53 + 0.65X$ using the least-squares method).

Figure 13.9—Median Fit Regression Line for Attitude/Achievement Bivariate Data

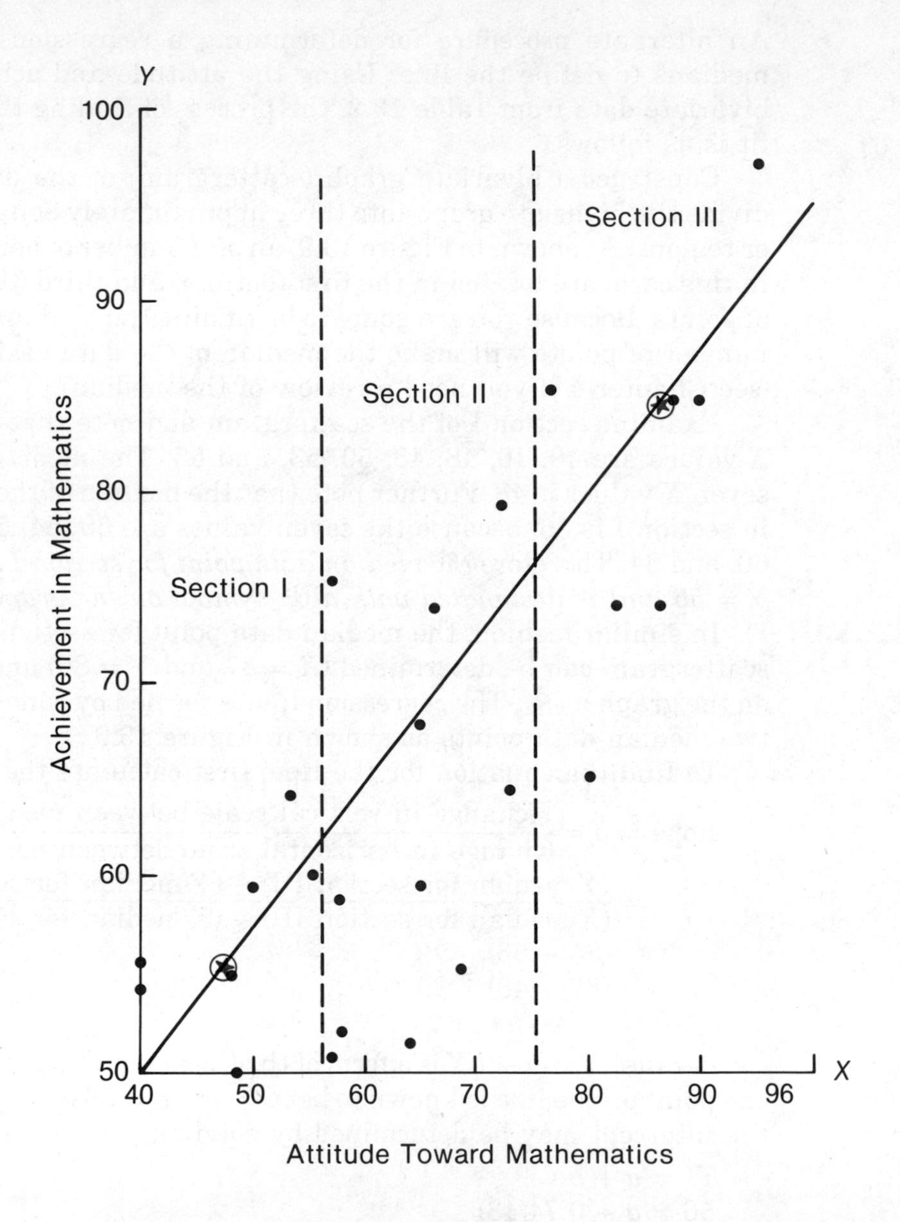

The median fit technique requires less computation than the least-squares method; however, the accuracy of predictions of the median fit line will generally be inferior to the least-squares line for the same bivariate data.

COMPUTER FOCUS 13—Regression

Purpose

This Computer Focus is an extension of the Computer Focus in Chapter 12. Therefore the purpose is an extension of the one for the previous chapter. This Computer Focus provides an opportunity to practice your program modification skills and your debugging abilities.

Problem

In the previous Computer Focus (12), the BASIC statements for computing ΣX, ΣY, Σx^2, Σy^2, and Σxy were used in the computation of the Pearson product-moment r. Recall that the regression coefficient $b = \frac{\Sigma xy}{\Sigma x^2}$ and $a = \overline{Y} - b\overline{X}$.

Design, code, and develop a BASIC program that first will use these statements to compute the means of two variables in a bivariate distribution and then will calculate the values to compute b (the slope) and a (the intercept) for a set of bivariate data. Have the program print the regression equation.

KEY TERMS AND NOTATION

Beta (β) as a symbol for slope
Confidence band
Critical value of F
Degrees of freedom error (df_e)
Degrees of freedom regression (df_r)
Degrees of freedom total (df_t)
Dependent variable
Homoschedasticity
Independent variable
Intercept (a)
Least-squares criterion
Linear
Linear equation
Line of best fit
Mean square error (MS_e)
Mean square regression (MS_r)
Predicted value of Y (Y')
Regression coefficient
Slope (b and β)
Slope-intercept form of linear equation
Standard error of estimate (S_e)
Sum of squares error (SS_e)
Sum of squares regression (SS_r)
Sum of squares total (SS_t or Σy^2)

APPLICATION EXERCISES

13.1 Briefly define the following terms:
a. Y-intercept
b. slope
c. standard error of estimate
d. regression line
e. least-squares criterion
f. Y'

13.2 In the regression equation: $Y' = -9.67 + (-0.67)X$, what is the value of the slope (b) and the intercept (a)?

13.3 Write the regression equations for a line with intercepts and slopes as follows:
a. intercept = 3.00, slope = 0.50
b. intercept = −1.50, slope = 0.35
c. intercept = −4.20, slope = −1.60
d. intercept = 0, slope = 3

13.4 Construct a graph of the following regression equations by plotting three points that determine the line.
a. $Y' = 4.00 + 0.30X$
b. $Y' = 0 + 1X$ or $Y' = X$
c. $Y' = 5 + 0X$ or $Y' = 5$

13.5 Given the regression equation: $Y' = 0.4 + 0.25X$, find the predicted values of Y for the following X-values:
a. $X = 4$
b. $X = 100$
c. $X = 2.4$
d. $X = 0$
e. $X = -8$
f. $X = 7$

13.6 Given the following intermediate summary computations, compute a and b and write the regression equation.
a. $\Sigma x^2 = 40$; $\Sigma y^2 = 130.2$; $\Sigma xy = 30$; $\overline{X} = 3.00$; $\overline{Y} = 7.30$
b. $\Sigma x^2 = 218.1$; $\Sigma y^2 = 30.9$; $\Sigma xy = 60.7$; $\overline{X} = 4.1$; $\overline{Y} = 19.3$

13.7 Use the following bivariate distribution where X is the independent variable and Y is the dependent variable to answer the items that follow.

X	Y
7	3
13	6
2	2
4	5
15	14
10	10
19	8
28	19
26	15
22	17

a. Compute: ΣX; ΣY; ΣX^2; ΣY^2; ΣXY; $\overline{X}$; $\overline{Y}$
b. Using results from a., compute: Σx^2; Σy^2; and Σxy
c. From the results of a. and b., find:
   r (correlation coefficient)
   a (Y-intercept)
   b (slope)
d. From the regression equation derived in c., what is the value of Y' for $X = 20$?
e. State an appropriate null hypothesis for the problem and test for significance at the 0.05 level.
f. Find the standard error of estimate.
g. Sketch a regression line and a 68% confidence band on the graph.

13.8 For the following set of bivariate data:

Variable I	Variable II
5	8
3	6
4	4
7	9
2	5
6	5
4	7
5	3
8	9
6	4

a. Compute the correlation coefficient.

b. Designate variable II as dependent (Y) and variable I as independent (X). Compute a and b and write the regression equation.
c. Interchange the variables used in b. so that Y in b. becomes X and X in b. becomes Y—reverse the dependent and independent roles of the variables from b. Compute the new regression equation.
d. Based on your results in c., what is your conclusion about predicting Y from X and, conversely, predicting X from Y; that is, are the two regressions the same? Was the designation of independent and dependent variables an issue in a.?
e. On the same grid, draw the two regression equations derived in b. and c. What does the point at which the two intersect represent?

13.9 Using the results from Application Exercise 13.8 b., what is the probability that someone scoring 20 on the X-measure would score 15 or higher on the Y-measure? ($S_e = 2.05$.)

13.10 Suppose high school grade-point average (GPA) was used to predict college freshman grade-point average with the equation: $Y' = 0.04 + 0.85X$ (where X is high school GPA and Y is college freshman GPA). Assume that $S_e = 0.12$.
a. What college freshman GPA would be predicted for a student with a 2.8 high school GPA?
b. To belong to a certain student club, a freshman must have a GPA of 2.5 or higher. What is the probability that a student with a 2.8 high school GPA will remain academically eligible for the club?
c. Approximately 95% of the students who have 2.8 GPA's in high school are expected to have college freshman GPA's between what two points?

13.11 *Optional* Use the data in Table 13.2 and compute the value of $\Sigma e^2 = \Sigma(Y - Y')^2$ where Y' is calculated using the least-squares technique:

$$Y' = 23.53 + 0.65X.$$

Then find $\Sigma e^2 = \Sigma(Y - Y')^2$ where Y' is calculated using the median fit technique:

$$Y' = 20.48 + 0.74X.$$

How do the two values of Σe^2 compare, and what conclusion do you draw from these results?

14

t-Test

CHAPTER OBJECTIVES

Upon completion of the chapter, students will be able to:

1. Describe similarities and differences between *t*-distributions and the normal curve.
2. Distinguish between correlated and independent *t*-test models.
3. Perform the necessary procedural tasks for determining whether to use a separate variance or a pooled variance *t*-test model.
4. Compute the degrees of freedom for the three *t*-test formulas discussed in the chapter.
5. Calculate *t*-values using the appropriate *t*-test model when provided data.
6. Use a table of *t*-values to determine the critical value of *t* for the area(s) of rejection.
7. Interpret a test of the null hypothesis from a *t*-test.
8. Discuss "nonsignificant" results in an appropriate manner.

The health teacher at Highland High School, Miss Yeager, conducted an experiment to test a hypothesis about the relationship between physical fitness and self-esteem. The underlying rationale for her experiment was that exercise and proper diet would lead to higher levels of fitness, which in turn would result in increased self-confidence and well-being.

To test her hypothesis, Miss Yeager randomly divided the volunteers for the experiment into two groups. One group, called the experimental group, agreed to conform to a prescribed daily exercise and diet program. The other group, designated a control group, were given "health" pills and were told to continue their present life-style and maintain their daily routine except that they were to take three of the "health" pills per day, one after each meal. Although the fact was unknown to the control group, the "health" pills were placebos with no nutritional value.

The two groups remained on their treatment programs for 16 weeks under Miss Yeager's supervision. At the end of the project she administered an instrument designed to assess self-esteem for both the experimental and control groups. She was extremely eager to use these data to test her fitness/well-being hypothesis. To do this she needed to compare the self-esteem scores of the two groups to determine if one group (presumably the experimental group) had higher scores on the average than the other group. Miss Yeager used a statistical technique known as a *t-test* to help her make an objective comparison of the two groups.

Early in the history of inferential statistics, large samples were used and normal curve models served quite well. While the normal curve and its characteristics provide an adequate basis for probability statements with large samples, it is inappropriate for small-sample work, for example, when $N < 30$. A major problem for statisticians is that the standard error of the mean of a sampling distribution may be quite inaccurate when N is small. Specifically, when an estimate (statistic) of the population standard deviation (parameter) is used to determine the standard error of the mean, the estimator is only an estimate and will vary from sample to sample. While it is true that over a large number of samples the mean of the s-values will equal σ, for any one sample s will deviate from σ on a chance probability basis. The lack of accuracy in such cases may be of considerable consequence in hypothesis testing. The t-test model, however, is well suited for hypothesis testing when samples are relatively small and s is used as an estimate of σ.

Early in the 1900's, a young statistician named William S. Gossett, who was employed at Guiness' Brewery in Dublin, demonstrated that when the standard error of the mean is calculated using small samples, the sampling distribution of means is not a normal curve. With the help of a professor, Gossett was able to empirically derive general mathematical expressions for sampling statistics obtained from small samples. Because students were not permitted to publish scientific findings, Gossett, writing under the pseudonym "**Student**," published his work in 1908. Although his work was very important for small sample statistical work, his publication received very little attention until Sir Ronald Fisher included Gossett's work in what is considered to be the first textbook of modern statistics. The development of Student's *t*-distributions is considered one of the major breakthroughs from classical to modern statistical inference. Today, hypothesis tests using the *t*-probability distributions are still referred to as *Student's t-tests*.

The ***t*-distribution** is actually a family of distributions rather than just a single curve as was the case with the normal distribution. As sample sizes become fairly large (over 30), the appearance of the *t*-curve is much like the normal distribution. As N gets small, the curve remains symmetric about the mean, but the tails of the curve get higher and thus the probability statements are altered accordingly.

The family of *t*-curves vary according to sample size, and the shape of the *t*-distributions varies as a function of the degrees of freedom for the sample. Recall that the degrees of freedom (**df**) represent the number of independent pieces of information on which a sample statistic is based. Another way of conceptualizing df is N minus the number of parameters estimated in calculating a sample statistic. For example, in the chapter on correlation, the computation of r required the deviations from the two means $\overline{X}$ and $\overline{Y}$ (estimates of parameters). Thus when r is used as an estimate of the population rho, it loses two degrees of freedom (df $= N - 2$). The df for the null hypothesis test for which a *t*-distribution is appropriate can be readily determined. And a different *t*-distribution exists for every value of df.

A few selected *t*-curves are shown in Figure 14.1 according to their respective degrees of freedom. The procedure for finding degrees of freedom for a *t*-test will be discussed later. It will suffice for the time being to realize that df for a *t*-distribution are based on the sample size, N.

Figure 14.1—*t*-Distributions with 1, 5, and 20 Degrees of Freedom and Normal Distribution Comparison

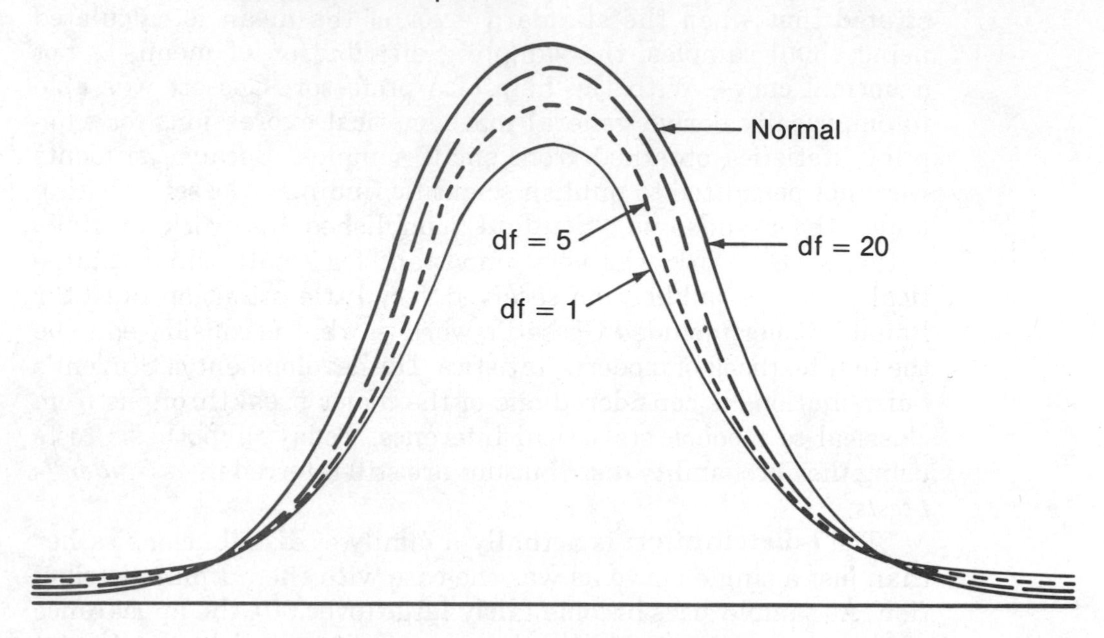

DIFFERENCES BETWEEN TWO SETS OF VALUES

The *t*-distribution has many roles in hypothesis testing. In this chapter you will study one of the primary uses of a *t*-test, which is testing for differences between two sets of data. In nearly all cases, one of two situations exists. First, two sets of data (different measurements with the same scaling) may be *generated by one group of subjects*. Such is the case when high school seniors take the SAT, which generates a *verbal* score and a *quantitative* score, both with the same scaling. In the context of a *t*-test, a potential question of interest might be whether there exists a "significant" difference between the two sets of scores.

A second common situation that generates two sets of data arises when *two groups of subjects* are measured with the same assessment device. For example, suppose one group of students earns aerobic fitness points by jogging according to some specific schedule, whereas another group earns points via standardized aerobic dance routines. After six weeks in the respective programs, the aerobic fitness of members of both groups is measured. Two sets of values are generated by two groups using the same measurement. Again, the question of concern would be whether or not the two sets of values are significantly different.

Factors Influencing the Differences Between Groups

Assume data have been collected on two groups of people, labeled group A and group B. Let the respective sample sizes of the two groups be denoted N_A and N_B. The means and standard deviations of the two groups are $\overline{X}_A$, $\overline{X}_B$, s_A, and s_B. Two of the many possible outcomes for the respective distributions are illustrated in Figure 14.2. Assume that the sample sizes are equal ($N_A = N_B$) and the standard deviations of the two sets of data are also equal ($s_A = s_B$).

Figure 14.2—Distributions Showing Differences between Means and Degree of Overlap

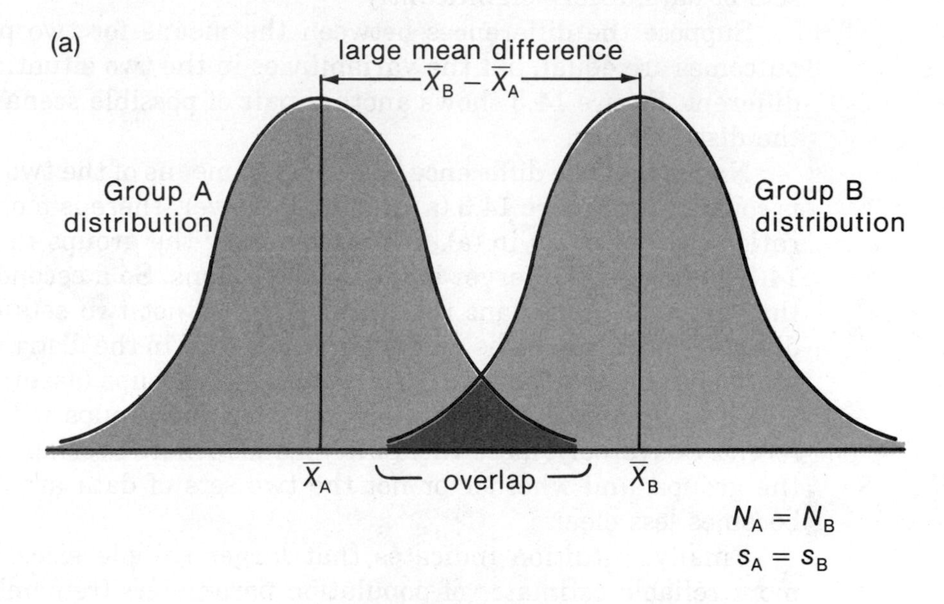

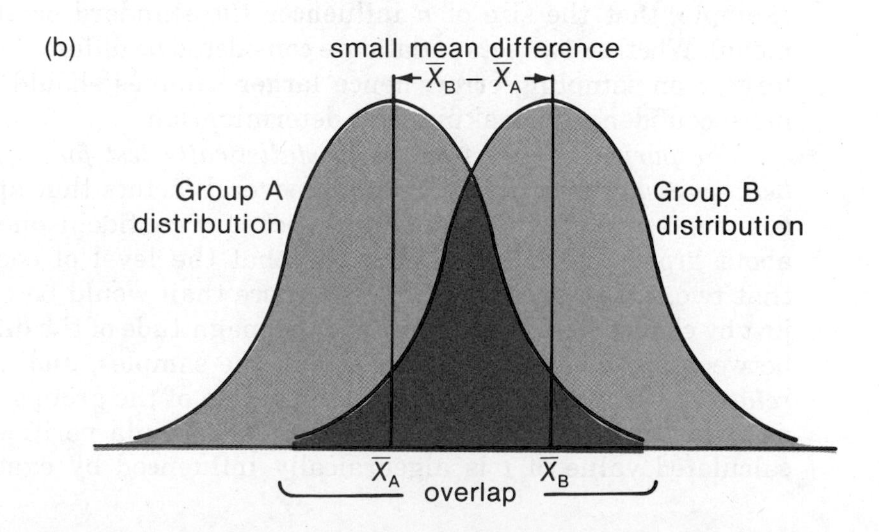

By observation and common sense, it is not difficult to determine which of the two outcomes (a or b) displays the greater separation between the groups. Clearly the outcome shown as (a) shows more separation (group difference). In general, the less overlap that exists between two sets, the more likely it is that the two data sets will be different. In the two cases shown in Figure 14.2, the extent of the overlap between the distributions is a function of the magnitude of the difference in the respective means. The correct conclusion from this example is that the size of the difference between the two means is a major factor in determining if the two sets of data differ "significantly."

Suppose the differences between the means for two possible outcomes are equal, but the variabilities in the two situations are different. Figure 14.3 shows another pair of possible scenarios for the distributions.

Notice that the difference between the means of the two groups is constant in Figure 14.3 (a) and (b). However, there is more separation (less overlap) in (a). Why? Obviously the groups in Figure 14.3 (b) have much larger standard deviations. So a second factor that plays an important role in whether or not two sets of data differ significantly is the amount of variability in the distributions. More specifically, the less variability in the groups (assuming all else is constant), the more likely it is that the groups will be different. Conversely, more variability leads to more overlap between the groups, and whether or not the two sets of data are distinct becomes less clear.

Finally, intuition indicates that larger sample sizes lead to more reliable estimates of population parameters (remember, for example, that the size of n influences the standard error of the mean). Whether two sets of data are considered as different depends largely on sampling error; hence larger samples should provide more confidence in making that determination.

The purpose of the t*-test is to statistically test for differences between two sets of data.* There are several factors that appeal to common sense that undoubtedly affect how confident one can be about group differences. It appears that the level of confidence that two sets of data are different (more than would be expected just by chance) is *directly related* to the magnitude of the difference between the means and to the size of the samples, and *inversely related* to the variability of the sample data of the groups. Indeed, as shown in Figure 14.4, a typical t-test formula verifies that a calculated value of t is algebraically influenced by exactly the

Figure 14.3—Two Distributions: One with Groups with Large Variability; One with Homogeneous Groups

(a) small variability

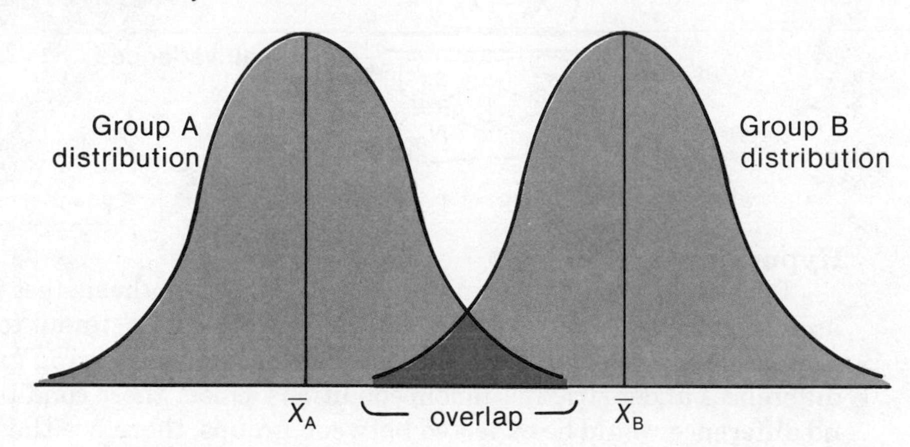

(b) large variability

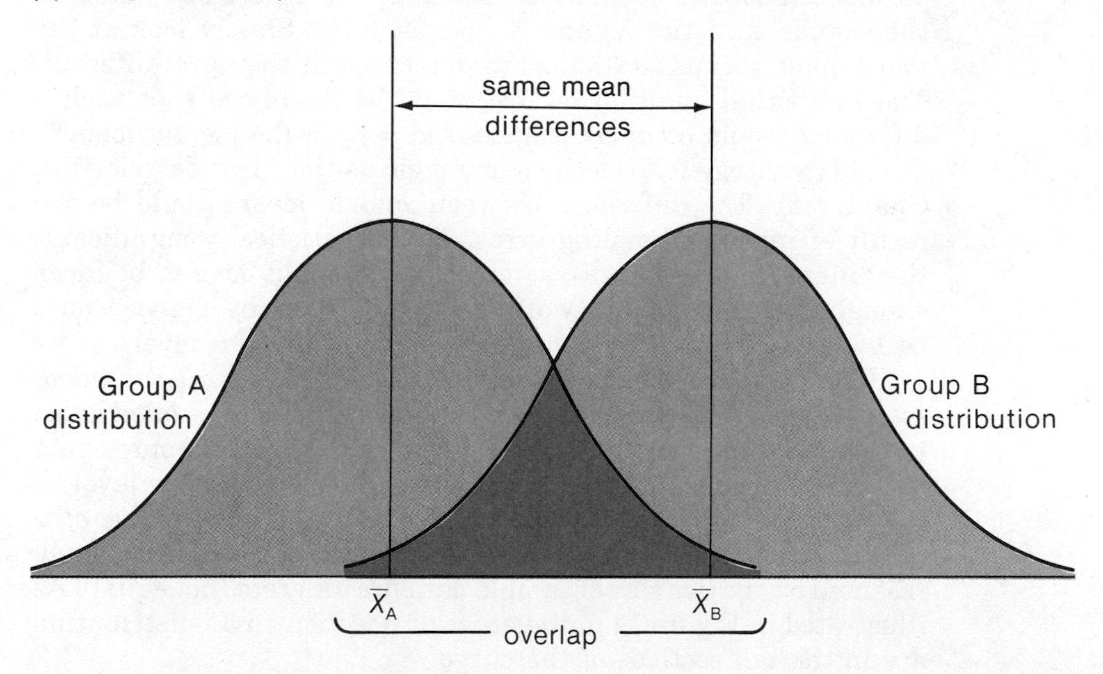

same factors. Notice, from an arithmetic standpoint, what happens to the magnitude of t as a function of (1) the size of the mean difference, (2) the size of sample variances, and (3) the number of subjects in the samples.

Figure 14.4— Factors that Tend to Increase the Size of *t*

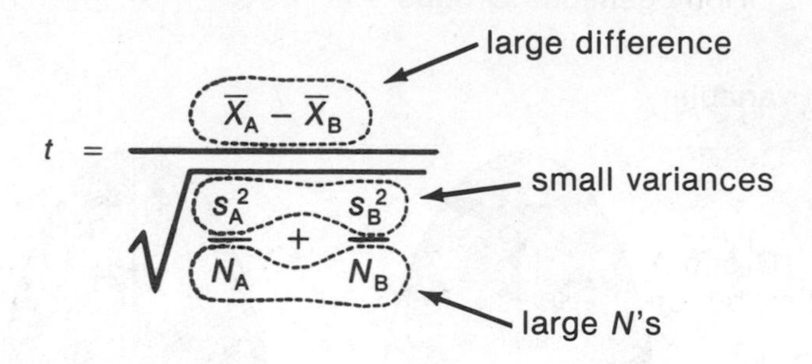

Hypothesis Testing

Because the only meaningful context for a hypothesis test is in an inferential statistics setting, the null hypothesis pertinent to the discussion is one of what would be the most likely occurring group differences under strictly random conditions. Under those conditions, no difference would be expected between groups; therefore the null hypothesis is $H_0{:}\mu_1 = \mu_2$ or, in another form, $H_0{:}\ \mu_1 - \mu_2 = 0$. The parameters for the population means, μ_1 and μ_2, are estimated by the sample statistics $\overline{X}_1$ and $\overline{X}_2$, respectively. Simply look at the two sample means (statistics) to determine if they are different. The inferential question is: "What is the likelihood that such a difference would occur by chance *if* $\mu_1 = \mu_2$ in the population(s)?"

This strategy follows the same logic as the *z*-test described in Chapter 11. The difference between sample means could be the result of random sampling error. To be statistically significant, the difference between two sample means would have to be large enough that the probability of such an occurrence by chance would be *less than* 5%, 1%, or some other predetermined α level.

Using sample statistics (as estimates of parameters), the *calculated value* of *t* is obtained using an equation. The *calculated value* is then compared to the *critical value* of *t* with the appropriate degrees of freedom (df) at the predetermined significance level. *If the calculated value is larger than the critical (tabled) value of* t, *the null hypothesis is rejected* and the means of the populations represented by the two sets of sample data are different (not equal). As illustrated in Figure 14.5, the areas of rejection on a *t*-distribution are in the tail sections of the curve.

Figure 14.6 displays critical values for the normal curve and *t*-curves with df = 10, 20, 100 for a two-tailed test of significance at $\alpha = 0.05$. Any calculated value of *t* falling in the areas of rejection would occur less than 5% ($p < 0.05$) of the time by chance. The

Figure 14.5—*t*-Distribution Areas of Rejection

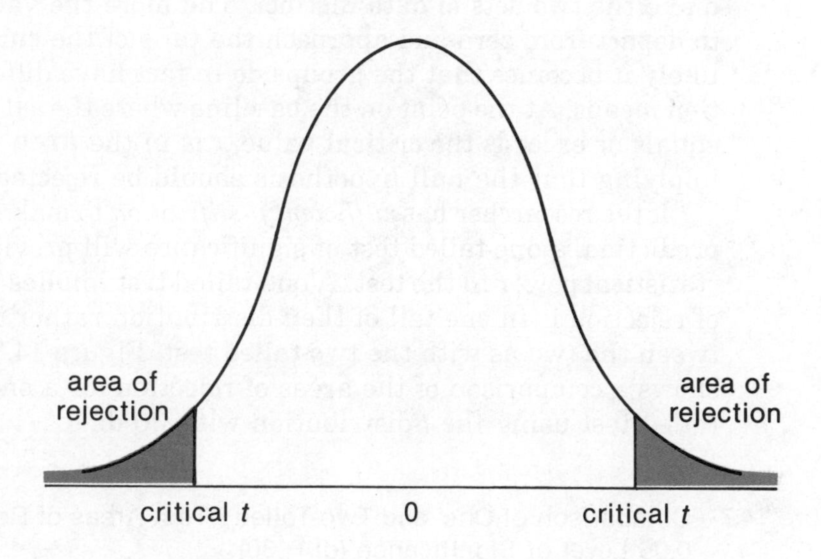

Figure 14.6—Critical Values for Normal Curve and *t*-Distributions with df = 10, 20, and 100 at the 0.05 Level of Significance (Two-Tailed Test) Areas of Rejection in Shaded Areas

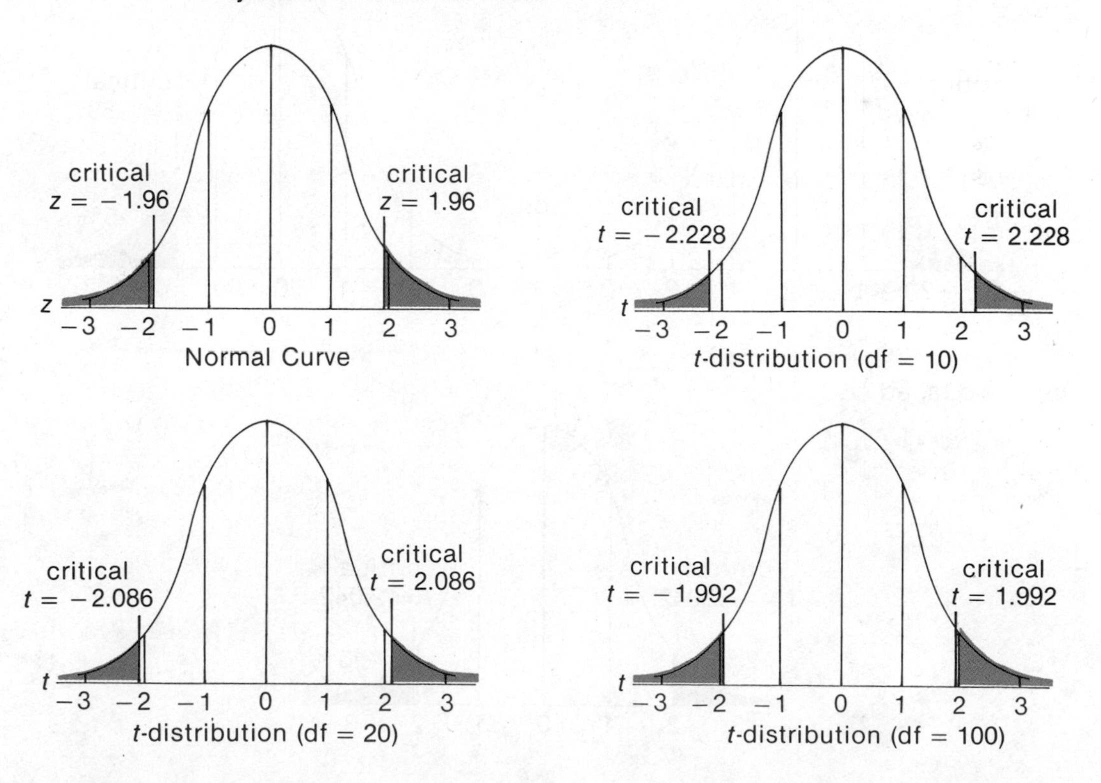

size of t is a function of the same distribution characteristics that make the two sets of data distinct. The more the value of t tends to depart from zero and approach the tails of the curve, the more likely it becomes that the groups do in fact have different population means. At the point on the baseline where the calculated value equals or exceeds the critical value, t is in the **area of rejection**, implying that the null hypothesis should be rejected.

If the researcher has *sufficient justification* to make a directional prediction, a one-tailed test of significance will provide additional statistical power to the test. A one-tailed test implies that the area of rejection is in one tail of the t-distribution rather than split between the two as with the two-tailed test. Figure 14.7 graphically shows a comparison of the areas of rejection for a one- and a two-tailed test using the t-distribution with 30 df.

Figure 14.7—Comparison of One- and Two-Tailed t-Test Areas of Rejection at the 0.05 Level of Significance (df = 30)

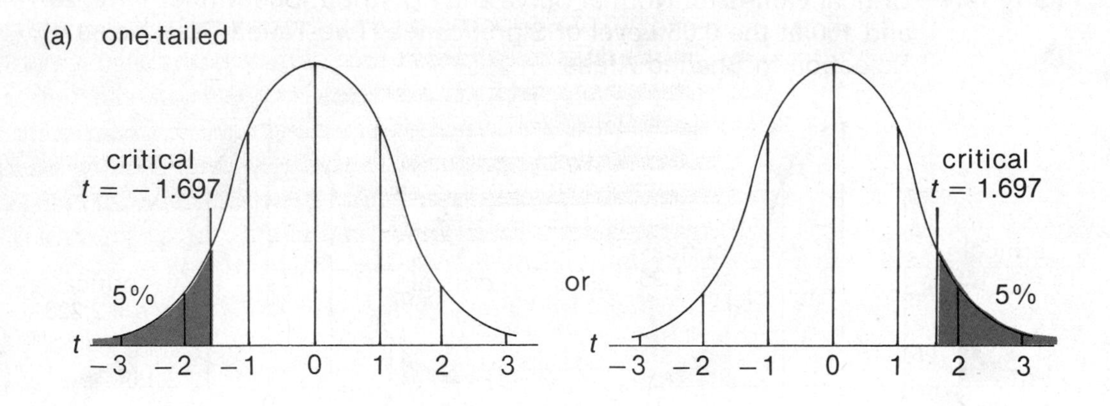

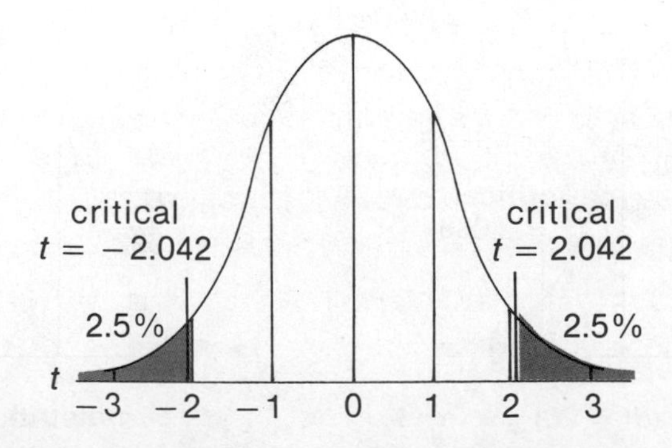

Appendix E provides critical values of t for both one- and two-tailed tests at the 0.01 and 0.05 levels of significance. As you may have guessed, finding the critical value of t is not much more complicated than finding a critical value of z from the normal curve. The format of Appendix E differs from the table of normal curve areas because the t-distribution takes on a different shape depending on the degrees of freedom associated with the problem. To circumvent the problem of creating a separate table for each possible value of df, Appendix E provides only six alpha levels. Because the t-distribution is symmetric about the mean of zero, negative values of t are not included and would be treated as though they were positive.

To use the table, three items of information are required. First, the degrees of freedom for the problem must be determined. Each equation for the t-test will be accompanied by the necessary expression for finding the df. Second, the predetermined value of α must be located at the top of the table. Finally, the researcher must locate the alpha level in the appropriate row at the top of the table depending on whether the test uses a one-tailed or a two-tailed area of rejection. The critical value of t is located at the intersection of the row (determined by df) and the column (determined by α and one- vs. two-tailed selection).

CORRELATED *t*-TEST

One commonly used model for the t-test is referred to as the **correlated *t*-test**. The purpose is to test for mean differences between two sets of data that have been collected on *one* group of subjects. In such a case the two sets of data will be correlated to some extent, and the correlated t-test model takes advantage of the relationship to provide more power to the statistical design.

Research Model

If two sets of data with the same score scaling are collected on one group of subjects, a correlated t-test is appropriate to analyze the differences between the means of the two sets. The following example illustrates such a design.

Miss Yeager, the high school health teacher introduced at the beginning of the chapter, was interested in starting an extracurricular fitness class for students. The class would meet before school three days a week. Her program consisted of one week of walking-jogging followed by six weeks of calisthenics and jogging. Miss Yeager was interested in obtaining some evidence on the effectiveness of her program for the 16 student participants. On the first day Miss Yeager measured each student's fitness level by counting heartbeat rate for a time interval of 25 seconds immediately after a ten-minute exercise session. In this case, the lower the score, the better the level of physical fitness. This first measurement is sometimes referred to as a pretest (because it was taken before the start of the program). At the completion of the seven-week program, she assessed the level of fitness again (called a posttest). Obviously the "treatment" was administered in the seven-week time interval between the pretest and the posttest.

Speaking from an inferential viewpoint, the two sets of data may represent different statistical populations: the pretest representing a population of students not involved in a fitness program; the posttest scores representing a population of students just completing a physical fitness program. The teacher will test H_0: $\mu_{pre} = \mu_{post}$ to see if the difference in $\overline{X}_{pre}$ and $\overline{X}_{post}$ is statistically significant. Miss Yeager's data for the 16 students are shown in Table 14.1.

Table 14.1 Fitness Data for Pretest and Posttest (Heartbeats in a 25-Second Time Interval)

Student #	Pretest Score X_1	Posttest Score X_2
1	26	20
2	28	21
3	20	15
4	24	22
5	25	25
6	18	15
7	26	26
8	25	20
9	28	29
10	24	20
11	21	15
12	25	27
13	29	24
14	17	12
15	19	20
16	26	20

Computational Procedures

Label the two sets of data Set 1 and Set 2.

The equation for computing a t-value for correlated data is:

$$t = \frac{\overline{X}_1 - \overline{X}_2}{\sqrt{\frac{s_1^2 + s_2^2 - 2r(s_1)(s_2)}{N}}}$$

where $\overline{X}_1$ and $\overline{X}_2$ are the respective group means,
s_1^2 and s_2^2 are the group variances,
s_1 and s_2 are the standard deviations,
r is the correlation between the two sets, and
N is the sample size (that is, the number of pairs of scores).

The equation requires the computation of summary statistics. As applied to Miss Yeager's problem, data from Table 14.1 are repeated along with the squares and cross products of the raw scores in Table 14.2. From the sums of the columns in Table 14.2, the means and deviation score sums of squares and cross products can be calculated. The calculations and formulas are as follows:

$$\overline{X}_1 = \frac{\Sigma X_1}{N} = \frac{381}{16} = \boxed{23.81}$$

$$\overline{X}_2 = \frac{\Sigma X_2}{N} = \frac{331}{16} = \boxed{20.69}$$

$$\Sigma x_1^2 = \Sigma X_1^2 - \frac{(\Sigma X_1)^2}{N}$$

$$= 9{,}279 - \frac{(381)^2}{16}$$

$$= 206.44$$

$$\Sigma x_2^2 = \Sigma X_2^2 - \frac{(\Sigma X_2)^2}{N}$$

$$= 7{,}191 - \frac{(331)^2}{16}$$

$$= 343.44$$

$$\Sigma x_1 x_2 = \Sigma X_1 X_2 - \frac{(\Sigma X_1)(\Sigma X_2)}{N}$$

$$= 8{,}089 - \frac{(381)(331)}{16}$$

$$= 207.06$$

Table 14.2 Fitness Data for Pretest and Posttest with Summary Computations

Student #	Pretest Score X_1	X_1^2	Posttest Score X_2	X_2^2	X_1X_2
1	26	676	20	400	520
2	28	784	21	441	588
3	20	400	15	225	300
4	24	576	22	484	528
5	25	625	25	625	625
6	18	324	15	225	270
7	26	676	26	676	676
8	25	625	20	400	500
9	28	784	29	841	812
10	24	576	20	400	480
11	21	441	15	225	315
12	25	625	27	729	675
13	29	841	24	576	696
14	17	289	12	144	204
15	19	361	20	400	380
16	26	676	20	400	520
Σ	381	9,279	331	7,191	8,089

$$\overline{X}_1 = 23.81 \qquad \overline{X}_2 = 20.69$$

$$\Sigma x_1^2 = 9{,}279 - \frac{(381)^2}{16} = 206.44$$

$$\Sigma x_2^2 = 7{,}191 - \frac{(331)^2}{16} = 343.44$$

$$\Sigma x_1 x_2 = 8{,}089 - \frac{(381)(331)}{16} = 207.06$$

$$s_1^2 = \frac{206.44}{15} = 13.76$$

$$s_2^2 = \frac{343.44}{15} = 22.90$$

$$s_1 = 3.71$$

$$s_2 = 4.79$$

$$r = \frac{207.06}{\sqrt{(206.44)(343.44)}} = 0.78$$

From the preceding results, the variances, standard deviations, and r can be computed as shown.

$$s_1^2 = \frac{\Sigma x_1^2}{N-1} = \frac{206.44}{15} = \boxed{13.76}$$

$$s_2^2 = \frac{\Sigma x_2^2}{N-1} = \frac{343.44}{15} = \boxed{22.90}$$

$$s_1 = \sqrt{\frac{\Sigma x_1^2}{N-1}} = \sqrt{13.76} = \boxed{3.71}$$

$$s_2 = \sqrt{\frac{\Sigma x_2^2}{N-1}} = \sqrt{22.90} = \boxed{4.79}$$

$$r = \frac{\Sigma x_1 x_2}{\sqrt{(\Sigma x_1^2)(\Sigma x_2^2)}} = \frac{207.06}{\sqrt{(206.44)(343.44)}} = \boxed{0.78}$$

The intermediate results from the preceding calculations that are highlighted by rectangles are now substituted into the formula for the correlated t-test.

$$\begin{aligned} t &= \frac{\overline{X}_1 - \overline{X}_2}{\sqrt{\dfrac{s_1^2 + s_2^2 - 2r(s_1)(s_2)}{N}}} \\ &= \frac{23.81 - 20.69}{\sqrt{\dfrac{13.76 + 22.90 - (2)(0.78)(3.71)(4.79)}{16}}} \\ &= \frac{3.12}{\sqrt{\dfrac{36.66 - 27.72}{16}}} \\ &= \frac{3.12}{\sqrt{0.56}} \\ &= \frac{3.12}{0.75} \\ &= 4.16 \quad \text{(calculated value of } t\text{)} \end{aligned}$$

Calculator keystrokes for the mean (see Chapter 4), the variance and standard deviation (see Chapter 5), and r (see Chapter 12) have already been presented. Given these statistics, the correlated t-test value can be computed with calculator keystrokes as shown in Figure 14.8.

The number of degrees of freedom for the correlated t-test problem is $N - 1$; or, in the case of Miss Yeager's data, df = 16 − 1 or 15.

Miss Yeager decided to test the null hypothesis at the 5% level of significance. She is entitled to use a one-tailed test on the basis of the following rationale. Logically, the exercise program would *not hinder* physical fitness; if anything, the program would either maintain the students' current fitness level or improve it. Thus a one-tailed test is justified.

Figure 14.8—Calculator Keystrokes

PROBLEM:

Solve correlated *t*-test:

$$t = \frac{23.81 - 20.69}{\sqrt{\dfrac{13.76 + 22.90 - (2)(0.78)(3.71)(4.79)}{16}}}$$

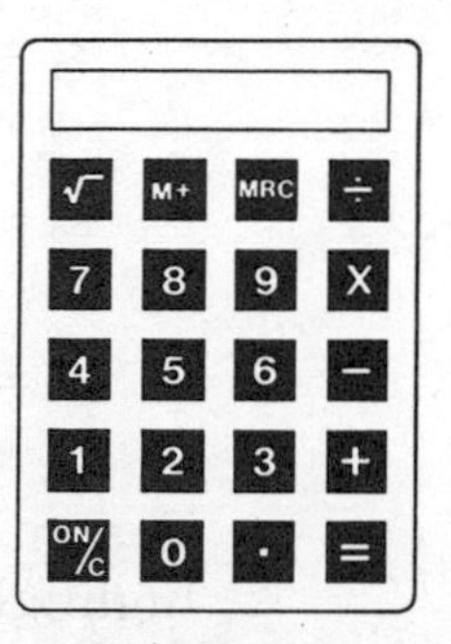

Procedure	Arithmetic Operation	Calculator Keystrokes	Display
1. Solve for value of denominator	**1.** $\sqrt{\dfrac{13.76 + 22.90 - 2(0.78)(3.71)(4.78)}{16}}$ = 0.7473869	**1.** 2 [×]	2
		0.78 [×]	1.56
		3.71 [×]	5.7876
		4.79 [=]	27.722604
		[M+]	27.722604
		13.76 [+]	13.76
		22.90 [−]	36.66
		[MRC]	27.722604
		[=] [÷]	8.937396
		16 [=]	0.5585872
		[√]	0.7473869
		[CLM]	0.7473869
		[M+]	0.7473869
2. Solve for value of numerator	**2.** 23.81 − 20.69 = 3.12	**2.** 23.81 [−]	23.81
		20.69 [=]	3.12
3. Divide numerator by denominator	**3.** 3.12 ÷ 0.7473869 = 4.17	**3.** [÷] [MRC]	0.7473869
		[=]	4.1745446

From Appendix E, the critical value of t with 15 df at $\alpha = 0.05$ for a one-tailed test is 1.753. Because the *calculated* t-*value* for the fitness data *exceeds* the *critical* t-*value* (4.16 > 1.753), the null hypothesis is rejected. Consequently, Miss Yeager is relatively confident that her training program has an effect on the measure of fitness for the population of students. In other words, a difference of 3.12 points or more between the means in this experiment would occur less than 5% of the time just by chance ($p < 0.05$). Hence the difference in this case is not considered a result of random chance but the result of the fitness program. Figure 14.9 graphically shows the statistical test with the calculated t-value falling in the area of rejection.

Figure 14.9—Graphic Illustration of *t*-Test Hypothesis Test for Fitness Data (One-Tailed Test at 0.05 Level)

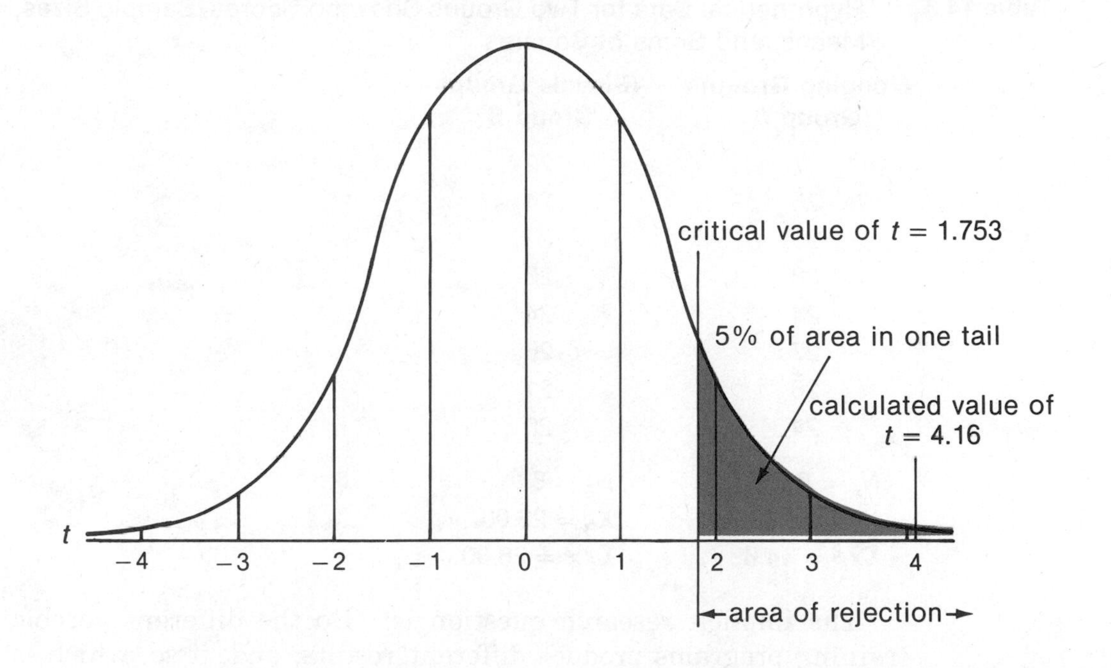

A glance at the sample means for the two sets of data reveals that the scores (heart rates) are *lower* at posttest time than at pretest time. Because this was the anticipated direction of the outcome, the test not only revealed a significant difference between the means, but confirmed the predicted direction of the change over the training period.

INDEPENDENT *t*-TEST

Suppose, instead of one physical fitness program, Miss Yeager was interested in the relative effectiveness of two aerobic training programs. If this issue was of interest, the *t*-test could also be used to test for group differences. However, in this case two groups (one for each of the two programs) would be assessed on aerobic fitness at the end of the program, and a test of group mean differences would be accomplished with a *t*-test.

Suppose Miss Yeager randomly divided the 16 participating students into two equal groups, $N_A = 8$; $N_B = 8$. Group A would train according to a prescribed jogging program for six weeks. Group B would exercise on bicycles according to a prescribed routine. The hypothetical fitness data collected at the conclusion of the six-week period are shown in Table 14.3.

Table 14.3 Hypothetical Data for Two Groups Showing Scores, Sample Sizes, Means, and Sums of Squares

(Jogging Group) Group A	(Bicycle Group) Group B
27	22
25	24
26	24
28	23
24	24
27	26
26	21
24	20
$N_A = 8$	$N_B = 8$
$\overline{X}_A = 25.88$	$\overline{X}_B = 23.00$
$\Sigma x_A^2 = 14.88$	$\Sigma x_B^2 = 26.00$

The implicit research question is: "Do the different aerobic training programs produce different results; and, if so, which is better?" From a statistical perspective, the question is: "Is the difference between the means of the two groups (2.88) large enough to suggest that this difference did not occur just by chance?" Because Miss Yeager is dealing with two distinct (independent) groups of subjects, an **independent *t*-test** is the appropriate technique to test the null hypothesis, H_0: $\mu_A = \mu_B$.

Determining Which Model to Use

There are two independent *t*-test formulas from which to choose. They are referred to as: (1) the **pooled variance** formula and (2) the **separate variance** model. The selection process depends first on whether the samples are from populations with *equal variability* and second on whether the sample sizes (N_A and N_B) are equal. The rules can be summarized as follows:

1. If the population variances are not equal and the sample sizes are unequal, that is, $\sigma_A^2 \neq \sigma_B^2$, and $N_A \neq N_B$ (where $\neq$ means "not equal to"), use the separate variance model.
2. If $\sigma_A^2 = \sigma_B^2$, use the pooled variance model.
3. If $\sigma_A^2 \neq \sigma_B^2$ and $N_A = N_B$, use either formula. The degrees of freedom will be adjusted.

Testing for Homogeneity of Variance. As previously mentioned, each of the two groups has unbiased estimates of the population variances. The groups or sets of data are considered samples, and the issue is whether the populations represented by the two samples have equal variances; that is, whether the populations have **homogeneity of variance**. Consequently, a test of H_0: $\sigma_A^2 = \sigma_B^2$ must be performed before a decision is made about which *t*-test formula to use.

Although the *F*-statistic will be used extensively in the next chapter, its use will be demonstrated to determine if two populations have different variances. The ***F*-test** (*F*-ratio) is simply the larger variance divided by the smaller variance: $F = (\text{larger } s^2) \div (\text{smaller } s^2)$. In the example, data for finding the sample variances are taken from Table 14.3; the results are:

$$s_A^2 = \frac{14.88}{7} = 2.13;$$

$$s_B^2 = \frac{26.00}{7} = 3.71.$$

Therefore

$$F = \frac{s_B^2}{s_A^2} \quad \text{(larger variance in numerator)}$$

$$= \frac{3.71}{2.13}$$

$$= 1.74 \quad \text{(calculated value of } F\text{)}.$$

The calculated value must now be compared to the critical value (at the 0.05 level). To determine the critical value, the degrees of freedom for both groups are computed as follows:

$df_A = N_A - 1 = 8 - 1 = 7$
$df_B = N_B - 1 = 8 - 1 = 7.$

Therefore 7 df are associated with the numerator (group B), and 7 df are associated with the denominator (group A) of the *F*-ratio. With the aid of Appendix F we find that 7 df in the numerator will define a column; 7 df in the denominator will define a row. The intersection of this row and column reveals two numbers. The smaller number is the critical *F*-value for the 0.05 level of significance, and the larger number is the critical *F*-value for a 1% significance test.

For the example, a calculated *F*-value with 7 df in the numerator and 7 df in the denominator would have to exceed 3.79 to be statistically significant at the 0.05 level. Because our calculated value (1.74) for Miss Yeager's data was less than the critical (tabled) value, the null hypothesis $H_0: \sigma_A^2 = \sigma_B^2$ would not be rejected. Hence you would consider the equality of the population variances tenable, $\sigma_A^2 = \sigma_B^2$. Therefore the pooled variance *t*-test formula would be appropriate.

If the test of equality of the variances had revealed a significant difference, then the sample sizes would have been compared. In this case, $N_A = N_B$, and either *t*-test equation could have been used. If $\sigma_A^2 \neq \sigma_B^2$ and if $N_A \neq N_B$, then the separate variance formula would be employed as the *t*-test equation.

COMPUTATIONAL MODELS

Pooled Variance Model

Return to Miss Yeager's data shown in Table 14.3. It has already been determined that the pooled variance formula should be used to analyze the difference between the means.

The pooled variance formula is:

$$t = \frac{\overline{X}_A - \overline{X}_B}{\sqrt{\left(\frac{\Sigma x_A^2 + \Sigma x_B^2}{N_A + N_B - 2}\right)\left(\frac{1}{N_A} + \frac{1}{N_B}\right)}}$$

where $\overline{X}_A$ and $\overline{X}_B$ are the respective sample means,
Σx_A^2 and Σx_B^2 are the respective sums of squares, and
N_A and N_B are the sample sizes.

By substituting the values from Table 14.3 into the pooled variance formula, we find that the calculated value of t is:

$$t = \frac{25.88 - 23.00}{\sqrt{\left(\frac{14.88 + 26.00}{8 + 8 - 2}\right)\left(\frac{1}{8} + \frac{1}{8}\right)}}$$
$$= \frac{2.88}{\sqrt{\left(\frac{40.88}{14}\right)\left(\frac{2}{8}\right)}}$$
$$= \frac{2.88}{\sqrt{(2.92)(0.25)}}$$
$$= \frac{2.88}{\sqrt{0.73}}$$
$$= \frac{2.88}{0.85}$$
$$= 3.39 \quad \text{(calculated } t\text{-value)}$$

The calculator keystrokes for the pooled variance model t-test are shown in Figure 14.10.

For the pooled variance formula, the df $= N_A + N_B - 2$; or, in the present case, $8 + 8 - 2 = 14$ degrees of freedom. The critical value of t with 14 df at the 0.05 level of significance for a two-tailed test is 2.145 (Appendix E). The null hypothesis ($\mu_A = \mu_B$) is rejected. Miss Yeager can be reasonably confident that the two aerobic treatments do have different outcomes in the population of students. And by examining the means we see that method A (jogging) resulted in higher scores on the average than did method B (bicycling). The difference between the groups was probably too large to have been the result of chance; therefore Miss Yeager has a high degree of confidence about the relative merits of the training programs with respect to measurable fitness.

Separate Variance Model

The equation for the t-test using the separate variance model is:

$$t = \frac{\overline{X}_A - \overline{X}_B}{\sqrt{\frac{s_A^2}{N_A} + \frac{s_B^2}{N_B}}}$$

where $\overline{X}_A$ and $\overline{X}_B$ are the respective sample means,
s_A^2 and s_B^2 are the sample variances, and
N_A and N_B are the sample sizes.

Figure 14.10—Calculator Keystrokes

PROBLEM:

Solve pooled variance *t*-test:

$$t = \frac{25.88 - 23.00}{\sqrt{\left(\frac{14.88 + 26.00}{8 + 8 - 2}\right)\left(\frac{1}{8} + \frac{1}{8}\right)}}$$

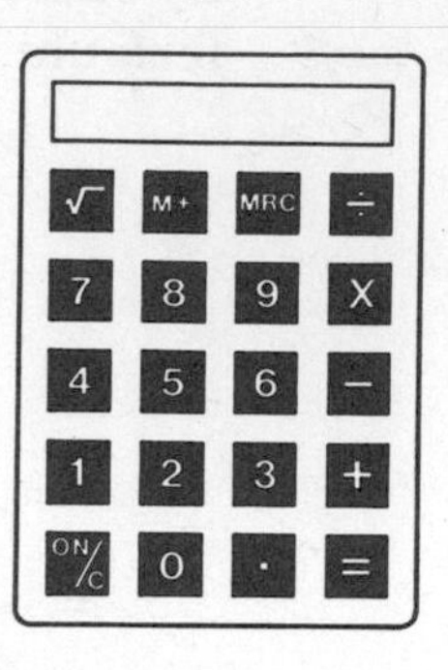

Procedure	Arithmetic Operation	Calculator Keystrokes	Display
1. Solve for value of denominator	**1.** $\sqrt{\left(\frac{14.88 + 26.00}{8 + 8 - 2}\right)\left(\frac{1}{8} + \frac{1}{8}\right)}$	**1.** 8 [+] 8 [−] 2 [=]	14
	$= \sqrt{\left(\frac{40.88}{14}\right)\left(\frac{2}{8}\right)}$	[M+]	14
	$= \sqrt{\frac{81.76}{112}}$	14.88 [+]	14.88
	$= \sqrt{0.73}$	26.00 [=]	40.88
	= 0.8544003	[÷] [MRC] [=]	2.92
		[CLM]	2.92
		[M+]	2.92
		2 [÷] 8 [=]	0.25
		[×] [MRC]	2.92
		[=]	0.73
		[√]	0.8544003
		[CLM]	0.8544003
		[M+]	0.8544003
2. Solve for value of numerator	**2.** 25.88 − 23.00 = 2.88	**2.** 25.88 [−]	25.88
		23 [=]	2.88
3. Divide numerator by denominator	**3.** 2.88 ÷ 0.8544003 = 3.37	**3.** [÷] [MRC]	0.8544003
		[=]	3.3707853

The computed value of t is found by substituting the values in the formula and performing the necessary arithmetic. The critical (tabled) value of t is found using Appendix E as follows:

1. Find critical value for $df_A = N_A - 1$.
2. Find critical value for $df_B = N_B - 1$.
3. Average the two t-values. That average is the critical value of t for the separate variance formula.

For the sake of illustration, suppose Miss Yeager's data yielded the following summary statistics:

Group A	Group B
$\overline{X}_A = 25$	$\overline{X}_B = 20$
$s_A^2 = 36$	$s_B^2 = 10$
$N_A = 25$	$N_B = 15$
$df_A = 24$	$df_B = 14$

The variances would be tested for equality as follows:

$$F = \frac{36}{10} = 3.6 \quad \text{(calculated value)}$$

The critical value of $F = 2.35$ (24 df in numerator, 14 df in denominator). As a result of the F-test, you conclude that

$$\sigma_A^2 \neq \sigma_B^2.$$

By observation, $N_A \neq N_B$. Therefore use the separate variance t-test model.

$$
\begin{aligned}
t &= \frac{\overline{X}_A - \overline{X}_B}{\sqrt{\frac{s_A^2}{N_A} + \frac{s_B^2}{N_B}}} \\
&= \frac{25 - 20}{\sqrt{\frac{36}{25} + \frac{10}{15}}} \\
&= \frac{5}{\sqrt{1.44 + 0.67}} \\
&= 3.45 \quad \text{(calculated } t\text{-value)}
\end{aligned}
$$

Figure 14.11 illustrates calculator keystrokes for the separate variance t-test for the data.

To determine the critical value of t using Appendix E:

$df_A = 24$; critical t @ 0.05 for two-tailed test = 2.064
$df_B = 14$; critical t @ 0.05 for two-tailed test = 2.145.

Therefore the critical $t = \frac{2.064 + 2.145}{2} = 2.10$.

The calculated t-value (3.45) exceeds the critical value (2.10) and therefore falls in an area of rejection. Thus on a two-tailed test of significance at the $\alpha = 0.05$ level, H_0: $\mu_A = \mu_B$ would be *rejected.* Group A scored significantly higher than group B based on estimates of the population means.

To summarize the procedures, the flow diagram in Figure 14.12 provides a guide to aid in deciding which t-model to use. The figure provides the computational formula for t as well as showing how to determine the degrees of freedom for each model.

Figure 14.11—Calculator Keystrokes

PROBLEM:

Solve:

$$t = \frac{25 - 20}{\sqrt{\frac{36}{25} + \frac{10}{15}}}$$

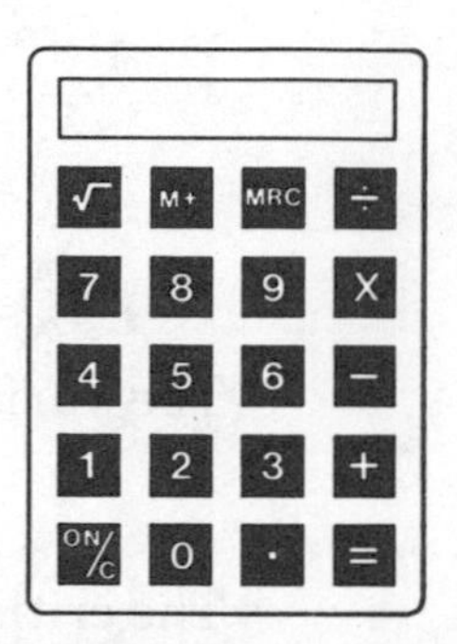

Procedure	Arithmetic Operation	Calculator Keystrokes	Display
1. Solve for value of denominator	**1.** $\sqrt{\frac{36}{25} + \frac{10}{15}}$	**1.** 36 [÷]	36
	$= \sqrt{2.1066666}$	25 [=]	1.44
	$= 1.451436$	[M+]	1.44
		10 [÷]	10
		15 [=]	0.6666666
		[+] [MRC]	1.44
		[=]	2.1066666
		[CLM]	2.1066666
		[√]	1.451436
		[M+]	1.451436
2. Solve for value of numerator	**2.** 25 − 20 = 5	**2.** 25 [−]	25
		20 [=]	5
3. Divide numerator by denominator	**3.** 5 ÷ 1.451436 = 3.45	**3.** [÷]	5
		[MRC]	1.451436
		[=]	3.4448642

Figure 14.12— Model for *t*-Test

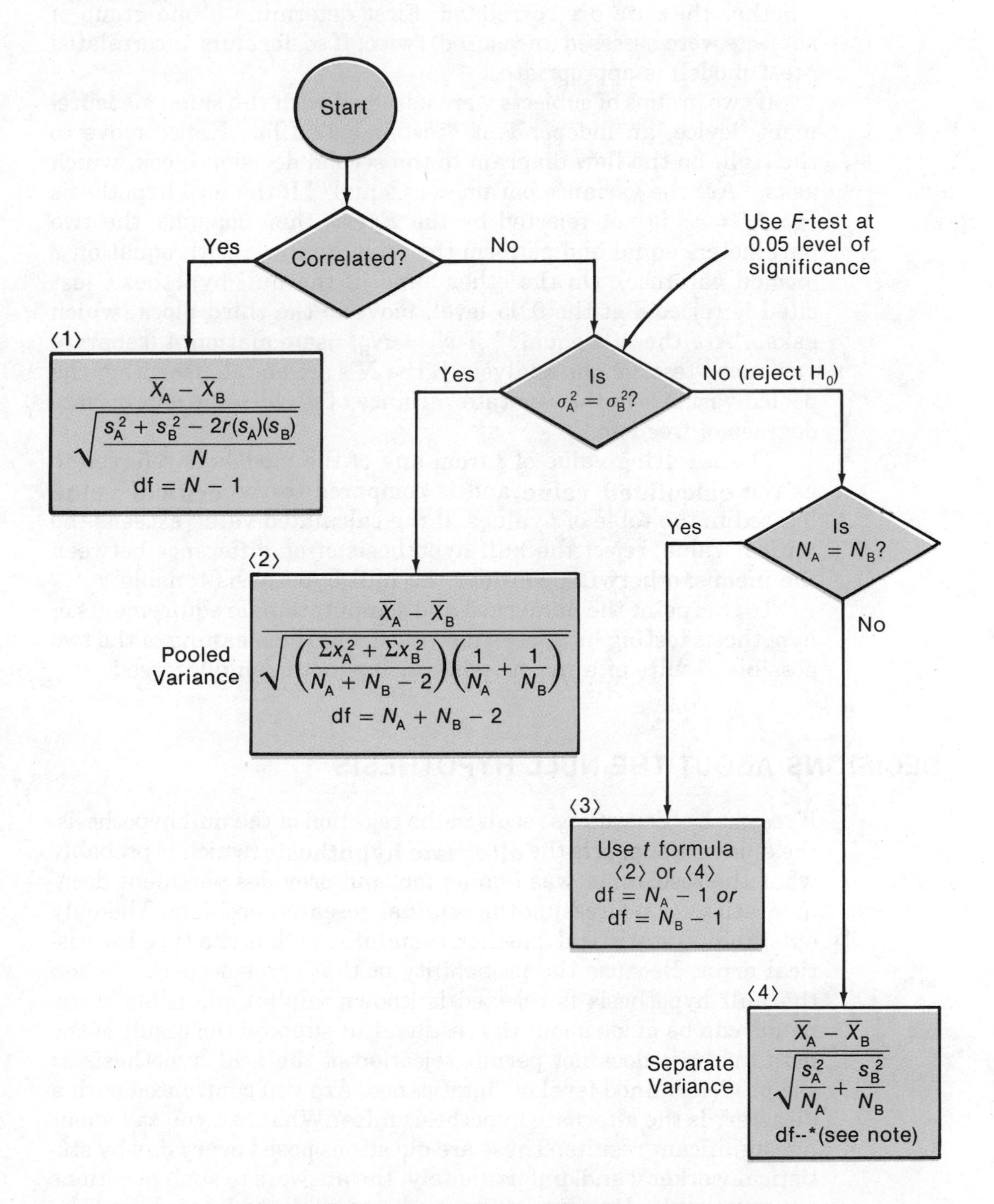

*If formula 4 is used, average tabled *t*-value for both df = $N_A - 1$ and df = $N_B - 1$.

Starting at the top of the flow diagram, the first question is whether the data are correlated. First determine if one group of subjects were assessed (measured) twice. If so, formula 1 (correlated *t*-test model) is appropriate.

If two groups of subjects were assessed with the same measurement device, an independent *t*-test is called for. Hence move to the right on the flow diagram to the second decision block, which asks, "Are the variance parameters equal?" If the null hypothesis H_0: $\sigma_A^2 = \sigma_B^2$ is not rejected by the *F*-test, then consider the two parameters equal and perform the *t*-test analysis with equation 2 (pooled variance). On the other hand, if the null hypothesis just cited is rejected at the 0.05 level, move to the third block, which asks, "Are the *N*'s equal?" If $N_A \neq N_B$, use equation 4 (separate variance *t*-test) for the analysis. If the *N*'s are equal, use either the pooled variance or the separate variance *t*-test with the appropriate degrees of freedom.

The resulting value of *t* from any of the models is referred to as the **calculated value** and is compared to the **critical value** located in the table of *t*-values. If the calculated value exceeds the critical value, reject the null hypothesis of no difference between the means; otherwise, consider the null hypothesis tenable.

To this point the numerical and computational requirements of hypothesis testing have been discussed, but the meaning of the two possible results of a hypothesis test has not been addressed.

DECISIONS ABOUT THE NULL HYPOTHESIS

If research observations result in the rejection of the null hypothesis, the evidence supports the **alternate hypothesis** (which is probably what the researcher was hoping for) and provides pertinent documentation for addressing the original research problem. The only outstanding statistical question remaining is that of a type I statistical error. Because the probability of that error occurring when the null hypothesis is rejected is known (alpha), plausible statements can be made about the results. But suppose the result of the data analysis does not permit rejection of the null hypothesis at the predetermined level of significance. Are you confronted with a disaster? Is the alternate hypothesis false? What can you say about nonsignificant results? These are questions posed every day by statistical workers; and, unfortunately, the answers to such questions are not simple. However, they can be sensibly addressed, and the failure to reject the null hypothesis should not be equated with experimental (or experimenter) failure.

The examination of what to say about a nonsignificant result begins with clarifying some relevant language and terminology. You may believe that slight differences in the language used in hypothesis testing are nothing more than trivial word games. Such a conjecture definitely is not true. Precise language permits a reasonable discussion about null hypothesis tests that do not lead to the rejection of the null hypothesis. Follow this discussion carefully. Start at the point at which a null hypothesis is tested. In traditional statistical hypothesis testing, two outcomes are possible. Either the null hypothesis is *rejected* at the specified significance level, or else one *fails to reject* the null hypothesis.

A statistically significant result is implied by a rejected null hypothesis, and a straightforward and widely used phrase is ". . . the null hypothesis was rejected at the 0.05 level of significance." On the other hand, the statistical test may not lead to the rejection of the null hypothesis. In such a case the investigator states that the statistical test **"failed to reject"** the null hypothesis. Notice that the phrase "failed to reject" is used and that *it does not mean* "accept." The two-pronged lesson is that: (1) "failed to reject" is not equivalent to "accept"; and (2) "failed to reject," which can be translated as "retain H_0 as plausible," is the appropriate language for describing nonsignificant statistical tests.

Why all the fuss about whether we accept a null hypothesis or fail to reject it? Because there is a huge difference between the respective meanings. Basically the null hypothesis is a conjectural statement of an expected outcome under random conditions. If you say that you accept the null hypothesis, you are implying that you believe that nothing more than random events are reflected in the different levels of your independent variable. Conversely, when you say that the test *failed to reject* the null hypothesis, at least two other possibilities are implied. One implication is that the evidence is insufficient to support the alternate hypothesis. A second implication is that the issue of anticipated (nonrandom) experimental effects remains essentially unresolved. Failure to reject the null hypothesis does not rule out the possibility that the alternate hypothesis is true. Within this context, you fail to confirm the alternative. If you "accept" the null hypothesis, you are in essence "rejecting" the alternate or research hypothesis, which is misleading at best.

To reiterate, virtually all null hypotheses may be rejected at *some level* of statistical significance. By prespecifying the alpha level, you have provided a guard against the commission of a type I error only. So, by failing to reject a null hypothesis, you are implying that to do otherwise would constitute an unacceptable risk of a type I error. This discussion can be interpreted by these two statements:

1. Rejection of the null hypothesis enhances the credibility of the alternate hypothesis, whereas
2. Failure to reject the null hypothesis does not provide confirmation of the alternate hypothesis, nor does it provide cause for rejecting the alternative.

There is no cause for discouragement when your criterion for statistical significance is not satisfied. At best you have the option of deferring any final judgment until further evidence is available. At worst you may have to relinquish your research hypothesis for the time being. In either case the possibility remains that the alternate hypothesis is true but that the evidence to support it was simply insufficient.

Therefore several points should be emphasized. First, adopt an attitude that your investigation was a productive enterprise and don't apologize for "negative" results. If rejection of the null hypothesis had been a predetermined outcome, there would have been no justification for conducting the investigation. Second, emphasize that although the null hypothesis was not rejected, the truth of the alternate (research) hypothesis is still in question—the study did not resolve the issue. Finally, *do not* attempt to confirm the alternate hypothesis by using phrases such as "the results *approached* statistical significance"; "there was a *trend*, although not statistically significant, in the predicted direction"; or "the difference, while not significant at the predetermined alpha level of 0.01, did reach significance at the 0.05 level; therefore . . ." Such factors are inappropriate in reporting statistical results.

An old statistics professor used the analogy of striking a match to statistical significance. The match is not *nearly* lit, *tending toward* being lit, or *kind of* lit—it is either lit or unlit. The lesson is that a statistical test performed at a specified level of probability is either significant or not significant. Even though statisticians are not eager to interpret nonsupporting (inconclusive) results, they must be committed to reporting the state of affairs as it is revealed in the data. Be suspicious of statistical reports that violate this dictum.

COMPUTER FOCUS 14—Correlated *t*-Test

Purpose

The purpose of this Computer Focus is to enable you to use some of the programming expertise that you have enhanced during this course. In previous Computer Focus sections you have developed or studied programs and algorithms that contain the necessary computational routines for conducting a correlated *t*-test. More specifically, you have programmed routines for calculating the component parts of the correlated *t*-test equation. These include computing the means (see Computer Focus 4); the sums of squares, variances, and standard deviations (see Computer Focus 5); and Pearson *r*'s (see Computer Focus 11).

With some planning and design work, your knowledge could be used to develop and program a correlated *t*-test routine with the following specifications. The INPUT of the program would include: (1) *N*, the number of cases in the bivariate data set, and (2) the bivariate data itself. The OUTPUT should print the means, standard deviations (sample statistics), computed *t*-value, and degrees of freedom for the correlated *t*-test problem.

Problem

The task is to design (flowchart), write (code), and debug (implement) a BASIC program for computing a correlated *t*-test. Verify the accuracy of the program by analyzing data with known results.

KEY TERMS AND NOTATION

Area(s) of rejection
Calculated *t*-value
Correlated *t*-test
Critical *t*-value
df
F-test
"Fail to reject the null hypothesis"
Homogeneity of variance
Independent *t*-test
Pooled variance *t*-test
Separate variance *t*-test
"Student"
t-distribution

APPLICATION EXERCISES

14.1 Use Appendix F to determine if the *variances* of group 1 and group 2 distributions differ significantly at the 0.05 level.

a. $s_1^2 = 24$ $s_2^2 = 96$
   $N_1 = 15$ $N_2 = 25$

b. $s_1^2 = 87$ $s_2^2 = 42$
   $N_1 = 12$ $N_2 = 20$

c. $\Sigma x_1^2 = 884$ $\Sigma x_2^2 = 1{,}062$
   $N_1 = 13$ $N_2 = 19$

d. $\Sigma X_1 = 135$ $\Sigma X_2 = 387$
   $\Sigma X_1^2 = 2{,}471$ $\Sigma X_2^2 = 15{,}545$
   $N_1 = 8$ $N_2 = 10$

14.2 From the results of Application Exercise 14.1, decide whether a pooled variance or a separate variance *t*-test model would be more appropriate.

14.3 If $\overline{X}_A = 50$, $\overline{X}_B = 46$, $s_A^2 = 90$, $s_B^2 = 20$, $N_A = 9$, and $N_B = 5$:

a. Compute the *t*-value using the separate variance *t*-test formula.

b. Symbolically state the implied null hypothesis.

c. Using Appendix E, what is the appropriate decision about the null hypothesis at $\alpha = 0.05$ on a two-tailed test?

14.4 Compute the value of the pooled variance *t*-test using the following summary statistics:

$\Sigma x_1^2 = 216$ $\Sigma x_2^2 = 304$
$\overline{X}_1 = 150$ $\overline{X}_2 = 160$
$N_1 = 13$ $N_2 = 17$

Using a one-tailed test at the 0.05 level of significance, do the groups differ significantly?

14.5 Given the following summary data, test the null hypothesis at the 0.01 level of significance (two-tailed test) using the correlated *t*-test.

$N = N_1 = N_2 = 100$ $r = 0.41$
$s_1^2 = 64.0$ $s_2^2 = 49.0$
$\overline{X}_1 = 46.4$ $\overline{X}_2 = 42.5$

14.6 A group of students in an English as a Second Language (ESL) class were administered two IQ tests. One was based on language and purported to measure "verbal IQ." The other required problem solving with numbers, patterns, and puzzle-like exercises with very little verbal content—it measured "performance IQ." Scores on the two tests were as follows:

Verbal IQ	Performance IQ
90	92
105	115
95	98
93	103
92	99
96	95

Use the appropriate *t*-test model to test the null hypothesis that verbal IQ and performance IQ are the same for the population of ESL students. Use a one-tailed test at the 0.05 level of significance.

14.7 The varsity men's basketball coach, Coach Hanson, randomly divided his team into two groups. For six weeks group A practiced basketball free throws by taking as many practice shots as possible from the free-throw line for 20 minutes each day. Group B used a "visual-imagery" technique for the same length of time. Group B mentally concentrated on technique and making the shot for one minute before each actual practice shot. After six weeks each group shot 20 free throws with the following results (number of successful shots):

Group A	Group B
12	14
9	10
9	11
14	13
11	17
11	15
12	13
10	13
10	14
8	13
9	
11	

Use the appropriate *t*-test to decide if the data support one method over the other at the 0.05 level (two-tailed test).

14.8 The vocational agriculture class in Woodrow Wilson High School tested the relative effectiveness of two diets on beef cattle with the cooperation of a local feed-lot owner. They randomly divided 20 head of cattle into two groups and standardized the respective diets for a period of one year. Weight gains (in pounds) are as follows:

Diet A	Diet B
152	154
136	130
163	160
150	120
145	136
150	152
165	125
155	140
128	125
149	142

At the 0.05 level of significance (two-tailed test), does the evidence support a decision that one diet is more effective than the other with respect to weight gain in the cattle?

14.9 One psychological theory posits that first-born children will have higher motivation to achieve in school than later-born children in the same family. Pairs of children from ten families were administered an academic motivation assessment instrument. The resulting data scaled as *T*-scores were:

Family	First-Born	Later-Born
1	42	41
2	56	52
3	49	48
4	60	64
5	62	59
6	42	48
7	40	38
8	36	36
9	52	49
10	65	67

Using a correlated *t*-test at the 0.05 level of significance, do these data support the psychological theory?

14.10 At the beginning of this chapter, Miss Yeager had divided her exercise study participants into two groups. Summary statistics on a measure of self-esteem are as follows:

Experimental Group (Exercise & Diet)	Control Group (Placebo Pill)
$\overline{X} = 80$	$\overline{X} = 70$
$\Sigma x^2 = 1{,}216$	$\Sigma x^2 = 864$
$N = 20$	$N = 25$

Determine which of the two independent *t*-test equations should be used to test the null hypothesis. Then use the results of your *t*-test to decide if Miss Yeager's experimental treatment was effective in enhancing self-esteem.

15

Analysis of Variance

CHAPTER OBJECTIVES

Upon completion of the chapter, students will be able to:

1. Compare and contrast the purposes of analysis of variance and independent ***t***-tests.
2. Discuss the rationale for analysis of variance.
3. Compute the necessary values to complete an analysis of variance summary table from given data sets.
4. Determine when an ***a posteriori*** test is appropriately applied.
5. Interpret a statistically significant ***F***-value from an analysis of variance or a ***post hoc*** test.
6. Use Appendix F to determine the critical value of ***F*** for a particular problem.
7. Describe the two components of variability in an analysis of variance.

Ms. Heinrich wanted to determine the best way to present an important aspect of cell structure to her biology class. She had three choices: (1) she could lecture (method A), (2) she could use a visual photographic slide approach (method B), or (3) she could set up laboratory experiments using microscopes that allow students to learn via experimentation (method C). She decided to conduct an experiment with her 30 biology students. The students were randomly assigned to three instructional groups—10 to method A, 10 to method B, and 10 to method C. At the conclusion of the two-week unit on cell structure, the students were tested on the material. The scores (percentage correct) that served as the achievement criterion are shown in Table 15.1.

Table 15.1 Test Results from Ms. Heinrich's Study

Method A	Method B	Method C
80	94	75
76	82	70
75	82	82
72	80	80
84	79	72
79	93	74
75	90	80
70	85	71
81	83	68
82	85	73

The question of concern for Ms. Heinrich is whether the three methods of instruction have different effects on student achievement. Statistically speaking, she wants to know if the differences among the three groups are apt to be a function of chance or if something other than chance (for example, method of instruction) is probably responsible for the observed differences. Sound familiar?

DIFFERENCES AMONG TWO OR MORE GROUPS

In the last chapter the *t*-test was proposed as a way to test for significant differences between two sets of sample data. Ms. Heinrich's experiment involves more than two sets of data; consequently, she needs a method similar to a *t*-test that will enable her to test for differences among three groups. Such problems are generally approached with a technique called **analysis of variance** (which will be referred to as **ANOVA**).

The astute reader may wonder if three applications of an independent *t*-test would be appropriate—one to test for differences between method A and method B, one for method A vs. method C, and one for method B vs. method C. In other words, why couldn't Ms. Heinrich analyze her data two sets at a time using a *t*-test? The adjective "independent," as it applies to a type of *t*-test, requires among other conditions that estimates of certain parameters be truly independent. In practice, this means that repeated *t*-tests should not be used with the same data over and over again. To do so is "cheating probability" and increases the chances of committing a type I statistical error to a level well above the predetermined alpha. ANOVA procedures simultaneously test for differences among two or more sets of data, thereby avoiding problems encountered by using multiple *t*-tests.

The purpose of ANOVA is quite similar to that of the independent *t*-test, except ANOVA designs may include more than two sets of data. If differences between two independent sets of data are to be tested, the statistician may choose between an independent *t*-test or an ANOVA—the results will be identical. If a design includes more than two groups, say *k* groups, the null hypothesis tested by ANOVA is:

$$H_0: \mu_1 = \mu_2 = \mu_3 = \ldots = \mu_k$$

All possible pairs of means do not have to be significantly different for the null hypothesis to be rejected. The alternate hypothesis is that *at least one* pair of means will be significantly different. In the case of the three samples of biology students, the null hypothesis is:

$$H_0: \mu_A = \mu_B = \mu_C$$

where μ_A = the population mean test score of biology students taught by the lecture method,

μ_B = the population mean test score of biology students taught with visual imagery, and

μ_C = the population mean test score of biology students using the laboratory.

RATIONALE FOR ANOVA

The name of the technique (analysis of variance) considered in the present chapter is not obviously linked to the statement of the null hypothesis just given. It appears that the method should be called "analysis of means," because the ultimate aim is to test for mean differences. In this section that apparent contradiction will be resolved. You will learn that a null hypothesis about means is actually approached through analyzing components of variance.

Partitioning Variance

Because you will be dealing with variability (variance), here is a preview of the rationale for ANOVA using the data from Ms. Heinrich's class (Table 15.1).

1. If all students ($N = 30$) are combined into one (total) group, individuals' scores will vary about the grand mean. The term "grand mean" is used here to describe the arithmetic mean of the total group, because later when this group is divided into three separate groups, each will have its own mean. This variability about the grand mean may be expressed in terms of standard deviation (s), variance (s^2), or sum of squares of the deviation scores (Σx^2). The notation for variability of the combined group will be subscripted with a T to indicate **total variance**.
2. When the total group of 30 students is divided into three groups, the total variance is also divided. Some of the variance accompanies group A, some goes with group B, and some with group C. This is referred to as **within-group variance**. It is a measure of dispersion of group members' scores about the respective group means. Within-group variance is also known as **error variance** in statistical literature. Recall that "error" refers to unexplained variance; it does not mean that a mistake has been made.
3. Part of the total variance (of the combined group) was a result of the fact that the three individual group means deviated from the grand mean ($\overline{X}_T$). Such variability is known as **between-groups variance**. Between-groups variance is frequently called **treatment variance** by statisticians to indicate that it results from different treatments or methods of manipulation of the groups.
4. Eventually the variance *between* the groups will be compared to variance *within* each group to form the statistical test using efficient computational equations. Computations that follow illustrate how the total variance is distributed (partitioned).

Table 15.2 displays Ms. Heinrich's data along with the intermediate computations for finding the sum of squares for the combined group.

Table 15.2 Test Results from Ms. Heinrich's Study with Summary Preliminary Calculations

	Method A	X^2	Method B	X^2	Method C	X^2
	80	6,400	94	8,836	75	5,625
	76	5,776	82	6,724	70	4,900
	75	5,625	82	6,724	82	6,724
	72	5,184	80	6,400	80	6,400
	84	7,056	79	6,241	72	5,184
	79	6,241	93	8,649	74	5,476
	75	5,625	90	8,100	80	6,400
	70	4,900	85	7,225	71	5,041
	81	6,561	83	6,889	68	4,624
	82	6,724	85	7,225	73	5,329
Sums	774	60,092	853	73,013	745	55,703
Means	77.4		85.3		74.5	

Totals $\Sigma X = 2{,}372$ $\Sigma X^2 = 188{,}808$ $N = 30$

Grand mean of all 30 scores = 79.07

Total Variability. To illustrate the partitioning of the combined group's variability, the summary data from Table 15.2 are used to compute the total sums of squares designated SS_T. As a reminder, the **sum of squares (SS)** for any set of numbers is:

$$\Sigma x^2 = \Sigma X^2 - \frac{(\Sigma X)^2}{N}.$$

Thus the SS_T is calculated in the usual way considering all 30 scores as one combined group.

$$SS_T = 188{,}808 - \frac{(2{,}372)^2}{30}$$
$$= 1{,}261.87$$

Now you will see how that variability (SS_T) is partitioned into two components.

Within-Group Variability. Within-group variability arises from the fact that individuals differ within their own respective groups. This is one of life's interesting aspects: people differ. So variation is expected simply because people, their attributes, and in this case their scores do vary within groups. Considering Ms. Heinrich's three groups individually, the sum of squares can be computed for each group as follows:

$$\text{Group A:}\quad \Sigma x_A^2 = 60{,}092 - \frac{(774)^2}{10} = 184.4$$

$$\text{Group B:}\quad \Sigma x_B^2 = 73{,}013 - \frac{(853)^2}{10} = 252.1$$

$$\text{Group C:}\quad \Sigma x_C^2 = 55{,}703 - \frac{(745)^2}{10} = 200.5$$

Summing the individual groups' sums of squares yields the sum of squares *within* groups (SS_W).

$$SS_W = (184.4 + 252.1 + 200.5) = 637.$$

Between-Groups Variability. The means of groups A, B, and C shown in Table 15.2 are 77.4, 85.3, and 74.5, respectively. Part of the total variability (SS_T) results from differences between the individual group means and the grand mean for the total group. Suppose, for illustration purposes, each student's score is replaced with the mean of the particular group to which the student belongs. Table 15.3 displays such a data arrangement.

Table 15.3 Group Means Substituted for Individual Scores Eliminating Within-Group Variation

	Method A	Method B	Method C
	77.4	85.3	74.5
	77.4	85.3	74.5
	77.4	85.3	74.5
	77.4	85.3	74.5
	77.4	85.3	74.5
	77.4	85.3	74.5
	77.4	85.3	74.5
	77.4	85.3	74.5
	77.4	85.3	74.5
	77.4	85.3	74.5
Sum X =	774	853	745
Sum X^2 =	59,907.6	72,760.9	55,502.5

Notice that in Table 15.3 *the within-group variability has been eliminated* from the data. By observation you can see that variability does still exist between the groups. More specifically, each group differs from the overall grand mean (79.07). The sum of squares between groups (SS_B) can be readily calculated because the within-group variability has been eliminated. From Table 15.3,

$\Sigma X = 774 + 853 + 745 = 2{,}372$; and
$\Sigma X^2 = 59{,}907.6 + 72{,}760.9 + 55{,}502.5 = 188{,}171$.

Therefore

$$SS_B = 188{,}171 - \frac{(2{,}372)^2}{30} = 624.8667.$$

Now, notice a fundamental relationship:

$$SS_T = SS_B + SS_W$$

that is, $1{,}261.8667 = 624.8667 + 637$. All of the overall variability (SS_T) has been divided into between-groups (SS_B) and within-group (SS_W) sums of squares.

Comparing Between-Groups and Within-Group Variability

The distribution of the three groups shown in Figure 15.1 depicts how the components of variance provide information about the differences among the groups.

By contrasting the two situations in Figure 15.1, you can see that (a) has larger between-groups variability and smaller within-group variability than (b). At a glance it appears from the display that you would expect the three groups in (a) to be more distinct (less overlap) because of the two components of variation. The effect of large between-groups variation is that the groups show more separation. Similarly, small within-group variation tends to create more separation among the groups.

In the final analysis, *if the between-groups variability is considerably larger than the within-group variability, a distinct difference among the groups would be expected.* ANOVA makes use of such a rationale by comparing the between-groups variance to the within-group variance. In the previous chapter you learned that variances can be compared with the aid of the F-distribution to determine if one variance is significantly larger than another. Consequently, using an F-test to determine whether the between-groups variance is larger than the within-group variance constitutes the method known as analysis of variance. If the between-groups variance is significantly larger, the null hypothesis of no differences among the population means is rejected.

Figure 15.1—Comparison of Between-Groups and Within-Group Components of Variance

(a)

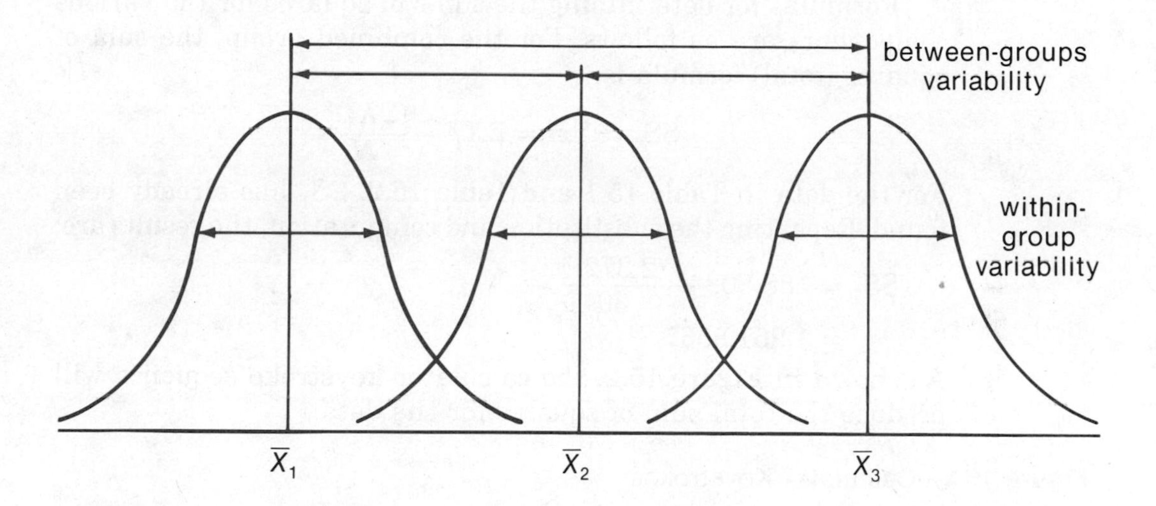

(b)

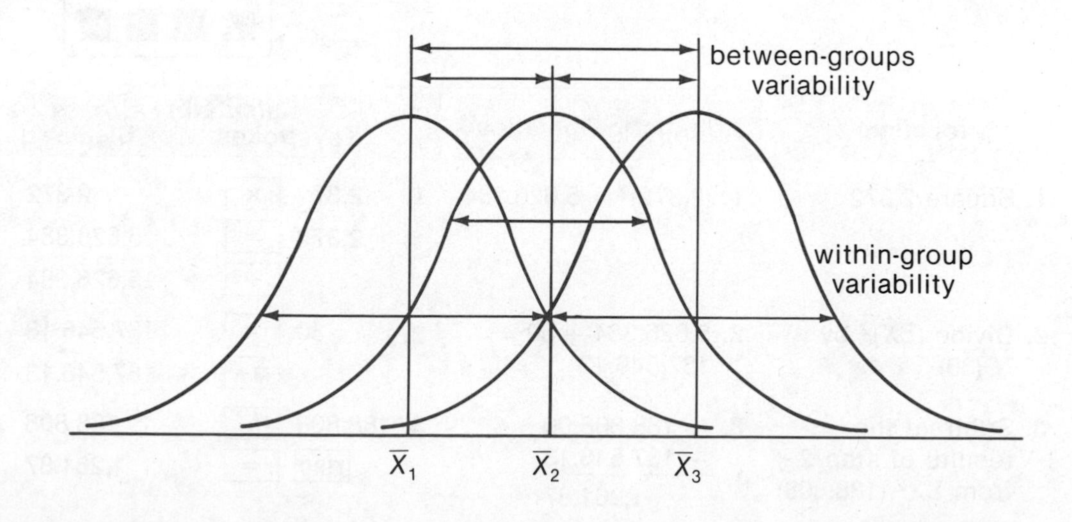

COMPUTATIONAL PROCEDURES

Although the methods described in the previous section could be used to compute ANOVA, more efficient means are available that reduce the computational chores.

Equations

Formulas for determining the sums of squares for the various components are as follows. For the combined group, the sum of squares (total) formula is:

$$SS_T = \Sigma x^2 = \Sigma X_T^2 - \frac{(\Sigma X_T)^2}{N}$$

For the data in Table 15.1 and Table 15.2, SS_T has already been found. Repeating the substitution and computation, the results are:

$$SS_T = 188{,}808 - \frac{(2{,}372)^2}{30}$$
$$= 1{,}261.8667$$

As shown in Figure 15.2, the calculator keystroke sequence will produce the total sum of squares for the data.

Figure 15.2—Calculator Keystrokes

PROBLEM:

Solve:

$$SS_T = 188{,}808 - \frac{(2{,}372)^2}{30}$$

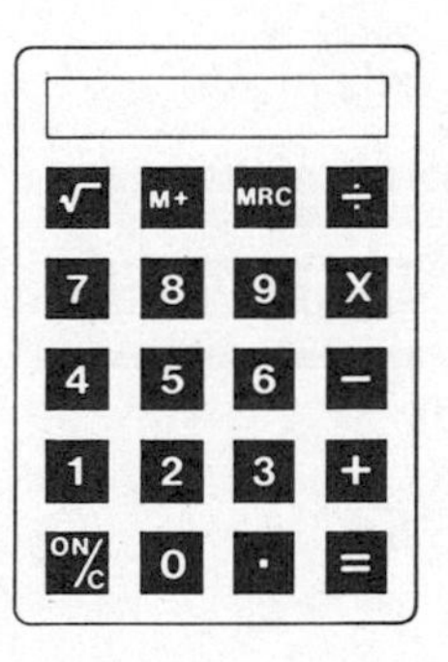

Procedure	Arithmetic Operation	Calculator Keystrokes	Display
1. Square 2,372	1. $(2{,}372)^2 = 5{,}626{,}384$	1. 2,372 [×]	2,372
		2,372 [=]	5,626,384
		[÷]	5,626,384
2. Divide $(\Sigma X)^2$ by N (30)	2. 5,626,384 ÷ 30 = 187,546.13	2. 30 [=]	187,546.13
		[M+]	187,546.13
3. Subtract the results of step 2 from ΣX^2 (188,808)	3. 188,808.00 − 187,546.13 = 1,261.87	3. 188,808 [−]	188,808
		[MRC] [=]	1,261.87

The sum of squares (between groups) is found by:

$$SS_B = \frac{(\Sigma X_A)^2}{n_A} + \frac{(\Sigma X_B)^2}{n_B} + \frac{(\Sigma X_C)^2}{n_C} - \frac{(\Sigma X_T)^2}{N}$$

For Ms. Heinrich's data, the between-groups sum of squares is:

$$SS_B = \frac{(774)^2}{10} + \frac{(853)^2}{10} + \frac{(745)^2}{10} - \frac{(2{,}372)^2}{30} = 624.8667$$

The calculator keystrokes are illustrated in Figure 15.3.

Recall that $SS_T = SS_B + SS_W$. This implies that the within-group sum of squares can be found by subtracting SS_B from SS_T:

$$SS_W = SS_T - SS_B = 1{,}261.8667 - 624.8667 = 637.$$

Although the sums of squares provide an indication of variability, the *F-test* uses a probability distribution based on the ratio of two *variances*. Remember these two important points: (1) both the between-groups and within-group variances need to be computed, and (2) in the context of ANOVA, variance is referred to as **mean square (MS)**.

Earlier in the text, variance of a sample was defined as sums of squares divided by degrees of freedom. In ANOVA procedures, degrees of freedom are associated with each source of variance. Thus $\mathbf{df_B}$ represents **degrees of freedom between groups**; $\mathbf{df_W}$ indicates **degrees of freedom within groups**; and $\mathbf{df_T}$ denotes **degrees of freedom for the total**. The degrees of freedom (df) for the total and the two components of variance are:

(total) $df_T = N - 1$ (where N = total number of subjects),
(between) $df_B = k - 1$ (where k = number of groups), and
(within) $df_W = N - k$ or $df_T - df_B$ (where n and k are as defined above).

The df for each source of variance for the biology class experiment are:

$$df_T = 30 - 1 = 29$$
$$df_B = 3 - 1 = 2$$
$$df_W = 30 - 3 = 27.$$

To this point we have computed the sums of squares and the degrees of freedom. These are summarized as follows (rounded to two decimal places, which is usually sufficient):

Source of Variance	df	SS
Between Groups	2	624.87
Within Group	27	637.00
Total	29	1,261.87

Figure 15.3—Calculator Keystrokes

PROBLEM:

Solve: SS (between groups)

$$SS_B = \frac{(774)^2}{10} + \frac{(853)^2}{10} + \frac{(745)^2}{10} - \frac{(2,372)^2}{30}$$

Procedure	Arithmetic Operation	Calculator Keystrokes	Display
1. Evaluate $\frac{(774)^2}{10}$ and add to memory	1. 774 × 774 = 599,076; 599,076 ÷ 10 = 59,907.6	1. 774 [×]	774
		774 [÷]	599,076
		10 [=]	59,907.6
		[M+]	59,907.6
2. Evaluate $\frac{(853)^2}{10}$ and add to memory	2. 853 × 853 = 727,609; 727,609 ÷ 10 = 72,760.9	2. 853 [×]	853
		853 [÷]	727,609
		10 [=]	72,760.9
		[M+]	72,760.9
3. Evaluate $\frac{(745)^2}{10}$ and add to memory	3. 745 × 745 = 555,025; 555,025 ÷ 10 = 55,502.5	3. 745 [×]	745
		745 [÷]	555,025
		10 [=]	55,502.5
		[M+]	55,502.5
4. Evaluate $\frac{(2,372)^2}{30}$ and subtract from memory	4. 2,372 × 2,372 = 5,626,384; 5,626,384 ÷ 30 = 187,546.13	4. 2,372 [×]	2,372
		2,372 [÷]	5,626,384
		30 [=]	187,546.13
		[M−]	187,546.13
	188,171.00 − 187,546.13 = 624.87	[MRC]	624.87

To find the variance for each component (called "mean square" from here on), simply divide the sums of squares by the respective degrees of freedom. The mean squares (MS) are:

$$MS_B = \frac{624.87}{2} = 312.44$$

$$MS_W = \frac{637.00}{27} = 23.59.$$

Because we are interested in a comparison of MS_B to MS_W, the total mean square (MS_T) is not required. Now the summary table appears as:

Source of Variance	df	SS	MS
Between Groups	2	624.87	312.44
Within Group	27	637.00	23.59
Total	29	1,261.87	

In the previous chapter you learned that *a ratio comparison of two variances (MS) yields a calculated F-value*. Therefore for our data: $F = \frac{312.44}{23.59} = 13.24.$

ANOVA Summary Table

The computational efforts are now complete and are shown in an ANOVA summary table:

Source of Variance	df	SS	MS	*F*
Between Groups	2	624.87	312.44	13.24
Within Group	27	637.00	23.59	
Total	29	1,261.87		

An **ANOVA summary table** is usually included when ANOVA results are reported. A summary of the computational formulas in the respective cells is provided in Table 15.4.

Table 15.4 ANOVA Summary Table with Computational Formulas

Source	df	SS	MS	*F*
Between Groups	$k - 1$	$\frac{(\Sigma X_A)^2}{n_A} + \frac{(\Sigma X_B)^2}{n_B} + \ldots - \frac{(\Sigma X_T)^2}{N}$	$\frac{SS_B}{df_B}$	$\frac{MS_B}{MS_W}$
Within Group	$N - k$	$SS_T - SS_B$	$\frac{SS_W}{df_W}$	
Total	$N - 1$	$\Sigma X_T^2 - \frac{(\Sigma X_T)^2}{N}$		

INTERPRETATION OF THE *F*-VALUE

Using the methods described in the previous section, the *calculated value* of *F* was determined. As with other inferential hypothesis testing techniques, the calculated *F*-value is compared to a critical (tabled) *F*-value. *If the calculated* F*-value exceeds the critical value, the null hypothesis is rejected.* If the calculated *F*-value is less than the critical value, the null hypothesis is not rejected.

The critical value of *F* is a point that defines the area of rejection for an ***F*-distribution**. As was the case for a *t*-distribution, the shape of the *F*-distribution changes according to the degrees of freedom associated with the numerator and denominator of the ***F*-ratio**. Notice in Figure 15.4 that an ANOVA (*F*-test) uses only the right tail of the distribution as an area of rejection. This is because the concern is only with how many times larger the between-groups variance is than the within-group variance.

Figure 15.4—*F*-Distribution Showing Critical *F*-Value and Area of Rejection

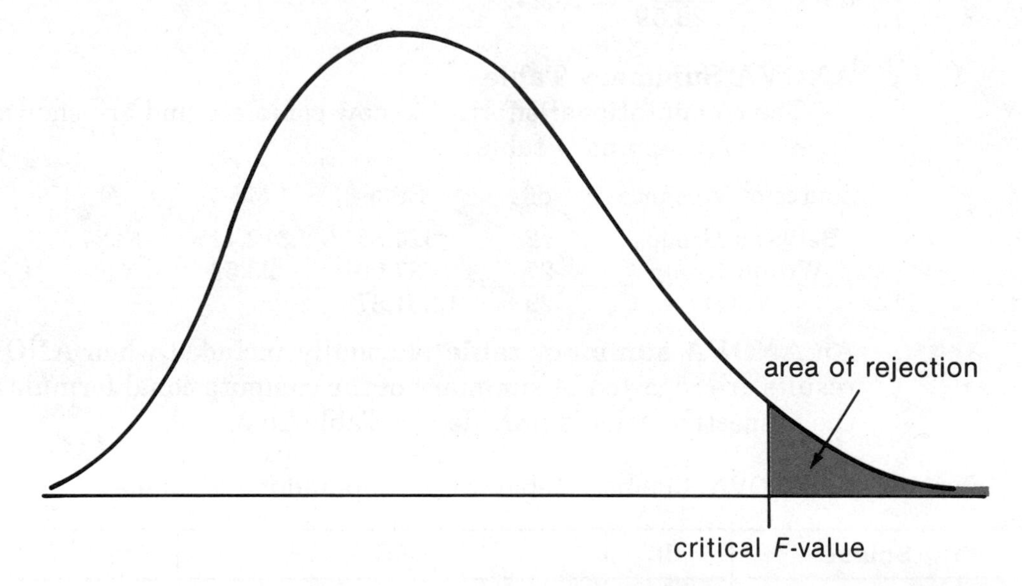

The shape of an *F*-curve is skewed in a positive direction, and the area of rejection is located in the part of the curve with the gradual slope.

For our purposes the area of rejection will be either 1% or 5% of the area under the curve. Critical values of *F* are found in Appendix F. To review locating critical values, find the df for the numerator (between groups) in the top row. Then locate df for the denominator (within group) in the first column. The two values in the table where the particular row and column intersect are critical values of *F*: the top number for the 0.05 level of significance; the bottom (bold print) number for the 0.01 level of significance.

Returning to the ANOVA data from the example, the calculated value of *F* was 13.24. The ANOVA summary table also provides the numerator degrees of freedom ($df_B = 2$) and the denominator degrees of freedom ($df_W = 27$) for determining the critical value of *F*. Using an alpha level of 0.05, the critical value of *F* from Appendix F is 3.35. Figure 15.5 graphically shows that the calculated value of *F* (13.24) is larger than the critical (tabled) value and thus falls in the area of rejection.

Figure 15.5—Critical and Calculated Values on the *F*-Distribution with 2 and 27 df

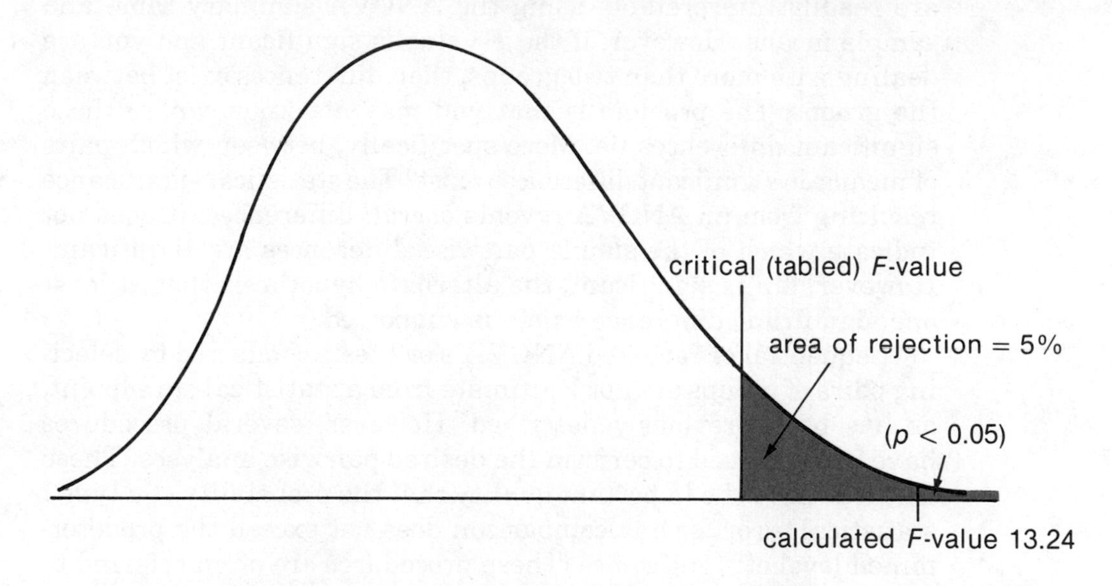

Therefore the null hypothesis:

$$H_0: \mu_A = \mu_B = \mu_C$$

is rejected. Ms. Heinrich now has evidence that the three instructional methods are not equally effective.

Interestingly, although she knows at this point that the means are significantly ($P < 0.05$) different, it is not obvious how the groups compare because the group means were not computed when the ANOVA was performed. The calculations involved only variances; therefore for interpretation purposes she would compute arithmetic means for the groups to determine their relative standings on the test score results. The means are provided as descriptive statistics in Table 15.2: $\overline{X}_A = 77.4$, $\overline{X}_B = 85.3$, and $\overline{X}_C = 74.5$. The respective group means together with an ANOVA summary table seem to end the analysis and provide all the information needed for interpreting the results. But do they? What happens if F is significant?

WHAT TO DO IF *F* IS SIGNIFICANT—MULTIPLE COMPARISON TEST

ANOVA is a powerful statistical tool for testing differences between means of two or more groups. If, in an ANOVA problem, F is not significant, or if only two groups are being considered, the results are readily interpretable using the ANOVA summary table and sample means. However, if the F-value is significant and you are dealing with more than two groups, then differences exist between the groups. The problem is that you may not know where these significant differences lie. More specifically, between which pairs of means do significant differences exist? The statistical significance resulting from an ANOVA reveals *overall* differences; it does not indicate which of the simple pairwise differences are significant. However, if F is significant, the alternate hypothesis that at least one significant difference exists is supported.

Sequential or repeated ANOVA's or t-tests conducted by selecting pairs of groups are not legitimate from a statistical standpoint, as has been previously described. However, several procedures have been devised to perform the desired pairwise analyses. These methods have the important quality that the probability of a type I statistical error for any comparison does not exceed the predetermined level of significance. These procedures are often referred to as ***a posteriori*** or ***post hoc* tests**. Their purpose is to make simple pairwise comparisons between groups *after* an ANOVA has resulted in a statistically significant overall result.

One of the more popular *a posteriori* tests is a method developed by a statistician named Scheffé. For each pairwise comparison, the **Scheffé procedure** calls for the computation of an F-value using the formula:

$$F = \frac{(\overline{X}_1 - \overline{X}_2)^2}{MS_W\left(\frac{1}{n_1} + \frac{1}{n_2}\right)(k-1)}$$

The calculated F-value has $(k - 1)$ degrees of freedom in the numerator and $(N - k)$ degrees of freedom in the denominator just as in the overall ANOVA. If the calculated values are greater than the tabled values of F from Appendix F, the two groups are significantly different; that is, the null hypothesis is rejected.

In general, the number of possible comparisons that can be made with k groups is $\frac{1}{2}(k)(k - 1)$. In our example with three groups $(k = 3)$ of biology students, the number of possible comparisons can be computed: $\frac{1}{2}(3)(2) = 3$. A simple listing verifies the number of pairs that are to be compared:

group A vs. group B,
group A vs. group C, and
group B vs. group C.

Now apply the Scheffé method to the numerical example. From previous computations you know that the necessary data for the problem are:

$$\overline{X}_A = 77.4$$
$$\overline{X}_B = 85.3$$
$$\overline{X}_C = 74.5$$
$$n_A = n_B = n_C = 10$$
$$MS_W = 23.59$$
$$k = 3.$$

The critical F-value obtained from Appendix F at alpha = 0.05 with 2 df in the numerator and 27 df in the denominator (commonly shortened to the notation df = 2,27) is 3.35; consequently, any calculated value of F greater than 3.35 from the Scheffé procedure will indicate a significant $(p < 0.05)$ difference between the pairs of means.

Comparing the means of group A and group B yields:

$$F = \frac{(77.4 - 85.3)^2}{23.59\left(\frac{1}{10} + \frac{1}{10}\right)(3 - 1)}$$
$$= \frac{62.41}{(23.59)(0.2)(2)}$$
$$= \frac{62.41}{9.436}$$
$$= 6.61 \quad (\textit{reject } H_0\text{: } \mu_A = \mu_B).$$

The calculator keystrokes for Scheffé's *a posteriori* test are shown in Figure 15.6.

Comparing the means for group A and group C results in another F-value (note that if the n's are equal, the denominator remains constant for all comparisons).

$$F = \frac{(77.4 - 74.5)^2}{9.436}$$
$$= \frac{8.41}{9.436}$$
$$= 0.89 \quad (\textit{not significant}; \text{H}_0 \text{ is } \textit{tenable})$$

Figure 15.6—Calculator Keystrokes

PROBLEM:

Solve:

$$F = \frac{(77.4 - 85.3)^2}{23.59\left(\frac{1}{10} + \frac{1}{10}\right)(3 - 1)}$$

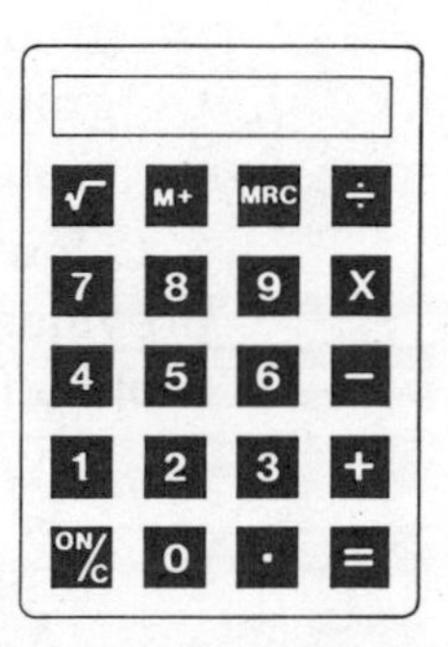

Procedure	Arithmetic Operation	Calculator Keystrokes		Display
1. Calculate value of denominator	**1.** $(23.59)(\frac{1}{10} + \frac{1}{10})(3 - 1)$	**1.** 1	÷	1
	$= (23.59)(0.2)(2)$	10	=	0.1
	$= (23.59)(0.4)$		M+	0.1
	$= 9.436$	1	÷	1
		10	=	0.1
			M+	0.1
		23.59	×	23.59
		MRC	×	4.718
		2	=	9.436
2. Calculate value of numerator	**2.** 77.4	**2.** 77.4	−	77.4
	− 85.3	85.3	×	−7.9
	−7.9	−7.9	=	62.41
	$(-7.9)^2 = 62.41$			
3. Divide numerator by denominator	**3.** $62.41 \div 9.436 = 6.61$	**3.**	÷	62.41
		9.436	=	6.6140313

Finally, comparing the means for group B and group C yields the following results:

$$F = \frac{(85.3 - 74.5)^2}{9.436}$$
$$= \frac{116.84}{9.436}$$
$$= 12.36 \quad (\textit{reject } H_0\text{: } \mu_A = \mu_C)$$

The results can now be interpreted. Group B had a significantly ($p < 0.05$) *higher* test score mean than group A. Likewise, group B scored significantly ($p < 0.05$) *higher* than group C. The data do not provide evidence of a difference between group A and group C ($p > 0.05$). Consequently, method B (visual imagery via slides) was apparently the best instructional method in Ms. Heinrich's experiment.

Note that whenever an overall null hypothesis is rejected by an ANOVA, it is reasonable to assume that at least one mean differs from another mean. If the sample n's are equal, one is justified in concluding that a significant difference between the largest mean and the smallest mean exists on the basis of the ANOVA. *Post hoc* tests are not as powerful as ANOVA. Therefore it is not uncommon for an investigator to find ANOVA showing a significant F-value but none of the *a posteriori* comparisons revealing pairwise differences. In such a case the extreme (largest) difference is designated as significant on the basis of the original ANOVA result.

COMPUTER FOCUS 15—Testing a Computer-Generated Approximation to the Normal Curve: An Application of ANOVA and the *F*-Ratio

Purpose

This Computer Focus involves randomly generating an important distribution (the normal curve) using a computer. You will get a "feel" for how well approximations, which are usually more efficient to generate, compare to exact solutions under various conditions of sample size.

Purpose

The mathematician S. M. Ulam, while working at the Los Alamos Scientific Laboratory in New Mexico in the mid-1940's, formalized a set of procedures for studying and modeling problems with randomly generated simulations. For decades these methods, known as Monte Carlo methods, have been refined and applied in a variety of fields of research. The issue of the degree of accuracy of randomly generated approximations or simulations is an important one. The object of this Computer Focus is to test an approximation to the normal probability curve on two critical characteristics: (1) mean and (2) variance.

The two routines shown in this Computer Focus are algorithms to generate normally distributed values. The first one produces a *satisfactory approximation* to the normal curve, whereas the second algorithm produces *exact* values (under the assumption that the random number distribution [RND] function generates independent random numbers). The approximation routine needs to be executed *N* times to produce *N* values; the second one produces two values for each execution. Both routines should be located in a loop. Because the routines are not complete programs, you will need to design a structure for looping through the routines and for storing the output to complete the problem of this Computer Focus.

Program 1

```
100 REM APPROXIMATION TO NORMAL DISTRIBUTION
    OF Z-SCORES

105 LET J = 0

110       FOR I = 1 TO 12

115       J = J + RND(1)

120       NEXT I

125 Z = J - 6

130 REM OUTPUT STANDARD NORMAL Z-SCORE
    VARIABLE--Z
```

Program 2 (Called the Box-Muller Method)

```
100 REM  EXACT NORMAL DISTRIBUTION OF Z-SCORES

105 A = 6.283185308

110 B = RND(1)

115 C = RND(1)

120 B = -2*LOG(B)

125 B = SQR(B)

130 C = A*B

135 Z1 = B*SIN(C)

140 Z2 = B*COS(C)

145 REM  OUTPUT STANDARD NORMAL Z-SCORES -- Z1
    AND Z2
```

Two necessary although insufficient conditions for distributions to be considered statistically equivalent are that their means and variances, respectively, are not different. You now know that ANOVA will test for inequality of means and that the *F*-ratio is used to compare two variances (see Chapter 14 in addition to the current chapter). Your problem is to determine if the mean and variance of a distribution of "approximate" values differ from those in a distribution of "exact" values. You will need to use the routines to generate two sets of data. Then apply ANOVA procedures to test for mean differences. Finally, use the *F*-ratio as illustrated in the *t*-test chapter to determine if the variances are different. Try the experiment with different sample sizes. To reduce the amount of computation you must complete, have the routines accumulate ΣX and ΣX^2. Draw your conclusion about the suitability of the approximation routine.

KEY TERMS AND NOTATION

Analysis of variance (ANOVA)
Between-groups variance
df_B (degrees of freedom between groups)
df_T (degrees of freedom total)
df_W (degrees of freedom within group)
Error variance
F-distribution
F-ratio
MS
Mean square
post hoc test
Scheffé procedure
Sum of squares (SS)
Summary table for ANOVA
Total variance
Treatment variance
Within-group variance

APPLICATION EXERCISES

15.1 Complete the following ANOVA summary tables and use Appendix F to determine whether or not the null hypothesis in each case would be rejected at the 0.05 level of significance.

a.

Source	df	SS	MS	*F*
Between	4	84	21.00	?
Within	45	210	4.67	
Total	49	294		

b.

Source	df	SS	MS	*F*
Between	2	42	?	?
Within	24	218	?	
Total	26	260		

c.

Source	df	SS	MS	*F*
Between	2	?	?	?
Within	12	?	18	
Total	?	300		

15.2 Complete the following summary table and answer the questions that follow.

Source	df	SS	MS	*F*
Between	5	?	?	?
Within	?	?	40	
Total	20	100		

a. What is the value of *F*?
b. Is *F* significant at the 0.01 level?
c. How many groups were involved in the analysis?
d. How many individual subjects were in the total group?

15.3 Suppose that Major Brunner, recruiting officer for the region, wanted to increase high school vocational counselors' interest in military job and career potential. Major Brunner designed an experiment involving three randomly selected groups of counselors. The data that follow represent the results of the hypothetical experiment. Group I received printed materials describing military occupations of potential interest to their counselees. Group II systematically met once a week with military recruiters for the purpose of discussing military occupational opportunities. Group III participated in on-site visits to military installations to observe vocational training. After a period of time, an attitude checklist was administered to the counselors to assess how the counselors felt about recommending military careers to their students. The higher the score, the more positive the attitude toward military occupations. Perform an ANOVA at the 0.05 level of significance and, if appropriate, perform an *a posteriori* comparison and interpret the results of the experiment for Major Brunner.

Group		
I	II	III
2	8	14
4	10	16
6	12	18
8	14	20
10	16	22
12	7	11
5	9	7
7	4	9

15.4 Compute an analysis of variance on the following data. Complete an ANOVA summary table and perform a *post hoc* test on the results. ($\alpha = 0.05$.)

Group			
A	B	C	D
14	17	14	8
12	15	12	6
10	12	12	5
10	9	11	4
9	9	11	2
6	7	10	2
6	7	10	2

15.5 Following are ANOVA summary tables, group means, and sample sizes. In each case perform an *a posteriori* test and interpret the results at the 0.05 level of significance.

a. ANOVA Summary

Source	df	SS	MS	*F*
Between	2	313.8	156.9	30.58
Within	27	138.6	5.13	
Total	29			

	Group A	Group B	Group C
Mean	17.3	22.4	25.1
n	10	10	10

b. ANOVA Summary

Source	df	SS	MS	*F*
Between	2	70	35	14
Within	12	30	2.5	
Total	14	100		

	Group A	Group B	Group C
Mean	3	4	8
n	5	5	5

c. ANOVA Summary

Source	df	SS	MS	*F*
Between	2	76	38	4.75
Within	15	120	8	
Total	17	196		

	Group A	Group B	Group C
Mean	9	10	16
n	6	6	6

15.6 Several members of the varsity golf team were discussing the relative merits of various brands of golf balls. There was disagreement about which brand gives the most distance with a drive from the tee box. The team members had narrowed the field to four brands: brand W, brand R, brand H, and brand T. As one skilled in statistical application to practical problems, they turned to you for your advice regarding a statistically based experiment that would resolve the argument about which brand (if any) would carry the longest distance off a drive. Describe your experiment and the statistical procedures for analyzing the data collected from the experiment.

16

Chi Square

CHAPTER OBJECTIVES

Upon completion of the chapter, students will be able to:

1. Distinguish between observed and expected frequencies.
2. Compute a chi square given an observed and expected frequency distribution.
3. Determine expected frequencies for a one- or two-variable chi square problem when provided observed frequencies.
4. Identify the critical value of a chi square given an alpha level and degrees of freedom.
5. State and test a null hypothesis with an appropriate chi square technique.
6. Specify the conditions under which Yates' correction factor is used.

To devise a strategy for his sprinters, Mr. White, the high school track coach, collected data on the 200-meter dash events in the state. The race begins around a curve and finishes on the straight part of the track. Runners are "staggered" at the beginning so that each runner will eventually run the same distance. Coach White has collected data on the 200-meter winners for an entire season by noting the lane in which the winner ran. Lane 1 is on the inside of the curve; lane 8 is on the outside of the track. For the fifty-six 200-meter dashes, the winning lanes and the number of winners were as follows:

Lane Number	1	2	3	4	5	6	7	8
Number of Winners	5	6	4	7	7	9	8	10

The question Coach White wants answered is: "Do the results indicate that the number of winners is equally distributed across the eight lanes?" The data in this case are counts or frequencies, not test scores or measured variables. Therefore inferential statistical methods described to this point are not appropriate for addressing the coach's question. However, one statistical technique that will test a null hypothesis relevant to the question is known as *chi square* (pronounced "ki" as in the word "kite").

The chi square tests described in this chapter are only two of literally dozens of applications that are based on a very important sampling distribution called the chi square (χ^2) distribution. Chi square statistical tests are part of a group of techniques known as *nonparametric* techniques. **Nonparametric** procedures generally require only a few assumptions about the population parameters and can be used with numerical data scaled on less than an interval level and with frequency counts on one or two nominal variables. Note that Coach White's data consist of frequencies on a nominal (lane position) variable. Recall that nominal variables only provide names or labels; they do not provide quantitative information. In contrast, *parametric* techniques such as the Pearson r, regression, t-test, and analysis of variance require that at least one of the variables (usually the dependent variable) be a measured variable on a continuous interval or ratio scale.

From a computational perspective, you will notice that for the first time in this text, a value of the sum of squares (Σx^2) will not be calculated. This is because the mean of the distribution, and therefore the deviation from the mean, is not useful with nominal data. Consequently, the chi square tests as well as other nonparametric statistical tests found in more advanced texts are generally easier to calculate than parametric tests such as the t-test or ANOVA.

In general, the **chi square test** is applicable in determining if an *observed frequency* distribution or an observed joint frequency distribution differs from an *expected* distribution. The **observed frequency** distribution is the one obtained via a counting process, whereas the **expected frequency** distribution is a theoretical distribution based either on an *a priori* expectation or on random occurrences.

The chi square random variable is:

$$\chi^2 = \sum \frac{(O - E)^2}{E}$$

where O is the observed frequency and E is the expected frequency.

(An exception to this formula, called Yates' correction, will be discussed later.)

The strategy for solving chi square problems in general is not much different from procedural solutions for other hypothesis tests. The χ^2 is a test of the null hypothesis that states that there is no difference between the observed frequency distribution and a theoretical expected frequency distribution. Briefly, the steps are:

1. determine the expected frequency distribution (E);
2. compute the value of the chi square random variable using the formula;
3. determine the appropriate degrees of freedom (df) for the problem according to the chi square application requirements;
4. compare the calculated value of χ^2 with the critical value of χ^2 for the appropriate df and alpha level as with other probability distributions; and
5. interpret the results.

As with the *F*-distribution described in the previous chapter, the chi square distribution actually forms a family of curves depending on the degrees of freedom. As shown in Figure 16.1, the area of rejection is in the tail in the direction of the skew. Because the chi square distribution changes shape with varying degrees of freedom, only three levels of significance for χ^2 critical values are reported with the respective degrees of freedom in Appendix G.

Figure 16.1—Distributions of Chi Square at df = 5, 10, and 3 and Areas of Rejection at 0.05 Level

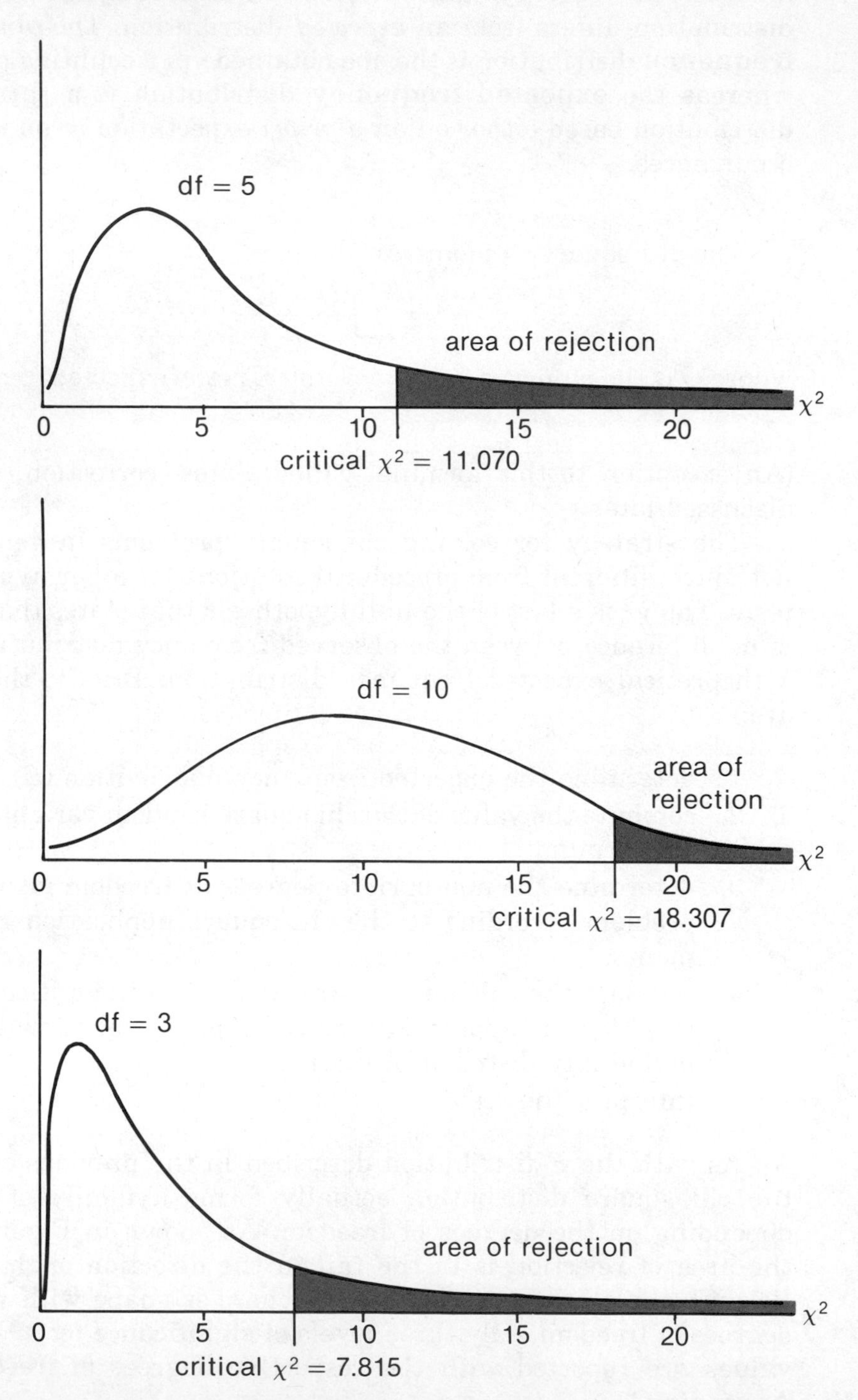

ONE-VARIABLE APPLICATIONS

In its most basic form as a test of *goodness of fit*, the chi square test can be used to test for differences between an observed distribution and an expected distribution with equiprobable events for each level of a nominal variable. The term **goodness of fit** refers to how well the observed distribution "fits" the theoretical expectations. Coach White's problem posed at the beginning of the chapter will illustrate this concept. Recall that the coach was interested in whether there were advantages to being assigned to certain lanes in the 200-meter dash. The expected distribution, if there are no advantages to having certain lanes, would be seven winners for each lane. That is, the coach would expect an equal number of winners for each lane. And, because the data are for 56 winners running in eight lanes, the expectation of the number of winners for each lane would be one eighth of 56, or 7. The original sample of data (observed) is repeated with the corresponding expected frequency distribution as follows:

Lane Number	1	2	3	4	5	6	7	8
Number of Winners	5	6	4	7	7	9	8	10
Expected Winners	7	7	7	7	7	7	7	7

To a statistician, Coach White's question might be translated to this form: "Does the observed distribution of winners differ from the expected distribution more than could be anticipated just by random chance, say 5% or less of the time?" The chi square goodness-of-fit test addresses this issue by testing the null hypothesis H_0: $O = E$ or H_0: $O - E = 0$, which states that there is no difference between the observed and expected frequencies in an inferential context.

The step-by-step solution following the strategy outlined earlier is as follows:

1. $E = \frac{56}{7}$ (equal number of winners for each lane was expected).

2. $$\chi^2 = \sum \frac{(O - E)^2}{E}$$

$$= \frac{(5-7)^2}{7} + \frac{(6-7)^2}{7} + \frac{(4-7)^2}{7} + \frac{(7-7)^2}{7} + \frac{(7-7)^2}{7} + \frac{(9-7)^2}{7} + \frac{(8-7)^2}{7} + \frac{(10-7)^2}{7}$$

$$= \frac{4 + 1 + 9 + 0 + 0 + 4 + 1 + 9}{7}$$

$$= \frac{28}{7} = 4.0$$

The calculator keystrokes for computing the chi square are shown in Figure 16.2.

Figure 16.2— Calculator Keystrokes

PROBLEM:

$$\chi^2 = \sum \frac{(O - E)^2}{E}$$

Note: Since the numerator is squared in each term, the absolute difference between *O* and *E* is used (no negatives).

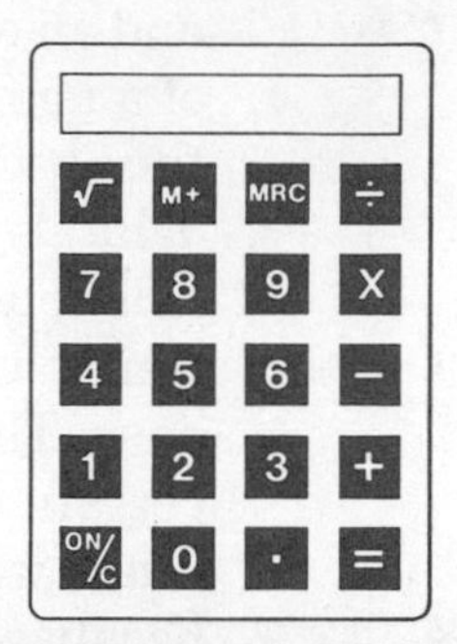

Procedure	Arithmetic Operation	Calculator Keystrokes	Display
1. Square each $\lvert O - E \rvert$ and divide each by E	1. $\frac{(5-7)^2}{7} = \frac{4}{7}$	1. 7 [−] 5 [=]	2
		[×] 2 [÷]	4
	$\frac{(6-7)^2}{7} = \frac{1}{7}$	7 [=] [M+]	0.5714285
	$\frac{(4-7)^2}{7} = \frac{9}{7}$	7 [−] 6 [=]	1
	$\frac{(7-7)^2}{7} = \frac{0}{7}$	[×] 1 [÷]	1
		7 [=] [M+]	0.1428571
	$\frac{(7-7)^2}{7} = \frac{0}{7}$	7 [−] 4 [=]	3
		[×] 3 [÷]	9
	$\frac{(9-7)^2}{7} = \frac{4}{7}$	7 [=] [M+]	1.2857142
	$\frac{(8-7)^2}{7} = \frac{1}{7}$	0 [M+]	0
		0 [M+]	0
	$\frac{(10-7)^2}{7} = \frac{9}{7}$	9 [−] 7 [=]	2
		[×] 2 [÷]	4
		7 [=] [M+]	0.5714285
		8 [−] 7 [=]	1
		[×] 1 [÷]	1
		7 [=] [M+]	0.1428571
		10 [−] 7 [=]	3
		[×] 3 [÷]	9
		7 [=] [M+]	1.2857142
2. Sum the terms. Results equal the calculated χ^2	2. $\chi^2 = \frac{4}{7} + \frac{1}{7} + \frac{9}{7} + \frac{0}{7} + \frac{0}{7} + \frac{4}{7} + \frac{1}{7} + \frac{9}{7} = 4.0$	2. [MRC]	3.9999996

3. For one variable problems such as the present illustration, the degrees of freedom are one less than the number of cells (levels of the variable, or lanes in this case); that is, if k = number of cells, then df = $k - 1$. Therefore, for the present problem, df = 8 − 1 = 7 degrees of freedom.
4. The critical value of χ^2 with 7 df at the 0.05 alpha level using Appendix G is 14.067, which exceeds the calculated value of 4.0.
5. The deviation between the distribution that Coach White observed and what he would expect is so small that such a discrepancy would occur more than 5% of the time just by chance, or $p > 0.05$. Therefore the coach has no evidence that certain lanes are advantageous—the null hypothesis is not rejected. Figure 16.3 shows the relationship between the calculated value and the critical value of χ^2. The calculated value would have to exceed 14.067 for the null hypothesis to be rejected.

Another example illustrates a slight variation on computing E. Suppose that the principal of Glennwood High, Mrs. Wilson, is

Figure 16.3—Chi Square Distribution (7 Degrees of Freedom)

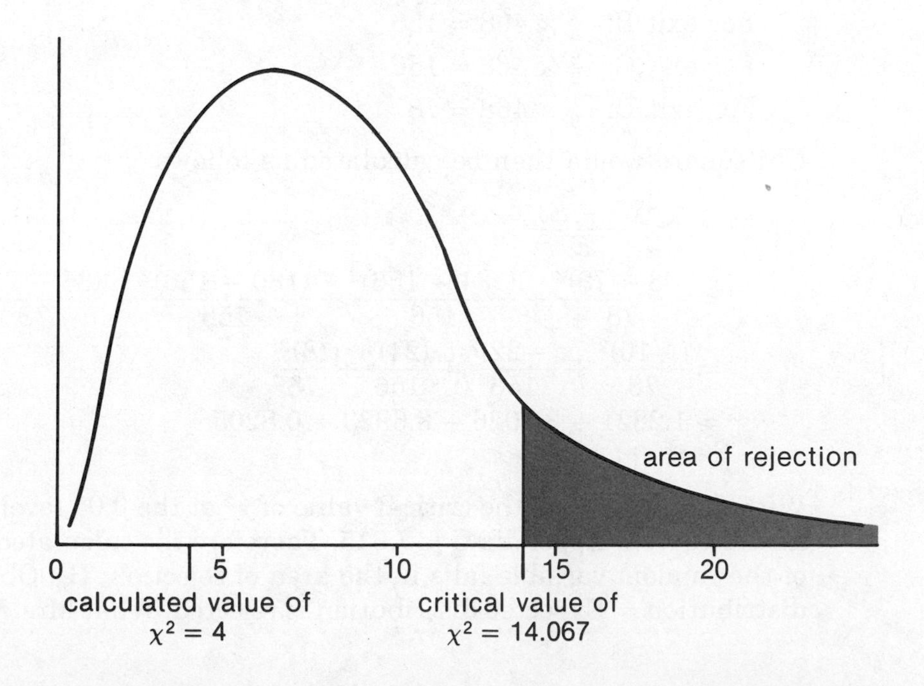

interested in the fire-drill evacuation from the school auditorium. The auditorium has four exits: exit A, exit B, exit C, and exit D. However, exits B and C are double-width doors, so Mrs. Wilson *expects* twice as many students to use exits B and C as use exits A and D. In other words, the most efficient evacuation through exits A, B, C, and D is in the respective ratio of 1:2:2:1.

The fire-drill procedure is activated while the student body is in the auditorium. Mrs. Wilson has teachers stationed at each of the four exits to count the number of students that pass through each door. The tabulated (observed) results were as follows:

Exit A	68
Exit B	134
Exit C	180
Exit D	86

Does this distribution differ from Mrs. Wilson's expectations more than would be expected by chance 5% of the time? In solving this problem, the expected number of students leaving each exit is not equal, because the doorways are not equal in size. Rather, the ratio of 1:2:2:1 would be used to compute *E*. That is, one-sixth of the 468 students would be expected to use exit A, two-sixths to use exit B, two-sixths to use exit C, and one-sixth to use exit D. Thus the expected frequencies would be calculated as:

For exit A: $\frac{1}{6} \times 468 = 78$

For exit B: $\frac{2}{6} \times 468 = 156$

For exit C: $\frac{2}{6} \times 468 = 156$

For exit D: $\frac{1}{6} \times 468 = 78$

Chi square would then be calculated as follows:

$$\begin{aligned}
\chi^2 &= \sum \frac{(O-E)^2}{E} \\
&= \frac{(68-78)^2}{78} + \frac{(134-156)^2}{156} + \frac{(180-156)^2}{156} + \frac{(86-78)^2}{78} \\
&= \frac{(-10)^2}{78} + \frac{(-22)^2}{156} + \frac{(24)^2}{156} + \frac{(8)^2}{78} \\
&= 1.2821 + 3.1026 + 3.6923 + 0.8205 \\
&= 8.90
\end{aligned}$$

With df $= 4 - 1 = 3$, the critical value of χ^2 at the 0.05 level of significance from Appendix G is 7.815. Therefore the calculated value of the random variable falls in the area of rejection; H_0: Observed distribution = Expected distribution is rejected. What Mrs. Wilson observed is departing from what she expected too much to be a function of chance as much as 5% of the time.

In examining the observed distribution and expected distribution for each cell, it becomes clear that according to Mrs. Wilson's expected distribution too few students were leaving the auditorium through exits A and B; too many through exits C and D. She now has an empirical basis for restructuring the fire-drill evacuation procedures from the auditorium.

TWO-VARIABLE APPLICATION

Chi square is also used to analyze the *joint frequency distribution* of two variables. Such an application is often called a **test of independence**, which, as the name implies is a test to determine if the levels of one variable are independent of the levels of the other. This application is identical to determining if the expected joint frequency distribution is significantly different from an observed joint frequency distribution.

To illustrate how the expected frequencies are computed, consider a generalized 3 by 3 *contingency table*. Recall that a **contingency table** is a table with rows and columns that shows the joint frequency distribution of two variables. The following table uses lowercase letters to represent the observed joint frequencies within the *cells* (a **cell** in this case being the intersection of a row and a column). Capital letters are used to designate the row and column marginals. **Marginals** are the sums of the frequencies in the rows and columns. *T* is the total number of frequencies in the table.

OBSERVED

				Row Marginals (Totals)
	a	b	c	L
	d	e	f	M
	g	h	i	N
Column Marginals (Totals)	P	Q	S	T

The corresponding expected frequencies are computed using the marginals from the observed frequency distribution, as follows:

EXPECTED

$\frac{LP}{T}$	$\frac{LQ}{T}$	$\frac{LS}{T}$
$\frac{MP}{T}$	$\frac{MQ}{T}$	$\frac{MS}{T}$
$\frac{NP}{T}$	$\frac{NQ}{T}$	$\frac{NS}{T}$

To obtain the expected frequency for the upper-left cell, L is multiplied by P and the product is divided by T. Similarly, the expected value for each cell is computed by the respective formulas, or you can multiply the respective row and column marginals for a cell and then divide by the total number of frequencies. As would be anticipated, this procedure results in theoretical joint frequencies that are proportional to both the respective row and the column in which the cell is located.

Consider two variables: grade level in high school and preference for a particular brand of computer. Suppose a survey asked students to identify their grade level and the brand of computer they preferred. The hypothetical observed results are as follows:

OBSERVED

	Preferred Computer Brand			
	A	B	C	Row Marginals
Sophomore	22	30	16	68
Junior	20	25	10	55
Senior	12	15	34	61
Column Marginals	54	70	60	Total = 184

Suppose the null hypothesis of no difference between the observed and expected frequencies will be tested at the 0.01 level of significance. The expected frequencies would be calculated as follows.

EXPECTED

	Preferred Computer Brand		
	Brand A	Brand B	Brand C
Sophomore	$\frac{(68)(54)}{184}$	$\frac{(68)(70)}{184}$	$\frac{(68)(60)}{184}$
Junior	$\frac{(55)(54)}{184}$	$\frac{(55)(70)}{184}$	$\frac{(55)(60)}{184}$
Senior	$\frac{(61)(54)}{184}$	$\frac{(61)(70)}{184}$	$\frac{(61)(60)}{184}$

When the terms are simplified, the expected frequencies are as follows:

	Preferred Computer Brand		
	Brand A	Brand B	Brand C
Sophomore	19.9565	25.8696	22.1739
Junior	16.1413	20.9239	17.9348
Senior	19.9022	23.2065	19.8913

Because the expected frequencies are only intermediate results to be used in subsequent calculations, it is advisable to carry to four or more decimal-place accuracy in the chi square computation. Just as with the one-variable chi square, a cell-by-cell calculation is necessary to get the cumulative contribution of each cell to the value of χ^2, as shown.

$$\begin{aligned}\chi^2 = &\frac{(22 - 19.9565)^2}{19.9565} + \frac{(30 - 25.8695)^2}{25.8695} + \frac{(16 - 22.1739)^2}{22.1739} \\ &+ \frac{(20 - 16.1413)^2}{16.1413} + \frac{(25 - 20.9239)^2}{20.9239} + \frac{(10 - 17.9348)^2}{17.9348} \\ &+ \frac{(12 - 17.9022)^2}{17.9022} + \frac{(15 - 23.2065)^2}{23.2065} + \frac{(34 - 19.8913)^2}{19.8913} \\ = &\ 0.2092 + 0.6595 + 1.7190 + 0.9225 + 0.7940 + 3.5106 \\ &+ 1.9459 + 2.9021 + 10.0072 \\ = &\ 22.67\end{aligned}$$

To determine the critical value of χ^2, the degrees of freedom for a contingency table must be computed by $(R - 1)(C - 1)$ where R is the number of rows and C is the number of columns in the table. For the present example, df = $(3 - 1)(3 - 1) = (2)(2) = 4$. The critical χ^2 with 4 df at the 0.01 level, as shown in Appendix G, is 13.277. Thus the calculated value, $\chi^2 = 22.67$, is statistically significant at the 0.01 level ($p < 0.01$).

In light of the evidence, it appears that computer preference *is not independent* of grade level. Or, to phrase the result in different terms, grade level *is related to* computer brand preference. More specifically, far more seniors selected brand C and slightly more sophomores and juniors preferred brands A and B than would be expected based on the observations.

YATES' CORRECTION

As mentioned earlier, the chi square formula may need modification under some conditions. The reason for this is that the chi square distribution is continuous while the levels of the variable or variables in a chi square table are generally discrete (discontinuous). **Yates' correction**, called the correction for discontinuity, compensates for error introduced when comparing a calculated chi square based on discrete data to continuous tabled (critical) values of chi square. This biased error causes the calculated value of chi square to be slightly inflated. Generally this error is negligible and can be ignored because it is usually of no consequence.

However, if one or both of the following conditions are present in a problem, an adjustment should be made to the chi square calculation formula. These conditions are:

1. df = 1, and/or
2. more than 20% of the cells have an expected frequency of less than 5.

The adjustment is called Yates' correction and is shown by:

$$\chi^2 = \sum \frac{(|O - E| - 0.5)^2}{E}$$

As an illustration of Yates' correction, consider a problem with df = 1. Suppose you flip a coin 100 times and get 42 heads and 58 tails. Use a chi square to test the fairness of the coin. That is, see if these data provide evidence that the coin is biased to yield either more heads or more tails than would be expected to occur by chance 5% of the time.

O	E	$(\|O - E\| - 0.5)$	$(\|O - E\| - 0.5)^2$
42	50	7.5	1.25
58	50	7.5	1.25
			$\chi^2 = 2.25$

As shown, χ^2 using Yates' correction is 2.25. This value is less than the critical value of 3.841. Therefore the evidence does not indicate an "unfair" coin; that is, a deviation this large from the 50-50 split would occur more than 5% of the time just by chance.

COMPUTER FOCUS 16—Testing Distributions with Chi Square: The Validity of Two Computer-Generated Distributions

Purpose

How close are pseudo-random number distributions to "real" random number distributions? That question probably cannot be answered directly; however, the distribution of random integers is known. That knowledge provides a hint about one method of testing the goodness of fit between the observed (pseudo-random integers) distributions and the expected (random integers) distributions. Therefore the purpose of this Computer Focus is to apply chi square to test such a comparison.

Background

In virtually all versions of the BASIC language, one can generate a random variable X, such that $0 < X < 1.0$, with a function RND(1). It follows that:

`10*RND(1)` results in random X: $0 < X < 10$

`100*RND(1)` results in random X: $0 < X < 100$

`1000*RND(1)` results in random X: $0 < X < 1{,}000$

and so on.

Further, INT(X), where X is any real number, truncates to the integer portion as follows:

`X = INT(8)` implies that $X = 8$,

`X = INT(8.26)` implies that $X = 8$,

`X = INT(8.97)` implies that $X = 8$.

Therefore INT(10*RND(1)) results in random X such that X is an integer and $0 \leq X \leq 9$, and INT(100*RND(1)) results in an integer $0 \leq X \leq 99$. Hence what happens if:

```
X = INT(RND(1)) + 1?

X = INT(10*RND(1)) + 1?

X = INT(100*RND(1)) + 1?

X = INT(100*RND(1)) + 50?
```

By varying the coefficient of the RND function and the value added to the INT function, you can vary the range of random integers. Computer-generated random numbers may be suspect because they are actually computed. Thus the two problems relate to the validity of computer generated distributions.

Problem 1

Write a BASIC program that will generate 500 "random" numbers from 1 to 10. Have the program accumulate the number of times each integer is generated. Apply a chi square goodness-of-fit test to determine whether or not the distribution is departing significantly from the expected random distribution. (See Application Exercise 16.10.)

Problem 2

In the previous Computer Focus you wrote a program that would generate an approximation of the normal curve. Using the same algorithm, have the computer generate a distribution of 500 numbers. Have the computer compute the mean and standard deviation of the numbers. Using Appendix C, you can determine how many values in a normal distribution would be expected to fall above one standard deviation above the mean, between the mean and one standard deviation above and below the mean, and below one standard deviation below the mean. Compare these expected values with the actual observed values using a chi square to determine the adequacy of the algorithm for generating a normal curve.

KEY TERMS AND NOTATION

Cell	Joint frequency
Chi square	Marginals
Contingency table	Nonparametric
Expected frequency	Observed frequency
Frequency	Test of independence
Goodness of fit	Yates' correction
Independence	

APPLICATION EXERCISES

16.1 Consider a chi square design with one variable and 10 cells.
a. How many degrees of freedom are associated with the problem?
b. Suppose the calculated value of chi square is 8.97. Is this significant at the 0.05 level?

16.2 If the calculated value of chi square is 10.93 for a one-variable design with six cells:
a. Is the null hypothesis rejected at the 0.05 level?
b. Is the null hypothesis rejected at the 0.01 level?

16.3 Find the critical (tabled) value of a chi square with 1 degree of freedom
a. at the 0.01 level.
b. at the 0.05 level.
c. Determine the square roots of the values of a. and b.
d. Compare the values obtained in c. with the 0.05 and 0.01 levels on a normal distribution of z-scores.
e. What does the relationship between chi square with df = 1 and z^2 appear to be?

16.4 A playing card is randomly selected from a well-shuffled stack of cards, the suit (diamond, spade, heart, or club) is noted, the card is replaced, and the experiment is repeated for 100 trials. The results are:

Diamonds	20
Spades	28
Hearts	25
Clubs	27

Does this distribution differ from what would be expected in a standard deck of playing cards at the 0.05 level of significance?

16.5 Eighty students were asked their preference about exams. Of the 80 responses, 18 preferred essay exams, 25 preferred matching items, 20 preferred multiple-choice tests, and 17 preferred short-answer items. Test at the 0.05 level the null hypothesis that the four types of exams are equally preferred in the population of students.

16.6 The junior class decided to sponsor a dance for the entire high school student body. They polled 120 students regarding preferences of nights on which to schedule the dance. The responses were tabulated as follows:

Preferred Night	Frequency
Friday	45
Saturday	50
Sunday	25

Test at the 0.05 level the hypothesis that the three nights are equally popular in the population.

16.7 On a true-false test of 100 items, a student gets 62 correct. Does this differ from what he or she would expect if the questions were answered randomly? Use Yates' correction and test at the 0.05 level of significance.

16.8 The enrollment of sophomores, juniors, and seniors in a particular school is in the ratio of 2:3:1. In a purportedly random sample of 180 students from the student body, the sample consisted of 63 sophomores, 84 juniors, and 33 seniors. Is there reason to doubt the randomness of the sampling procedure at the 0.05 level?

16.9 The ratio of hits to runs to errors by a high school baseball team for a season was 6:2:1. During the playoff tournament the team got 32 hits, scored 8 runs, and committed 5 errors. Using an alpha of 0.05, do these data indicate that the team played differently during the tournament than during the regular season?

16.10 A particular microcomputer was programmed to generate 100 random integers from 1 to 10. The distribution on the printout revealed the following frequencies:

Number	Frequency	Number	Frequency
1	12	6	8
2	11	7	12
3	8	8	9
4	15	9	10
5	7	10	8

Is there reason (at the 0.05 level) to doubt the validity of the random number algorithm used to generate the integers?

16.11 A 4 by 2 contingency table yields a chi square value of 5.68. Is this value significant at the 0.05 level?

16.12 A 6 by 4 contingency table of joint frequency has a chi square value of 30.62. Is this value significant at the 0.05 level?

16.13 Male and female students were assessed to determine their attitudes toward importance of the high school band. The results were as follows:

	Very Important	Moderately Important	Of Little Value
Male	30	20	5
Female	20	20	10

a. How many female students were involved in the sample?
b. How many males in the sample felt that the band was very important?
c. Compute the chi square value and determine if attitude toward the importance of the band is independent of student gender at the 0.05 level of significance.

16.14 Compute the value of the chi square and test the null hypothesis at the 0.05 level. Interpret each result.

a.

		School Level		
		Elementary	Junior High	Senior High
Grades Received	A or B	100	80	120
	C	130	90	110
	D or F	60	30	20

b.

Grades Received		Personality Type: Extroverted	Introverted
	A or B	60	40
	C	50	50
	D or F	50	46

c.

Gender		Prefer to Date Blondes?: Yes	Undecided	No
	Male	10	40	12
	Female	14	20	22

16.15 For the three chi square problems that follow, tell why Yates' correction should be used.

a.

	Left-Handed	Right-Handed
Observed	28	43

b.

Test Results		Answered First Item on Test: Right	Wrong
	Pass	16	2
	Fail	8	5

c.

Member of Student Politics Club		Preferred Political Party: Democrat	Republican	Independent	Undecided
	Yes	10	10	4	3
	No	8	9	2	2

16.16 The relationship between smoking and cancer was investigated for 100 recent deaths. The data were:

	Smoker	Nonsmoker
Cancer Present	34	15
Cancer Not Present	19	32

Use Yates' correction and determine if a relationship exists between the two variables at the 0.01 level of significance.

16.17 A school social worker studied the relationship between school absence and social class. Data were tabulated from school records for a one-semester period.

	Absence			
	None	1 Day	2 Days	3+ Days
Upper Class	20	16	10	4
Middle Class	12	9	5	3
Lower Class	8	5	4	3

Is social class related to absenteeism at the 0.05 level?

16.18 The curriculum director for a school district wished to know if students at various grade levels preferred different types of classes. She collected data from a sample of underclassmen (freshmen and sophomores) and upperclassmen (juniors and seniors):

	Preference	
	Lecture	Discussion
Underclassmen	85	90
Upperclassmen	97	95

Is preference for a particular type of class independent of grade level at the 0.05 level?

APPENDICES

APPENDIX A

Answers to Selected Application Exercises

CHAPTER 1

1.1 "Data is the plural form; "datum" is singular.

1.3 No; rank order (ordinal) data do not permit a determination of how much difference exists between two cases.

1.5 Inferential; only a portion (not the entire universe) of the many companies would be studied; an inference would be made about companies in general.

1.7 The Swiss cheese calorie *count* is more accurate than the *estimated* number of skin cancer incidences. Therefore, if "truthfulness" is equated with "accuracy," the first instance is probably closer to the truth. However, the projection is probably more important relative to public safety.

1.9 Sample

1.11 Summary descriptive statistics

CHAPTER 2

2.1 **Figure 1**—Importance of Extracurricular Activities

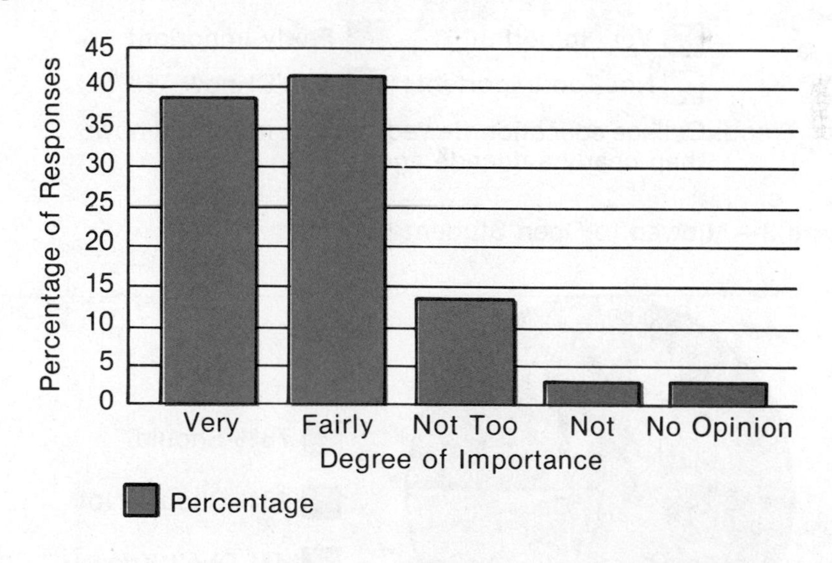

2.3 **Figure 2**—Importance of a College Education

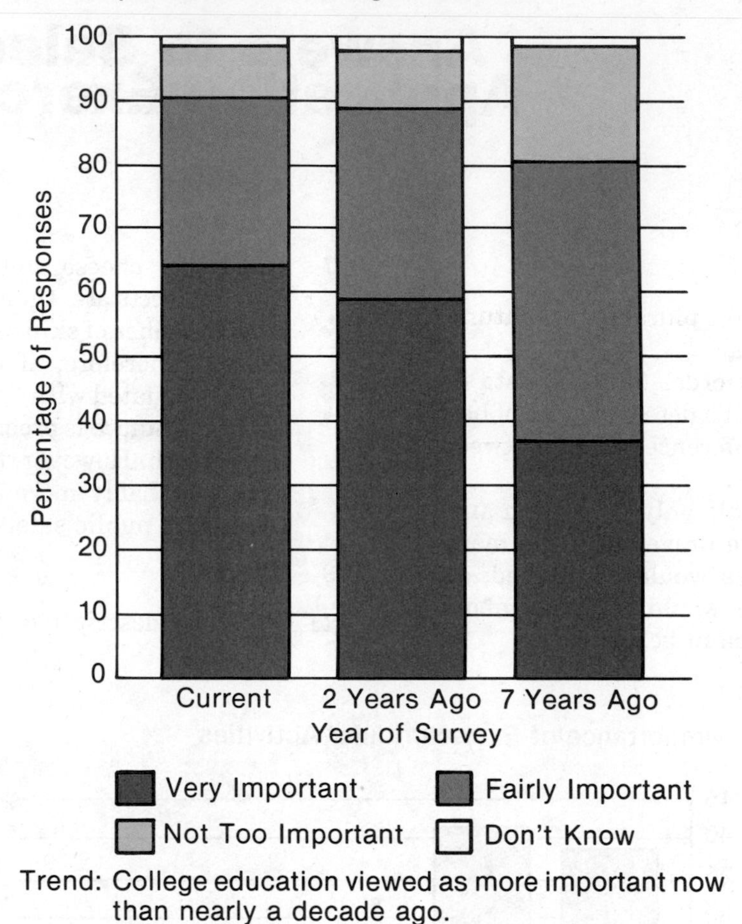

Trend: College education viewed as more important now than nearly a decade ago.

2.5 **Figure 3**—Allowed to Open Student Lockers

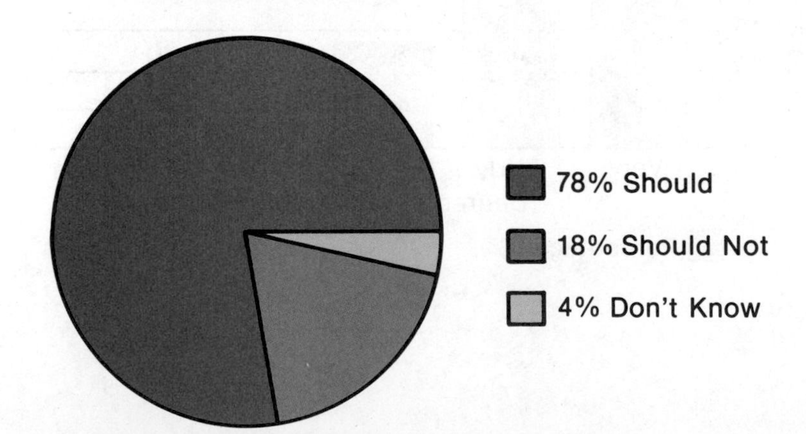

2.7 **Figure 4**—TV Viewing Time per Week

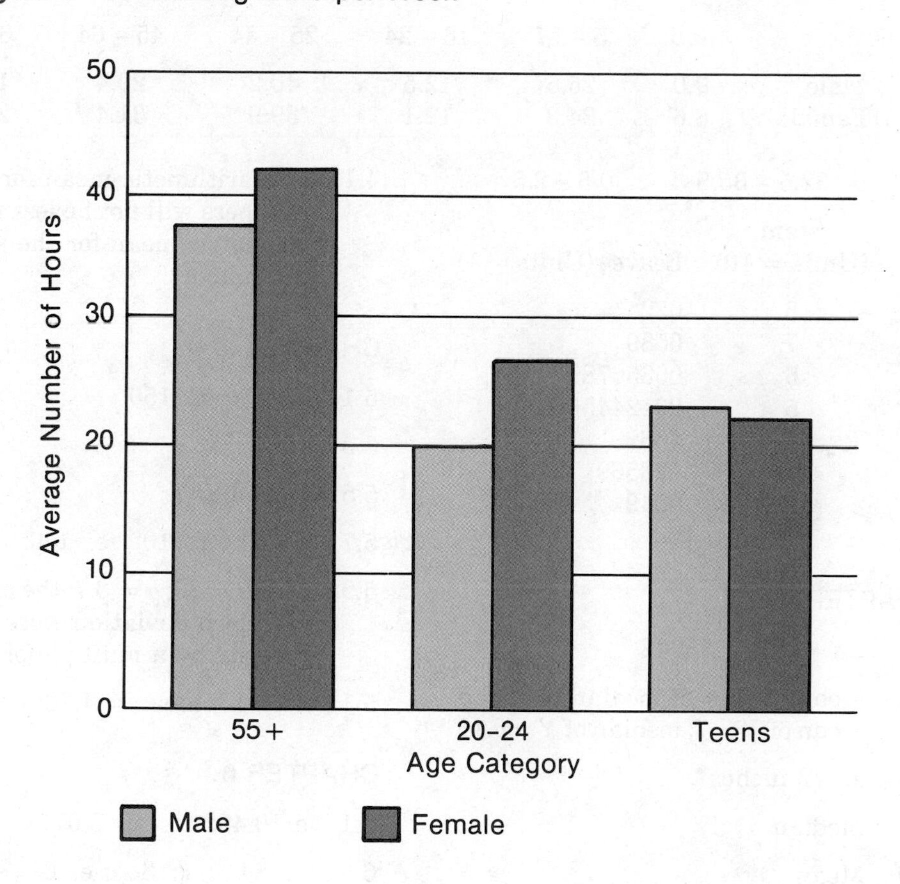

2.15 Grouped bar graph

CHAPTER 3

3.1

X	f	cf	rf	rcf
94	1	50	0.02	1.00
93	1	49	0.02	0.98
92	2	48	0.04	0.96
91	3	46	0.06	0.92
90	9	43	0.18	0.86
89	14	34	0.28	0.68
88	10	20	0.20	0.40
87	5	10	0.10	0.20
86	2	5	0.04	0.10
85	2	3	0.04	0.06
84	1	1	0.02	0.02

3.5

	Age Intervals					
	<5	5 – 17	18 – 24	25 – 44	45 – 64	65+
Male	9.0	25.5	12.5	40.2	29.4	13.7
Female	8.6	24.3	12.1	39.9	31.4	21.3

3.7 a. 32.5 – 35.5 c. 0.5 – 2.5

3.9

Stem (Units = 10)	Leaves (Units = 1)
8	022
7	0589
6	00356789
5	0012445689
4	4569
3	123569
2	0049
1	2

CHAPTER 4

4.1 $\Sigma X = 45$; $\Sigma Y = 42$

4.3 Mean of $X = 5$; median of $X = 5$
Mean of $Y = 7$; median of $Y = 6.15$

4.5 a. 72 inches

4.7 Median

4.9 Mean

4.11 a. Mean = 19.20
Median = 19
Mode = 18

4.13 Median because of extreme class size meeting in the gym

4.14 $\Sigma(X - 5) = 0$; the mean of the given set of numbers is 5. The sum of the deviation scores about the arithmetic mean is always equal to zero. The mean is the "center of gravity"; thus there is the same amount of deviation in a given set of numbers below the mean as there is above the mean.

4.16 22.22 miles per hour

4.18 The arithmetic mean for a set of numbers will not be less than the geometric mean for the same set of numbers.

CHAPTER 5

5.1 a. 2 c. 150

5.3 100

5.5 Variable Y

5.7 a. 10 c. 10 e. 1.43 g. 1.85

5.9 Mean = 20; $\sigma = 3.7$; the mean and standard deviation were both increased by a multiple of 2.

5.11 $N = 13$; mean = 4.23; $\sigma = 1.12$

CHAPTER 6

6.1 a. 0.4222 c. 0.0778

6.3 a. 84 c. 3 e. 6

6.5 a. 0 c. 1.50 e. −1.70

6.7 a. 216 c. 69

6.9 65

6.11 ±0.67

6.13 38

6.15 Sally

6.17 0.2266

6.21 $Q_1 = 90$; $Q_2 = 98.5$; $Q_3 = 109$
Interquartile range:
$I = Q_3 - Q_1 = 109 - 90 = 19$
Range of adjacent values:
$(1.5)(19) = 28.5$
Outliers: 60, 143

CHAPTER 7

7.1 $\frac{1}{52}$

7.3 $\frac{1}{1,200}$ if computer club exists; otherwise zero

7.5 $\frac{8}{25}$

7.7 $\frac{17}{25}$

7.9 $\frac{1}{8}$

7.11 a. $\frac{10}{19}$ b. 10 to 9

7.13 $\frac{21}{200} = 0.105 \approx 0.11$

7.14 c. 0.57

7.16 $\frac{2}{3}$

7.18 a. $\frac{7}{8}$ b. 1.00

7.20 a. 10 combinations
20 permutations
c. 70 combinations
1,680 permutations

7.22 0.10

7.24 Combination

7.26 0.29

7.28 Less than $\frac{225}{900}$ or ≤ 0.25

CHAPTER 8

8.1 $\frac{1}{16} + \frac{4}{16} + \frac{6}{16} + \frac{4}{16} + \frac{1}{16}$

8.3 $\frac{7}{128}$ or 0.055

8.5 a. $\frac{15}{64}$ b. $\frac{11}{32}$

8.7 a. 0.0250 c. 0.4430

8.9 a. 0.2514 c. 0.9078

8.11 a. 0.27 c. 0.18

8.13 a. 0.184 b. 0.18867

8.15 a. 4 b. 2

8.17 a. 0.0183 c. 0.1952

CHAPTER 9

9.3 No; all balloons did not have an equal opportunity of being selected, because some were too high (out of reach).

9.5 Randomly sample 322 fifth-grade students and administer a reading ability test.

9.7 No; probably not; volunteers cannot be assumed to be representative of any population except a population of volunteers, because the motivation that influenced them to volunteer makes them different from the remainder of the population.

9.9 One method would be to sample an appropriate number of pages at random and determine the average number of words per page on the random sample; then extrapolate (generalize) to the entire book.

CHAPTER 10

10.1 a. 5 c. 2 e. 1.25 g. 0.45

10.3 a. 50 b. 64

10.5 a. Standard error of the median by a factor of 1.25 times
b. Mean; less variation from sample to sample; more reliable

10.7 a. 95%
c. 95% confidence interval = {19.424 – 21.776}
99% confidence interval = {19.052 – 22.148}

10.9 a. 0.3446 or 34% b. 68%

CHAPTER 11

11.1 a. 1.64 (or 1.65) one-tailed test; 1.96 two-tailed test
c. 3.08 (or 3.09 or 3.10) one-tailed test; 3.30 two-tailed test

11.3 a. 1% b. 0

11.5 a. $p < 0.0001$
b. Reject; $p < 0.05$

11.7 Yes; $p < 0.01$

11.11 a. Larger N's imply smaller standard errors of the mean, which in turn imply larger z's
b. Larger z's imply smaller p's—more likely to reject the null hypothesis.

CHAPTER 12

12.1 a. − c. + e. +
g. − i. −

12.3 a. −1.00
b. All points are located on a line with a negative slope.

12.5 0.349 or 0.35 (or larger)

12.7 0.75

12.9 {0.20 – 0.64}

12.11 a. 0.94
c. $r^2 = 0.88$ or 88%

12.13 a. Rank order correlation = 0.89
b. Yes

CHAPTER 13

13.3 a. $Y' = 3.00 + 0.50X$
c. $Y' = -4.20 + (-1.60)X$

13.5 a. 1.4 c. 1 e. −1.6

13.7 a. $\Sigma X = 146$
$\Sigma Y = 99$
$\Sigma X^2 = 2{,}868$
$\Sigma Y^2 = 1{,}309$
$\Sigma XY = 1{,}881$
$\overline{X} = 14.6$
$\overline{Y} = 9.9$
c. $r = 0.89$; $a = 1.26$; $b = 0.59$

13.9 0.30

13.11 Least squares $\Sigma e^2 = 1{,}535.17$
Median fit $\Sigma e^2 = 1{,}791.466$
Hence the least-squares prediction is more accurate (that is, less squared error) than the median fit technique. The least-squares method minimizes the squared error.

CHAPTER 14

14.1 a. $F = 4.00$; $p < 0.05$; significantly different
c. $F = 1.25$; $p > 0.05$; not significantly different

14.3 a. $t = 1.07$
c. $p > 0.05$; no significant difference

14.5 $t = 4.76$; $p < 0.01$; reject the null hypothesis.

14.7 $F = 1.34$; therefore use pooled variance t-test. $t = 3.62$; group B outperformed group A at the 0.05 level of significance.

14.9 $t = 0.19$; df = 9; no significant difference; the theory is not supported.

CHAPTER 15

15.1 a. $F = 4.50$; reject null hypothesis ($p < 0.05$).
c. $df_T = 14$; $SS_B = 84$;
$SS_W = 216$; $MS_B = 42$
$F = 2.33$; $p > 0.05$; do not reject H_0.

15.3 $F = 6.94$; reject H_0 ($p < 0.05$).
Group I vs. group II
$F = 1.17$ ($p > 0.05$)
Group I vs. group III
$F = 6.87$ ($p < 0.05$)
Group II vs. group III
$F = 2.32$ ($p > 0.05$)
Group III outperformed group I; no other differences.

15.5 a. A vs. B; $F = 12.68$; $p < 0.05$
A vs. C; $F = 29.65$; $p < 0.05$
B vs. C; $F = 3.55$; $p < 0.05$
Group C was significantly higher than both group A and group B; group B was significantly higher than group A.
c. A vs. B; $F = 0.19$; $p > 0.05$
A vs. C; $F = 9.19$; $p < 0.05$
B vs. C; $F = 6.75$; $p < 0.05$
Group C was significantly higher than group A and group B; no other differences.

CHAPTER 16

16.1 a. 9 b. No; $p > 0.05$

16.3 a. 6.6349
c. 2.58
e. Chi square with 1 df = z^2

16.5 Chi square = 1.90; $p > 0.05$; it is plausible that the types of exams are equally preferred.

16.7 Chi square (with Yates' correction) = 5.29; $p < 0.05$; the answers were probably not selected randomly.

16.9 Chi square (with Yates' correction) = 0.35; $p > 0.05$; the team's performance was the same and not significantly different from the usual performance.

16.11 No; $p > 0.05$

16.13 a. 50
c. Chi square = 3.44; $p > 0.05$; attitude toward band is independent of (not related to) gender.

16.15 a. 1 df
c. More than 20% of E-cells have frequency < 5.

16.17 Chi square (with Yates' correction) = 0.37; $p > 0.05$; social class is independent of absenteeism.

APPENDIX B

$e^{-\lambda}$ Values

Values of $e^{-\lambda}$ for $0 \leq \lambda < 1.00$

Tenths (First Decimal)	Hundredths (Second Decimal)									
λ	0	1	2	3	4	5	6	7	8	9
0.0	1.00000	0.99005	0.98020	0.97045	0.96079	0.95123	0.94176	0.93239	0.92312	0.91393
0.1	0.90484	0.89583	0.88692	0.87810	0.86936	0.86071	0.85214	0.84366	0.83527	0.82696
0.2	0.81873	0.81058	0.80252	0.79453	0.78663	0.77880	0.77105	0.76338	0.75578	0.74826
0.3	0.74082	0.73345	0.72615	0.71892	0.71177	0.70469	0.69768	0.69073	0.68386	0.67706
0.4	0.67032	0.66365	0.65705	0.65051	0.64404	0.63763	0.63128	0.62500	0.61878	0.61263
0.5	0.60653	0.60050	0.59452	0.58861	0.58275	0.57695	0.57121	0.56553	0.55990	0.55433
0.6	0.54881	0.54335	0.53794	0.53259	0.52729	0.52205	0.51685	0.51171	0.50662	0.50158
0.7	0.49659	0.49164	0.48675	0.48191	0.47711	0.47237	0.46767	0.46301	0.45841	0.45384
0.8	0.44933	0.44486	0.44043	0.43605	0.43171	0.42741	0.42316	0.41895	0.41478	0.41066
0.9	0.40657	0.40252	0.39852	0.39455	0.39063	0.38674	0.38289	0.37908	0.37531	0.37158

Values of $e^{-\lambda}$ for $\lambda = 1, 2, 3, \ldots, 10$

λ	1	2	3	4	5	6	7	8	9	10
$e^{-\lambda}$	0.36788	0.13534	0.04979	0.01832	0.00674	0.00248	0.00091	0.00034	0.00012	0.00005

Note: By using the laws of exponents, other values of $e^{-\lambda}$ can be obtained as follows:

$$e^{-2.59} = [e^{(-2)}][e^{(-0.59)}] = (0.13534)(0.55433) = 0.07502$$

APPENDIX C
Normal Curve Areas

Because of the normal curve's symmetry about the mean of zero, only positive z-scores are shown in this appendix. Proportional areas defined by negative z-score values are the same as those defined by the positive values.

Column A. The proportion of the total area under the curve between the mean ($z = 0$) and $\pm z$.

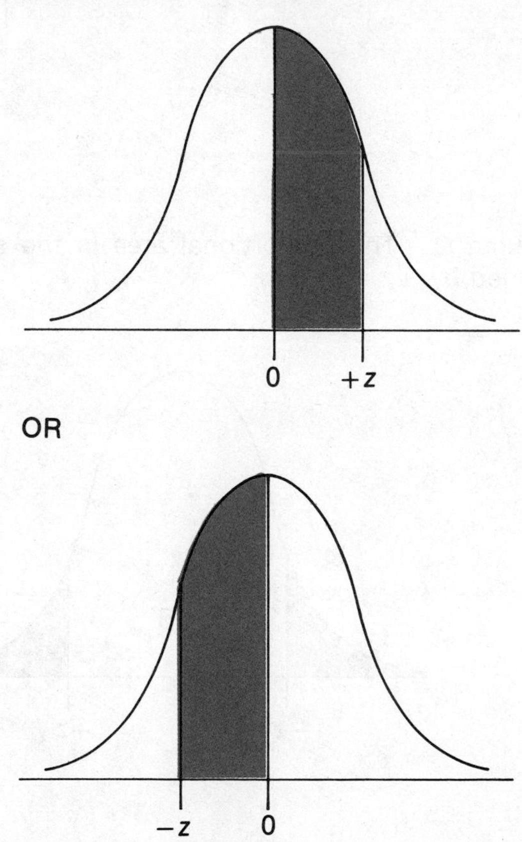

Column B. The proportion of the entire area under the curve in the larger part defined by $\pm z$.

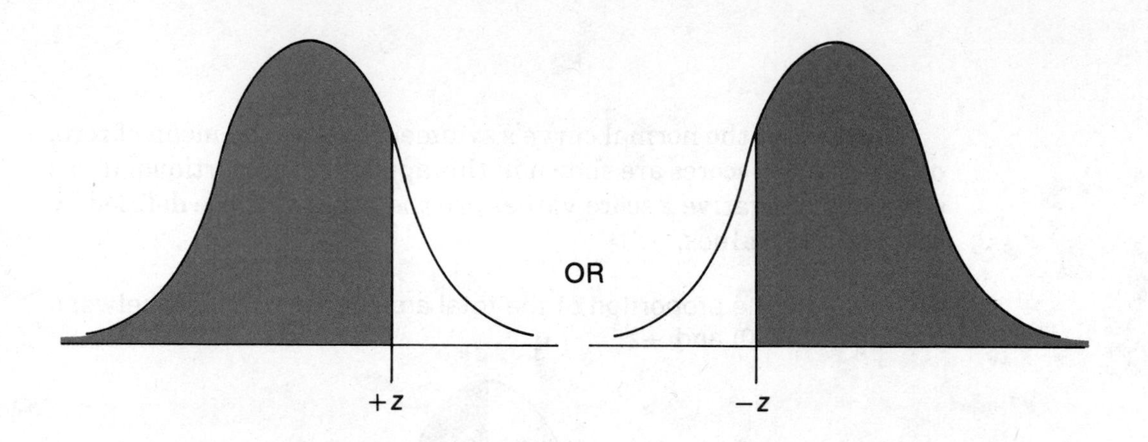

Column C. The proportional area in the smaller part of the curve defined by $\pm z$.

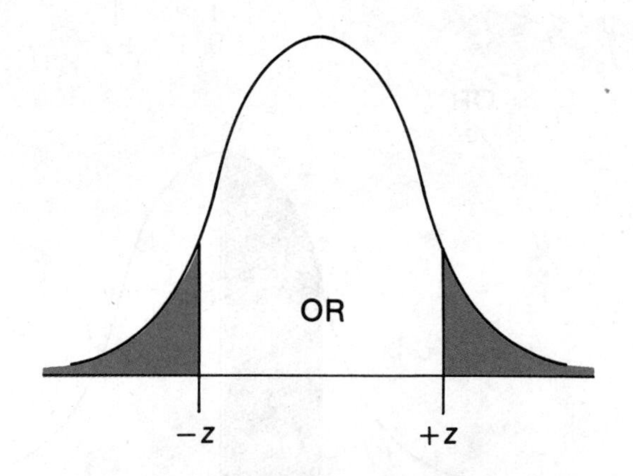

(1) z Standard Score ($\frac{x}{\sigma}$)	(2) A Area from Mean to $\frac{x}{\sigma}$	(3) B Area in Larger Portion	(4) C Area in Smaller Portion
0.00	.0000	.5000	.5000
0.01	.0040	.5040	.4960
0.02	.0080	.5080	.4920
0.03	.0120	.5120	.4880
0.04	.0160	.5160	.4840
0.05	.0199	.5199	.4801
0.06	.0239	.5239	.4761
0.07	.0279	.5279	.4721
0.08	.0319	.5319	.4681
0.09	.0359	.5359	.4641
0.10	.0398	.5398	.4602
0.11	.0438	.5438	.4562
0.12	.0478	.5478	.4522
0.13	.0517	.5517	.4483
0.14	.0557	.5557	.4443
0.15	.0596	.5596	.4404
0.16	.0636	.5636	.4364
0.17	.0675	.5675	.4325
0.18	.0714	.5714	.4286
0.19	.0753	.5753	.4247
0.20	.0793	.5793	.4207
0.21	.0832	.5832	.4168
0.22	.0871	.5871	.4129
0.23	.0910	.5910	.4090
0.24	.0948	.5948	.4052
0.25	.0987	.5987	.4013
0.26	.1026	.6026	.3974
0.27	.1064	.6064	.3936
0.28	.1103	.6103	.3897
0.29	.1141	.6141	.3859
0.30	.1179	.6179	.3821
0.31	.1217	.6217	.3783
0.32	.1255	.6255	.3745
0.33	.1293	.6293	.3707
0.34	.1331	.6331	.3669
0.35	.1368	.6368	.3632
0.36	.1406	.6406	.3594
0.37	.1443	.6443	.3557
0.38	.1480	.6480	.3520
0.39	.1517	.6517	.3483
0.40	.1554	.6554	.3446
0.41	.1591	.6591	.3409
0.42	.1628	.6628	.3372
0.43	.1664	.6664	.3336
0.44	.1700	.6700	.3300
0.45	.1736	.6736	.3264
0.46	.1772	.6772	.3228
0.47	.1808	.6808	.3192
0.48	.1844	.6844	.3156
0.49	.1879	.6879	.3121
0.50	.1915	.6915	.3085
0.51	.1950	.6950	.3050
0.52	.1985	.6985	.3015
0.53	.2019	.7019	.2981
0.54	.2054	.7054	.2946
0.55	.2088	.7088	.2912
0.56	.2123	.7123	.2877
0.57	.2157	.7157	.2843
0.58	.2190	.7190	.2810
0.59	.2224	.7224	.2776
0.60	.2257	.7257	.2743
0.61	.2291	.7291	.2709
0.62	.2324	.7324	.2676
0.63	.2357	.7357	.2643
0.64	.2389	.7389	.2611
0.65	.2422	.7422	.2578
0.66	.2454	.7454	.2546
0.67	.2486	.7486	.2514
0.68	.2517	.7517	.2483
0.69	.2549	.7549	.2451
0.70	.2580	.7580	.2420
0.71	.2611	.7611	.2389
0.72	.2642	.7642	.2358
0.73	.2673	.7673	.2327
0.74	.2704	.7704	.2296
0.75	.2734	.7734	.2266
0.76	.2764	.7764	.2236
0.77	.2794	.7794	.2206
0.78	.2823	.7823	.2177
0.79	.2852	.7852	.2148
0.80	.2881	.7881	.2119
0.81	.2910	.7910	.2090
0.82	.2939	.7939	.2061
0.83	.2967	.7967	.2033
0.84	.2995	.7995	.2005
0.85	.3023	.8023	.1977
0.86	.3051	.8051	.1949
0.87	.3078	.8078	.1922
0.88	.3106	.8106	.1894
0.89	.3133	.8133	.1867

(1) z Standard Score $\left(\frac{x}{\sigma}\right)$	(2) A Area from Mean to $\frac{x}{\sigma}$	(3) B Area in Larger Portion	(4) C Area in Smaller Portion	(1) z Standard Score $\left(\frac{x}{\sigma}\right)$	(2) A Area from Mean to $\frac{x}{\sigma}$	(3) B Area in Larger Portion	(4) C Area in Smaller Portion
0.90	.3159	.8159	.1841	1.35	.4115	.9115	.0885
0.91	.3186	.8186	.1814	1.36	.4131	.9131	.0869
0.92	.3212	.8212	.1788	1.37	.4147	.9147	.0853
0.93	.3238	.8238	.1762	1.38	.4162	.9162	.0838
0.94	.3264	.8264	.1736	1.39	.4177	.9177	.0823
0.95	.3289	.8289	.1711	1.40	.4192	.9192	.0808
0.96	.3315	.8315	.1685	1.41	.4207	.9207	.0793
0.97	.3340	.8340	.1660	1.42	.4222	.9222	.0778
0.98	.3365	.8365	.1635	1.43	.4236	.9236	.0764
0.99	.3389	.8389	.1611	1.44	.4251	.9251	.0749
1.00	.3413	.8413	.1587	1.45	.4265	.9265	.0735
1.01	.3438	.8438	.1562	1.46	.4279	.9279	.0721
1.02	.3461	.8461	.1539	1.47	.4292	.9292	.0708
1.03	.3485	.8485	.1515	1.48	.4306	.9306	.0694
1.04	.3508	.8508	.1492	1.49	.4319	.9319	.0681
1.05	.3531	.8531	.1469	1.50	.4332	.9332	.0668
1.06	.3554	.8554	.1446	1.51	.4345	.9345	.0655
1.07	.3577	.8577	.1423	1.52	.4357	.9357	.0643
1.08	.3599	.8599	.1401	1.53	.4370	.9370	.0630
1.09	.3621	.8621	.1379	1.54	.4382	.9382	.0618
1.10	.3643	.8643	.1357	1.55	.4394	.9394	.0606
1.11	.3665	.8665	.1335	1.56	.4406	.9406	.0594
1.12	.3686	.8686	.1314	1.57	.4418	.9418	.0582
1.13	.3708	.8708	.1292	1.58	.4429	.9429	.0571
1.14	.3729	.8729	.1271	1.59	.4441	.9441	.0559
1.15	.3749	.8749	.1251	1.60	.4452	.9452	.0548
1.16	.3770	.8770	.1230	1.61	.4463	.9463	.0537
1.17	.3790	.8790	.1210	1.62	.4474	.9474	.0526
1.18	.3810	.8810	.1190	1.63	.4484	.9484	.0516
1.19	.3830	.8830	.1170	1.64	.4495	.9495	.0505
1.20	.3849	.8849	.1151	1.65	.4505	.9505	.0495
1.21	.3869	.8869	.1131	1.66	.4515	.9515	.0485
1.22	.3888	.8888	.1112	1.67	.4525	.9525	.0475
1.23	.3907	.8907	.1093	1.68	.4535	.9535	.0465
1.24	.3925	.8925	.1075	1.69	.4545	.9545	.0455
1.25	.3944	.8944	.1056	1.70	.4554	.9554	.0446
1.26	.3962	.8962	.1038	1.71	.4564	.9564	.0436
1.27	.3980	.8980	.1020	1.72	.4573	.9573	.0427
1.28	.3997	.8997	.1003	1.73	.4582	.9582	.0418
1.29	.4015	.9015	.0985	1.74	.4591	.9591	.0409
1.30	.4032	.9032	.0968	1.75	.4599	.9599	.0401
1.31	.4049	.9049	.0951	1.76	.4608	.9608	.0392
1.32	.4066	.9066	.0934	1.77	.4616	.9616	.0384
1.33	.4082	.9082	.0918	1.78	.4625	.9625	.0375
1.34	.4099	.9099	.0901	1.79	.4633	.9633	.0367

(1) *z* Standard Score ($\frac{x}{\sigma}$)	(2) *A* Area from Mean to $\frac{x}{\sigma}$	(3) *B* Area in Larger Portion	(4) *C* Area in Smaller Portion
1.80	.4641	.9641	.0359
1.81	.4649	.9649	.0351
1.82	.4656	.9656	.0344
1.83	.4664	.9664	.0336
1.84	.4671	.9671	.0329
1.85	.4678	.9678	.0322
1.86	.4686	.9686	.0314
1.87	.4693	.9693	.0307
1.88	.4699	.9699	.0301
1.89	.4706	.9706	.0294
1.90	.4713	.9713	.0287
1.91	.4719	.9719	.0281
1.92	.4726	.9726	.0274
1.93	.4732	.9732	.0268
1.94	.4738	.9738	.0262
1.95	.4744	.9744	.0256
1.96	.4750	.9750	.0250
1.97	.4756	.9756	.0244
1.98	.4761	.9761	.0239
1.99	.4767	.9767	.0233
2.00	.4772	.9772	.0228
2.01	.4778	.9778	.0222
2.02	.4783	.9783	.0217
2.03	.4788	.9788	.0212
2.04	.4793	.9793	.0207
2.05	.4798	.9798	.0202
2.06	.4803	.9803	.0197
2.07	.4808	.9808	.0192
2.08	.4812	.9812	.0188
2.09	.4817	.9817	.0183
2.10	.4821	.9821	.0179
2.11	.4826	.9826	.0174
2.12	.4830	.9830	.0170
2.13	.4834	.9834	.0166
2.14	.4838	.9838	.0162
2.15	.4842	.9842	.0158
2.16	.4846	.9846	.0154
2.17	.4850	.9850	.0150
2.18	.4854	.9854	.0146
2.19	.4857	.9857	.0143
2.20	.4861	.9861	.0139
2.21	.4864	.9864	.0136
2.22	.4868	.9868	.0132
2.23	.4871	.9871	.0129
2.24	.4875	.9875	.0125

(1) *z* Standard Score ($\frac{x}{\sigma}$)	(2) *A* Area from Mean to $\frac{x}{\sigma}$	(3) *B* Area in Larger Portion	(4) *C* Area in Smaller Portion
2.25	.4878	.9878	.0122
2.26	.4881	.9881	.0119
2.27	.4884	.9884	.0116
2.28	.4887	.9887	.0113
2.29	.4890	.9890	.0110
2.30	.4893	.9893	.0107
2.31	.4896	.9896	.0104
2.32	.4898	.9898	.0102
2.33	.4901	.9901	.0099
2.34	.4904	.9904	.0096
2.35	.4906	.9906	.0094
2.36	.4909	.9909	.0091
2.37	.4911	.9911	.0089
2.38	.4913	.9913	.0087
2.39	.4916	.9916	.0084
2.40	.4918	.9918	.0082
2.41	.4920	.9920	.0080
2.42	.4922	.9922	.0078
2.43	.4925	.9925	.0075
2.44	.4927	.9927	.0073
2.45	.4929	.9929	.0071
2.46	.4931	.9931	.0069
2.47	.4932	.9932	.0068
2.48	.4934	.9934	.0066
2.49	.4936	.9936	.0064
2.50	.4938	.9938	.0062
2.51	.4940	.9940	.0060
2.52	.4941	.9941	.0059
2.53	.4943	.9943	.0057
2.54	.4945	.9945	.0055
2.55	.4946	.9946	.0054
2.56	.4948	.9948	.0052
2.57	.4949	.9949	.0051
2.58	.4951	.9951	.0049
2.59	.4952	.9952	.0048
2.60	.4953	.9953	.0047
2.61	.4955	.9955	.0045
2.62	.4956	.9956	.0044
2.63	.4957	.9957	.0043
2.64	.4959	.9959	.0041
2.65	.4960	.9960	.0040
2.66	.4961.	9961	.0039
2.67	.4962	.9962	.0038
2.68	.4963	.9963	.0037
2.69	.4964	.9964	.0036

(1) z Standard Score ($\frac{x}{\sigma}$)	(2) A Area from Mean to $\frac{x}{\sigma}$	(3) B Area in Larger Portion	(4) C Area in Smaller Portion	(1) z Standard Score ($\frac{x}{\sigma}$)	(2) A Area from Mean to $\frac{x}{\sigma}$	(3) B Area in Larger Portion	(4) C Area in Smaller Portion
2.70	.4965	.9965	.0035	3.00	.4987	.9987	.0013
2.71	.4966	.9966	.0034	3.01	.4987	.9987	.0013
2.72	.4967	.9967	.0033	3.02	.4987	.9987	.0013
2.73	.4968	.9968	.0032	3.03	.4988	.9988	.0012
2.74	.4969	.9969	.0031	3.04	.4988	.9988	.0012
2.75	.4970	.9970	.0030	3.05	.4989	.9989	.0011
2.76	.4971	.9971	.0029	3.06	.4989	.9989	.0011
2.77	.4972	.9972	.0028	3.07	.4989	.9989	.0011
2.78	.4973	.9973	.0027	3.08	.4990	.9990	.0010
2.79	.4974	.9974	.0026	3.09	.4990	.9990	.0010
2.80	.4974	.9974	.0026	3.10	.4990	.9990	.0010
2.81	.4975	.9975	.0025	3.11	.4991	.9991	.0009
2.82	.4976	.9976	.0024	3.12	.4991	.9991	.0009
2.83	.4977	.9977	.0023	3.13	.4991	.9991	.0009
2.84	.4977	.9977	.0023	3.14	.4992	.9992	.0008
2.85	.4978	.9978	.0022	3.15	.4992	.9992	.0008
2.86	.4979	.9979	.0021	3.16	.4992	.9992	.0008
2.87	.4979	.9979	.0021	3.17	.4992	.9992	.0008
2.88	.4980	.9980	.0020	3.18	.4993	.9993	.0007
2.89	.4981	.9981	.0019	3.19	.4993	.9993	.0007
2.90	.4981	.9981	.0019	3.20	.4993	.9993	.0007
2.91	.4982	.9982	.0018	3.21	.4993	.9993	.0007
2.92	.4982	.9982	.0018	3.22	.4994	.9994	.0006
2.93	.4983	.9983	.0017	3.23	.4994	.9994	.0006
2.94	.4984	.9984	.0016	3.24	.4994	.9994	.0006
2.95	.4984	.9984	.0016	3.30	.4995	.9995	.0005
2.96	.4985	.9985	.0015	3.40	.4997	.9997	.0003
2.97	.4985	.9985	.0015	3.50	.4998	.9998	.0002
2.98	.4986	.9986	.0014	3.60	.4998	.9998	.0002
2.99	.4986	.9986	.0014	3.70	.4999	.9999	.0001

Source: Appendix C is taken from Table II of Fisher & Yates': *Statistical Tables for Biological, Agricultural and Medical Research* published by Longman Group UK Ltd. London (previously published by Oliver and Boyd Ltd, Edinburgh) and by permission of the authors and publishers'.

APPENDIX D

Significance of *r*

Values of the Correlation Coefficient for Different Levels of Significance

df	*P* = .10	.05	.02	.01
1	.988	.997	.9995	.9999
2	.900	.950	.980	.990
3	.805	.878	.934	.959
4	.729	.811	.882	.917
5	.669	.754	.833	.874
6	.622	.707	.789	.834
7	.582	.666	.750	.798
8	.549	.632	.716	.765
9	.521	.602	.685	.735
10	.497	.576	.658	.708
11	.476	.553	.634	.684
12	.458	.532	.612	.661
13	.441	.514	.592	.641
14	.426	.497	.574	.623
15	.412	.482	.558	.606
16	.400	.468	.542	.590
17	.389	.456	.528	.575
18	.378	.444	.516	.561
19	.369	.433	.503	.549
20	.360	.423	.492	.537
21	.352	.413	.482	.526
22	.344	.404	.472	.515
23	.337	.396	.462	.505
24	.330	.388	.453	.496
25	.323	.381	.445	.487
26	.317	.374	.437	.479
27	.311	.367	.430	.471
28	.306	.361	.423	.463
29	.301	.355	.416	.456
30	.296	.349	.409	.449
35	.275	.325	.381	.418
40	.257	.304	.358	.393
45	.243	.288	.338	.372
50	.231	.273	.322	.354
60	.211	.250	.295	.325
70	.195	.232	.274	.302
80	.183	.217	.256	.283
90	.173	.205	.242	.267
100	.164	.195	.230	.254

Note: The probabilities given are for a two-tailed test of significance, that is with the sign of *r* ignored. For a one-tailed test of significance, the tabled probabilities should be halved.

Additional Values of r at the 5 and 1 Percent Levels of Significance

df	.05	.01
32	.339	.436
34	.329	.424
36	.320	.413
38	.312	.403
42	.297	.384
44	.291	.376
46	.284	.368
48	.279	.361
55	.261	.338
65	.241	.313
75	.224	.292
85	.211	.275
95	.200	.260
125	.174	.228
150	.159	.208
175	.148	.193
200	.138	.181
300	.113	.148
400	.098	.128
500	.088	.115
1,000	.062	.081

Source: Appendix D is taken from Table VI of Fisher & Yates': *Statistical Tables for Biological, Agricultural and Medical Research* published by Longman Group UK Ltd. London (previously published by Oliver and Boyd Ltd, Edinburgh) and by permission of the authors and publishers'.

APPENDIX E

t-Distribution Values

Selected alpha (significance) levels for one-tailed and two-tailed tests are shown in the top row and second row, respectively. The first column lists the degrees of freedom. The remainder of the table provides critical values of t corresponding to the respective df and alpha level.

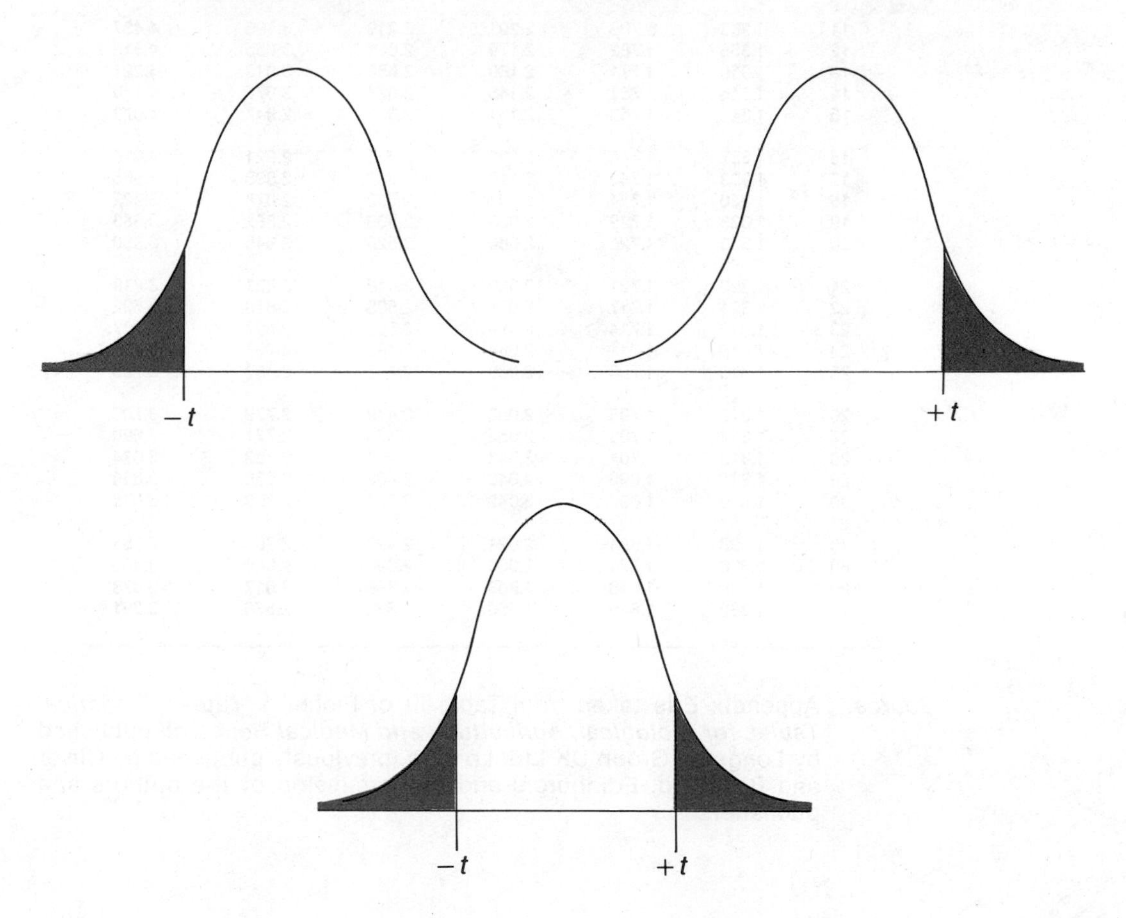

Distribution of t

	Level of significance for one-tailed test					
	.10	.05	.025	.01	.005	.0005
	Level of significance for two-tailed test					
df	.20	.10	.05	.02	.01	.001
1	3.078	6.314	12.706	31.821	63.657	636.619
2	1.886	2.920	4.303	6.965	9.925	31.598
3	1.638	2.353	3.182	4.541	5.841	12.941
4	1.533	2.132	2.776	3.747	4.604	8.610
5	1.476	2.015	2.571	3.365	4.032	6.859
6	1.440	1.943	2.447	3.143	3.707	5.959
7	1.415	1.895	2.365	2.998	3.499	5.405
8	1.397	1.860	2.306	2.896	3.355	5.041
9	1.383	1.833	2.262	2.821	3.250	4.781
10	1.372	1.812	2.228	2.764	3.169	4.587
11	1.363	1.796	2.201	2.718	3.106	4.437
12	1.356	1.782	2.179	2.681	3.055	4.318
13	1.350	1.771	2.160	2.650	3.012	4.221
14	1.345	1.761	2.145	2.624	3.977	4.140
15	1.341	1.753	2.131	2.602	2.947	4.073
16	1.337	1.746	2.120	2.583	2.921	4.015
17	1.333	1.740	2.110	2.567	2.898	3.965
18	1.330	1.734	2.101	2.552	2.878	3.922
19	1.328	1.729	2.093	2.539	2.861	3.883
20	1.325	1.725	2.086	2.528	2.845	3.850
21	1.323	1.721	2.080	2.518	2.831	3.819
22	1.321	1.717	2.074	2.508	2.819	3.792
23	1.319	1.714	2.069	2.500	2.807	3.767
24	1.318	1.711	2.064	2.492	2.797	3.745
25	1.316	1.708	2.060	2.485	2.787	3.725
26	1.315	1.706	2.056	2.479	2.779	3.707
27	1.314	1.703	2.052	2.473	2.771	3.690
28	1.313	1.701	2.048	2.467	2.763	3.674
29	1.311	1.699	2.045	2.462	2.756	3.659
30	1.310	1.697	2.042	2.457	2.750	3.646
40	1.303	1.684	2.021	2.423	2.704	3.551
60	1.296	1.671	2.000	2.390	2.660	3.460
120	1.289	1.658	1.980	2.358	2.617	3.373
∞	1.282	1.645	1.960	2.326	2.576	3.291

Source: Appendix E is taken from Table III of Fisher & Yates': *Statistical Tables for Biological, Agricultural and Medical Research* published by Longman Group UK Ltd. London (previously published by Oliver and Boyd Ltd, Edinburgh) and by permission of the authors and publishers'.

APPENDIX F

F-Distribution Values

To determine the critical *F*-value, locate df for the numerator (between groups) in the top row and df for the denominator (within group) in the left-hand column. The two values at the intersection of the respective row and column are critical values of *F* at 0.05 and 0.01.

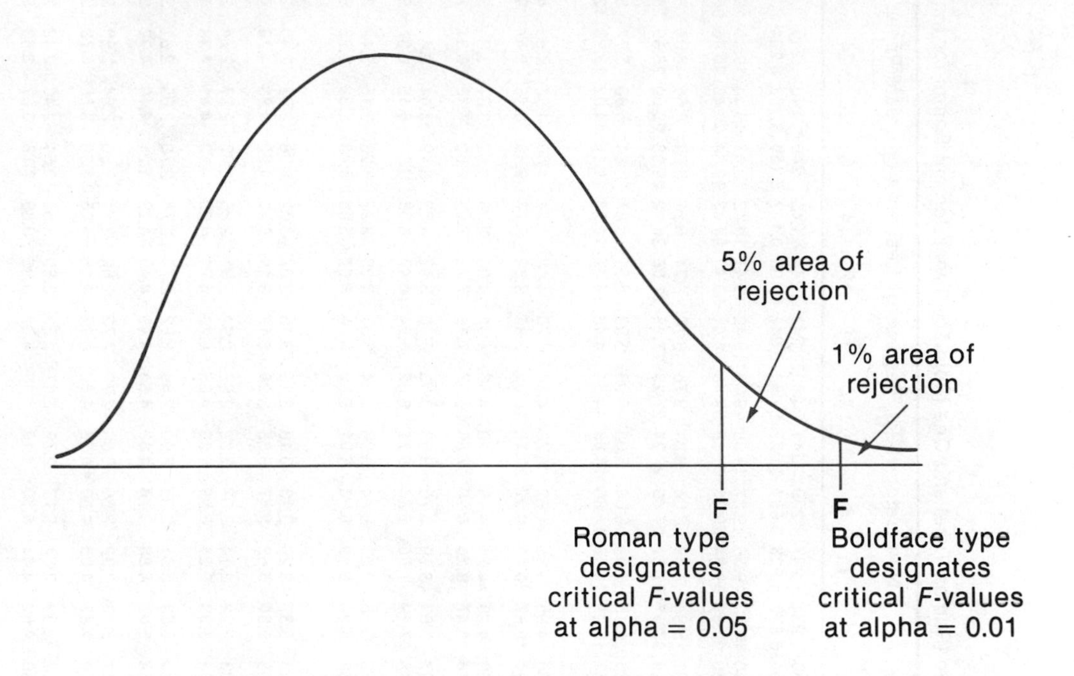

The Five (Roman Type) and One (Boldface Type) Percent Points for the Distribution of F

Numerator df—n_1 degrees of freedom (for greater mean square)

Denominator df

n_2	1	2	3	4	5	6	7	8	9	10	11	12	14	16	20	24	30	40	50	75	100	200	500	∞
1	161	200	216	225	230	234	237	239	241	242	243	244	245	246	248	249	250	251	252	253	253	254	254	254
	4,052	**4,999**	**5,403**	**5,625**	**5,764**	**5,859**	**5,928**	**5,981**	**6,022**	**6,056**	**6,082**	**6,106**	**6,142**	**6,169**	**6,208**	**6,234**	**6,258**	**6,286**	**6,302**	**6,323**	**6,334**	**6,352**	**6,361**	**6,366**
2	18.51	19.00	19.16	19.25	19.30	19.33	19.36	19.37	19.38	19.39	19.40	19.41	19.42	19.43	19.44	19.45	19.46	19.47	19.47	19.48	19.49	19.49	19.50	19.50
	98.49	**99.00**	**99.17**	**99.25**	**99.30**	**99.33**	**99.34**	**99.36**	**99.38**	**99.40**	**99.41**	**99.42**	**99.43**	**99.44**	**99.45**	**99.46**	**99.47**	**99.48**	**99.48**	**99.49**	**99.49**	**99.49**	**99.50**	**99.50**
3	10.13	9.55	9.28	9.12	9.01	8.94	8.88	8.84	8.81	8.78	8.76	8.74	8.71	8.69	8.66	8.64	8.62	8.60	8.58	8.57	8.56	8.54	8.54	8.53
	34.12	**30.82**	**29.46**	**28.71**	**28.24**	**27.91**	**27.67**	**27.49**	**27.34**	**27.23**	**27.13**	**27.05**	**26.92**	**26.83**	**26.69**	**26.60**	**26.50**	**26.41**	**26.35**	**26.27**	**26.23**	**26.18**	**26.14**	**26.12**
4	7.71	6.94	6.59	6.39	6.26	6.16	6.09	6.04	6.00	5.96	5.93	5.91	5.87	5.84	5.80	5.77	5.74	5.71	5.70	5.68	5.66	5.65	5.64	5.63
	21.20	**18.00**	**16.69**	**15.98**	**15.52**	**15.21**	**14.98**	**14.80**	**14.66**	**14.54**	**14.45**	**14.37**	**14.24**	**14.15**	**14.02**	**13.93**	**13.83**	**13.74**	**13.69**	**13.61**	**13.57**	**13.52**	**13.48**	**13.46**
5	6.61	5.79	5.41	5.19	5.05	4.95	4.88	4.82	4.78	4.74	4.70	4.68	4.64	4.60	4.56	4.53	4.50	4.46	4.44	4.42	4.40	4.38	4.37	4.36
	16.26	**13.27**	**12.06**	**11.39**	**10.97**	**10.67**	**10.45**	**10.27**	**10.15**	**10.05**	**9.96**	**9.89**	**9.77**	**9.68**	**9.55**	**9.47**	**9.38**	**9.29**	**9.24**	**9.17**	**9.13**	**9.07**	**9.04**	**9.02**
6	5.99	5.14	4.76	4.53	4.39	4.28	4.21	4.15	4.10	4.06	4.03	4.00	3.96	3.92	3.87	3.84	3.81	3.77	3.75	3.72	3.71	3.69	3.68	3.67
	13.74	**10.92**	**9.78**	**9.15**	**8.75**	**8.47**	**8.26**	**8.10**	**7.98**	**7.87**	**7.79**	**7.72**	**7.60**	**7.52**	**7.39**	**7.31**	**7.23**	**7.14**	**7.09**	**7.02**	**6.99**	**6.94**	**6.90**	**6.88**
7	5.59	4.74	4.35	4.12	3.97	3.87	3.79	3.73	3.68	3.63	3.60	3.57	3.52	3.49	3.44	3.41	3.38	3.34	3.32	3.29	3.28	3.25	3.24	3.23
	12.25	**9.55**	**8.45**	**7.85**	**7.46**	**7.19**	**7.00**	**6.84**	**6.71**	**6.62**	**6.54**	**6.47**	**6.35**	**6.27**	**6.15**	**6.07**	**5.98**	**5.90**	**5.85**	**5.78**	**5.75**	**5.70**	**5.67**	**5.65**
8	5.32	4.46	4.07	3.84	3.69	3.58	3.50	3.44	3.39	3.34	3.31	3.28	3.23	3.20	3.15	3.12	3.08	3.05	3.03	3.00	2.98	2.96	2.94	2.93
	11.26	**8.65**	**7.59**	**7.01**	**6.63**	**6.37**	**6.19**	**6.03**	**5.91**	**5.82**	**5.74**	**5.67**	**5.56**	**5.48**	**5.36**	**5.28**	**5.20**	**5.11**	**5.06**	**5.00**	**4.96**	**4.91**	**4.88**	**4.86**
9	5.12	4.26	3.86	3.63	3.48	3.37	3.29	3.23	3.18	3.13	3.10	3.07	3.02	2.98	2.93	2.90	2.86	2.82	2.80	2.77	2.76	2.73	2.72	2.71
	10.56	**8.02**	**6.99**	**6.42**	**6.06**	**5.80**	**5.62**	**5.47**	**5.35**	**5.26**	**5.18**	**5.11**	**5.00**	**4.92**	**4.80**	**4.73**	**4.64**	**4.56**	**4.51**	**4.45**	**4.41**	**4.36**	**4.33**	**4.31**
10	4.96	4.10	3.71	3.48	3.33	3.22	3.14	3.07	3.02	2.97	2.94	2.91	2.86	2.82	2.77	2.74	2.70	2.67	2.64	2.61	2.59	2.56	2.55	2.54
	10.04	**7.56**	**6.55**	**5.99**	**5.64**	**5.39**	**5.21**	**5.06**	**4.95**	**4.85**	**4.78**	**4.71**	**4.60**	**4.52**	**4.41**	**4.33**	**4.25**	**4.17**	**4.12**	**4.05**	**4.01**	**3.96**	**3.93**	**3.91**
11	4.84	3.98	3.59	3.36	3.20	3.09	3.01	2.95	2.90	2.86	2.82	2.79	2.74	2.70	2.65	2.61	2.57	2.53	2.50	2.47	2.45	2.42	2.41	2.40
	9.65	**7.20**	**6.22**	**5.67**	**5.32**	**5.07**	**4.88**	**4.74**	**4.63**	**4.54**	**4.46**	**4.40**	**4.29**	**4.21**	**4.10**	**4.02**	**3.94**	**3.86**	**3.80**	**3.74**	**3.70**	**3.66**	**3.62**	**3.60**
12	4.75	3.88	3.49	3.26	3.11	3.00	2.92	2.85	2.80	2.76	2.72	2.69	2.64	2.60	2.54	2.50	2.46	2.42	2.40	2.36	2.35	2.32	2.31	2.30
	9.33	**6.93**	**5.95**	**5.41**	**5.06**	**4.82**	**4.65**	**4.50**	**4.39**	**4.30**	**4.22**	**4.16**	**4.05**	**3.98**	**3.86**	**3.78**	**3.70**	**3.61**	**3.56**	**3.49**	**3.46**	**3.41**	**3.38**	**3.36**
13	4.67	3.80	3.41	3.18	3.02	2.92	2.84	2.77	2.72	2.67	2.63	2.60	2.55	2.51	2.46	2.42	2.38	2.34	2.32	2.28	2.26	2.24	2.22	2.21
	9.07	**6.70**	**5.74**	**5.20**	**4.86**	**4.62**	**4.44**	**4.30**	**4.19**	**4.10**	**4.02**	**3.96**	**3.85**	**3.78**	**3.67**	**3.59**	**3.51**	**3.42**	**3.37**	**3.30**	**3.27**	**3.21**	**3.18**	**3.16**

The Five (Roman Type) and One (Boldface Type) Percent Points for the Distribution of F

Denominator df n_2	Numerator df—n_1 degrees of freedom (for greater mean square) 1	2	3	4	5	6	7	8	9	10	11	12	14	16	20	24	30	40	50	75	100	200	500	∞
14	4.60	3.74	3.34	3.11	2.96	2.85	2.77	2.70	2.65	2.60	2.56	2.53	2.48	2.44	2.39	2.35	2.31	2.27	2.24	2.21	2.19	2.16	2.14	2.13
	8.86	**6.51**	**5.56**	**5.03**	**4.69**	**4.46**	**4.28**	**4.14**	**4.03**	**3.94**	**3.86**	**3.80**	**3.70**	**3.62**	**3.51**	**3.43**	**3.34**	**3.26**	**3.21**	**3.14**	**3.11**	**3.06**	**3.02**	**3.00**
15	4.54	3.68	3.29	3.06	2.90	2.79	2.70	2.64	2.59	2.55	2.51	2.48	2.43	2.39	2.33	2.29	2.25	2.21	2.18	2.15	2.12	2.10	2.08	2.07
	8.68	**6.36**	**5.42**	**4.89**	**4.56**	**4.32**	**4.14**	**4.00**	**3.89**	**3.80**	**3.73**	**3.67**	**3.56**	**3.48**	**3.36**	**3.29**	**3.20**	**3.12**	**3.07**	**3.00**	**2.97**	**2.92**	**2.89**	**2.87**
16	4.49	3.63	3.24	3.01	2.85	2.74	2.66	2.59	2.54	2.49	2.45	2.42	2.37	2.33	2.28	2.24	2.20	2.16	2.13	2.09	2.07	2.04	2.02	2.01
	8.53	**6.23**	**5.29**	**4.77**	**4.44**	**4.20**	**4.03**	**3.89**	**3.78**	**3.69**	**3.61**	**3.55**	**3.45**	**3.37**	**3.25**	**3.18**	**3.10**	**3.01**	**2.96**	**2.89**	**2.86**	**2.80**	**2.77**	**2.75**
17	4.45	3.59	3.20	2.96	2.81	2.70	2.62	2.55	2.50	2.45	2.41	2.38	2.33	2.29	2.23	2.19	2.15	2.11	2.08	2.04	2.02	1.99	1.97	1.96
	8.40	**6.11**	**5.18**	**4.67**	**4.34**	**4.10**	**3.93**	**3.79**	**3.68**	**3.59**	**3.52**	**3.45**	**3.35**	**3.27**	**3.16**	**3.08**	**3.00**	**2.92**	**2.86**	**2.79**	**2.76**	**2.70**	**2.67**	**2.65**
18	4.41	3.55	3.16	2.93	2.77	2.66	2.58	2.51	2.46	2.41	2.37	2.34	2.29	2.25	2.19	2.15	2.11	2.07	2.04	2.00	1.98	1.95	1.93	1.92
	8.28	**6.01**	**5.09**	**4.58**	**4.25**	**4.01**	**3.85**	**3.71**	**3.60**	**3.51**	**3.44**	**3.37**	**3.27**	**3.19**	**3.07**	**3.00**	**2.91**	**2.83**	**2.78**	**2.71**	**2.68**	**2.62**	**2.59**	**2.57**
19	4.38	3.52	3.13	2.90	2.74	2.63	2.55	2.48	2.43	2.38	2.34	2.31	2.26	2.21	2.15	2.11	2.07	2.02	2.00	1.96	1.94	1.91	1.90	1.88
	8.18	**5.93**	**5.01**	**4.50**	**4.17**	**3.94**	**3.77**	**3.63**	**3.52**	**3.43**	**3.36**	**3.30**	**3.19**	**3.12**	**3.00**	**2.92**	**2.84**	**2.76**	**2.70**	**2.63**	**2.60**	**2.54**	**2.51**	**2.49**
20	4.35	3.49	3.10	2.87	2.71	2.60	2.52	2.45	2.40	2.35	2.31	2.28	2.23	2.18	2.12	2.08	2.04	1.99	1.96	1.92	1.90	1.87	1.85	1.84
	8.10	**5.85**	**4.94**	**4.43**	**4.10**	**3.87**	**3.71**	**3.56**	**3.45**	**3.37**	**3.30**	**3.23**	**3.13**	**3.05**	**2.94**	**2.86**	**2.77**	**2.69**	**2.63**	**2.56**	**2.53**	**2.47**	**2.44**	**2.42**
21	4.32	3.47	3.07	2.84	2.68	2.57	2.49	2.42	2.37	2.32	2.28	2.25	2.20	2.15	2.09	2.05	2.00	1.96	1.93	1.89	1.87	1.84	1.82	1.81
	8.02	**5.78**	**4.87**	**4.37**	**4.04**	**3.81**	**3.65**	**3.51**	**3.40**	**3.31**	**3.24**	**3.17**	**3.07**	**2.99**	**2.88**	**2.80**	**2.72**	**2.63**	**2.58**	**2.51**	**2.47**	**2.42**	**2.38**	**2.36**
22	4.30	3.44	3.05	2.82	2.66	2.55	2.47	2.40	2.35	2.30	2.26	2.23	2.18	2.13	2.07	2.03	1.98	1.93	1.91	1.87	1.84	1.81	1.80	1.78
	7.94	**5.72**	**4.82**	**4.31**	**3.99**	**3.76**	**3.59**	**3.45**	**3.35**	**3.26**	**3.18**	**3.12**	**3.02**	**2.94**	**2.83**	**2.75**	**2.67**	**2.58**	**2.53**	**2.46**	**2.42**	**2.37**	**2.33**	**2.31**
23	4.28	3.42	3.03	2.80	2.64	2.53	2.45	2.38	2.32	2.28	2.24	2.20	2.14	2.10	2.04	2.00	1.96	1.91	1.88	1.84	1.82	1.79	1.77	1.76
	7.88	**5.66**	**4.76**	**4.26**	**3.94**	**3.71**	**3.54**	**3.41**	**3.30**	**3.21**	**3.14**	**3.07**	**2.97**	**2.89**	**2.78**	**2.70**	**2.62**	**2.53**	**2.48**	**2.41**	**2.37**	**2.32**	**2.28**	**2.26**
24	4.26	3.40	3.01	2.78	2.62	2.51	2.43	2.36	2.30	2.26	2.22	2.18	2.13	2.09	2.02	1.98	1.94	1.89	1.86	1.82	1.80	1.76	1.74	1.73
	7.82	**5.61**	**4.72**	**4.22**	**3.90**	**3.67**	**3.50**	**3.36**	**3.25**	**3.17**	**3.09**	**3.03**	**2.93**	**2.85**	**2.74**	**2.66**	**2.58**	**2.49**	**2.44**	**2.36**	**2.33**	**2.27**	**2.23**	**2.21**
25	4.24	3.38	2.99	2.76	2.60	2.49	2.41	2.34	2.28	2.24	2.20	2.16	2.11	2.06	2.00	1.96	1.92	1.87	1.84	1.80	1.77	1.74	1.72	1.71
	7.77	**5.57**	**4.68**	**4.18**	**3.86**	**3.63**	**3.46**	**3.32**	**3.21**	**3.13**	**3.05**	**2.99**	**2.89**	**2.81**	**2.70**	**2.62**	**2.54**	**2.45**	**2.40**	**2.32**	**2.29**	**2.23**	**2.19**	**2.17**
26	4.22	3.37	2.98	2.74	2.59	2.47	2.39	2.32	2.27	2.22	2.18	2.15	2.10	2.05	1.99	1.95	1.90	1.85	1.82	1.78	1.76	1.72	1.70	1.69
	7.72	**5.53**	**4.64**	**4.14**	**3.82**	**3.59**	**3.42**	**3.29**	**3.17**	**3.09**	**3.02**	**2.96**	**2.86**	**2.77**	**2.66**	**2.58**	**2.50**	**2.41**	**2.36**	**2.28**	**2.25**	**2.19**	**2.15**	**2.13**

The Five (Roman Type) and One (Boldface Type) Percent Points for the Distribution of F

Denominator df	Numerator df—n_1 degrees of freedom (for greater mean square)																							
n_2	1	2	3	4	5	6	7	8	9	10	11	12	14	16	20	24	30	40	50	75	100	200	500	∞
27	4.21	3.35	2.96	2.73	2.57	2.46	2.37	2.30	2.25	2.20	2.16	2.13	2.08	2.03	1.97	1.93	1.88	1.84	1.80	1.76	1.74	1.71	1.68	1.67
	7.68	**5.49**	**4.60**	**4.11**	**3.79**	**3.56**	**3.39**	**3.26**	**3.14**	**3.06**	**2.98**	**2.93**	**2.83**	**2.74**	**2.63**	**2.55**	**2.47**	**2.38**	**2.33**	**2.25**	**2.21**	**2.16**	**2.12**	**2.10**
28	4.20	3.34	2.95	2.71	2.56	2.44	2.36	2.29	2.24	2.19	2.15	2.12	2.06	2.02	1.96	1.91	1.87	1.81	1.78	1.75	1.72	1.69	1.67	1.65
	7.64	**5.45**	**4.57**	**4.07**	**3.76**	**3.53**	**3.36**	**3.23**	**3.11**	**3.03**	**2.95**	**2.90**	**2.80**	**2.71**	**2.60**	**2.52**	**2.44**	**2.35**	**2.30**	**2.22**	**2.18**	**2.13**	**2.09**	**2.06**
29	4.18	3.33	2.93	2.70	2.54	2.43	2.35	2.28	2.22	2.18	2.14	2.10	2.05	2.00	1.94	1.90	1.85	1.80	1.77	1.73	1.71	1.68	1.65	1.64
	7.60	**5.42**	**4.54**	**4.04**	**3.73**	**3.50**	**3.33**	**3.20**	**3.08**	**3.00**	**2.92**	**2.87**	**2.77**	**2.68**	**2.57**	**2.49**	**2.41**	**2.32**	**2.27**	**2.19**	**2.15**	**2.10**	**2.06**	**2.03**
30	4.17	3.32	2.92	2.69	2.53	2.42	2.34	2.27	2.21	2.16	2.12	2.09	2.04	1.99	1.93	1.89	1.84	1.79	1.76	1.72	1.69	1.66	1.64	1.62
	7.56	**5.39**	**4.51**	**4.02**	**3.70**	**3.47**	**3.30**	**3.17**	**3.06**	**2.98**	**2.90**	**2.84**	**2.74**	**2.66**	**2.55**	**2.47**	**2.38**	**2.29**	**2.24**	**2.16**	**2.13**	**2.07**	**2.03**	**2.01**
32	4.15	3.30	2.90	2.67	2.51	2.40	2.32	2.25	2.19	2.14	2.10	2.07	2.02	1.97	1.91	1.86	1.82	1.76	1.74	1.69	1.67	1.64	1.61	1.59
	7.50	**5.34**	**4.46**	**3.97**	**3.66**	**3.42**	**3.25**	**3.12**	**3.01**	**2.94**	**2.86**	**2.80**	**2.70**	**2.62**	**2.51**	**2.42**	**2.34**	**2.25**	**2.20**	**2.12**	**2.08**	**2.02**	**1.98**	**1.96**
34	4.13	3.28	2.88	2.65	2.49	2.38	2.30	2.23	2.17	2.12	2.08	2.05	2.00	1.95	1.89	1.84	1.80	1.74	1.71	1.67	1.64	1.61	1.59	1.57
	7.44	**5.29**	**4.42**	**3.93**	**3.61**	**3.38**	**3.21**	**3.08**	**2.97**	**2.89**	**2.82**	**2.76**	**2.66**	**2.58**	**2.47**	**2.38**	**2.30**	**2.21**	**2.15**	**2.08**	**2.04**	**1.98**	**1.94**	**1.91**
36	4.11	3.26	2.86	2.63	2.48	2.36	2.28	2.21	2.15	2.10	2.06	2.03	1.98	1.93	1.87	1.82	1.78	1.72	1.69	1.65	1.62	1.59	1.56	1.55
	7.39	**5.25**	**4.38**	**3.89**	**3.58**	**3.35**	**3.18**	**3.04**	**2.94**	**2.86**	**2.78**	**2.72**	**2.62**	**2.54**	**2.43**	**2.35**	**2.26**	**2.17**	**2.12**	**2.04**	**2.00**	**1.94**	**1.90**	**1.87**
38	4.10	3.25	2.85	2.62	2.46	2.35	2.26	2.19	2.14	2.09	2.05	2.02	1.96	1.92	1.85	1.80	1.76	1.71	1.67	1.63	1.60	1.57	1.54	1.53
	7.35	**5.21**	**4.34**	**3.86**	**3.54**	**3.32**	**3.15**	**3.02**	**2.91**	**2.82**	**2.75**	**2.69**	**2.59**	**2.51**	**2.40**	**2.32**	**2.22**	**2.14**	**2.08**	**2.00**	**1.97**	**1.90**	**1.86**	**1.84**
40	4.08	3.23	2.84	2.61	2.45	2.34	2.25	2.18	2.12	2.07	2.04	2.00	1.95	1.90	1.84	1.79	1.74	1.69	1.66	1.61	1.59	1.55	1.53	1.51
	7.31	**5.18**	**4.31**	**3.83**	**3.51**	**3.29**	**3.12**	**2.99**	**2.88**	**2.80**	**2.73**	**2.66**	**2.56**	**2.49**	**2.37**	**2.29**	**2.20**	**2.11**	**2.05**	**1.97**	**1.94**	**1.88**	**1.84**	**1.81**
42	4.07	3.22	2.83	2.59	2.44	2.32	2.24	2.17	2.11	2.06	2.02	1.99	1.94	1.89	1.82	1.78	1.73	1.68	1.64	1.60	1.57	1.54	1.51	1.49
	7.27	**5.15**	**4.29**	**3.80**	**3.49**	**3.26**	**3.10**	**2.96**	**2.86**	**2.77**	**2.70**	**2.64**	**2.54**	**2.46**	**2.35**	**2.26**	**2.17**	**2.08**	**2.02**	**1.94**	**1.91**	**1.85**	**1.80**	**1.78**
44	4.06	3.21	2.82	2.58	2.43	2.31	2.23	2.16	2.10	2.05	2.01	1.98	1.92	1.88	1.81	1.76	1.72	1.66	1.63	1.58	1.56	1.52	1.50	1.48
	7.24	**5.12**	**4.26**	**3.78**	**3.46**	**3.24**	**3.07**	**2.94**	**2.84**	**2.75**	**2.68**	**2.62**	**2.52**	**2.44**	**2.32**	**2.24**	**2.15**	**2.06**	**2.00**	**1.92**	**1.88**	**1.82**	**1.78**	**1.75**
46	4.05	3.20	2.81	2.57	2.42	2.30	2.21	2.14	2.09	2.04	2.00	1.97	1.91	1.87	1.80	1.75	1.71	1.65	1.62	1.57	1.54	1.51	1.48	1.46
	7.21	**5.10**	**4.24**	**3.76**	**3.44**	**3.22**	**3.05**	**2.92**	**2.82**	**2.73**	**2.66**	**2.60**	**2.50**	**2.42**	**2.30**	**2.22**	**2.13**	**2.04**	**1.98**	**1.90**	**1.86**	**1.80**	**1.76**	**1.72**
48	4.04	3.19	2.80	2.56	2.41	2.30	2.21	2.14	2.08	2.03	1.99	1.96	1.90	1.86	1.79	1.74	1.70	1.64	1.61	1.56	1.53	1.50	1.47	1.45
	7.19	**5.08**	**4.22**	**3.74**	**3.42**	**3.30**	**3.04**	**2.90**	**2.80**	**2.71**	**2.64**	**2.58**	**2.48**	**2.40**	**2.28**	**2.20**	**2.11**	**2.02**	**1.96**	**1.88**	**1.84**	**1.78**	**1.73**	**1.70**

The Five (Roman Type) and One (Boldface Type) Percent Points for the Distribution of *F*

	Numerator df—n_1 degrees of freedom (for greater mean square)																							
n_2	1	2	3	4	5	6	7	8	9	10	11	12	14	16	20	24	30	40	50	75	100	200	500	∞
50	4.03	3.18	2.79	2.56	2.40	2.29	2.20	2.13	2.07	2.02	1.98	1.95	1.90	1.85	1.78	1.74	1.69	1.63	1.60	1.55	1.52	1.48	1.46	1.44
	7.17	**5.06**	**4.20**	**3.72**	**3.41**	**3.18**	**3.02**	**2.88**	**2.78**	**2.70**	**2.62**	**2.56**	**2.46**	**2.39**	**2.26**	**2.18**	**2.10**	**2.00**	**1.94**	**1.86**	**1.82**	**1.76**	**1.71**	**1.68**
55	4.02	3.17	2.78	2.54	2.38	2.27	2.18	2.11	2.05	2.00	1.97	1.93	1.88	1.83	1.76	1.72	1.67	1.61	1.58	1.52	1.50	1.46	1.43	1.41
	7.12	**5.01**	**4.16**	**3.68**	**3.37**	**3.15**	**2.98**	**2.85**	**2.75**	**2.66**	**2.59**	**2.53**	**2.43**	**2.35**	**2.23**	**2.15**	**2.06**	**1.96**	**1.90**	**1.82**	**1.78**	**1.71**	**1.66**	**1.64**
60	4.00	3.15	2.76	2.52	2.37	2.25	2.17	2.10	2.04	1.99	1.95	1.92	1.86	1.81	1.75	1.70	1.65	1.59	1.56	1.50	1.48	1.44	1.41	1.39
	7.08	**4.98**	**4.13**	**3.65**	**3.34**	**3.12**	**2.95**	**2.82**	**2.72**	**2.63**	**2.56**	**2.50**	**2.40**	**2.32**	**2.20**	**2.12**	**2.03**	**1.93**	**1.87**	**1.79**	**1.74**	**1.68**	**1.63**	**1.60**
65	3.99	3.14	2.75	2.51	2.36	2.24	2.15	2.08	2.02	1.98	1.94	1.90	1.85	1.80	1.73	1.68	1.63	1.57	1.54	1.49	1.46	1.42	1.39	1.37
	7.04	**4.95**	**4.10**	**3.62**	**3.31**	**3.09**	**2.93**	**2.79**	**2.70**	**2.61**	**2.54**	**2.47**	**2.37**	**2.30**	**2.18**	**2.09**	**2.00**	**1.90**	**1.84**	**1.76**	**1.71**	**1.64**	**1.60**	**1.56**
70	3.98	3.13	2.74	2.50	2.35	2.23	2.14	2.07	2.01	1.97	1.93	1.89	1.84	1.79	1.72	1.67	1.62	1.56	1.53	1.47	1.45	1.40	1.37	1.35
	7.01	**4.92**	**4.08**	**3.60**	**3.29**	**3.07**	**2.91**	**2.77**	**2.67**	**2.59**	**2.51**	**2.45**	**2.35**	**2.28**	**2.15**	**2.07**	**1.98**	**1.88**	**1.82**	**1.74**	**1.69**	**1.62**	**1.56**	**1.53**
80	3.96	3.11	2.72	2.48	2.33	2.21	2.12	2.05	1.99	1.95	1.91	1.88	1.82	1.77	1.70	1.65	1.60	1.54	1.51	1.45	1.42	1.38	1.35	1.32
	6.96	**4.88**	**4.04**	**3.56**	**3.25**	**3.04**	**2.87**	**2.74**	**2.64**	**2.55**	**2.48**	**2.41**	**2.32**	**2.24**	**2.11**	**2.03**	**1.94**	**1.84**	**1.78**	**1.70**	**1.65**	**1.57**	**1.52**	**1.49**
100	3.94	3.09	2.70	2.46	2.30	2.19	2.10	2.03	1.97	1.92	1.88	1.85	1.79	1.75	1.68	1.63	1.57	1.51	1.48	1.42	1.39	1.34	1.30	1.28
	6.90	**4.82**	**3.98**	**3.51**	**3.20**	**2.99**	**2.82**	**2.69**	**2.59**	**2.51**	**2.43**	**2.36**	**2.26**	**2.19**	**2.06**	**1.98**	**1.89**	**1.79**	**1.73**	**1.64**	**1.59**	**1.51**	**1.46**	**1.43**
125	3.92	3.07	2.68	2.44	2.29	2.17	2.08	2.01	1.95	1.90	1.86	1.83	1.77	1.72	1.65	1.60	1.55	1.49	1.45	1.39	1.36	1.31	1.27	1.25
	6.84	**4.78**	**3.94**	**3.47**	**3.17**	**2.95**	**2.79**	**2.65**	**2.56**	**2.47**	**2.40**	**2.33**	**2.23**	**2.15**	**2.03**	**1.94**	**1.85**	**1.75**	**1.68**	**1.59**	**1.54**	**1.46**	**1.40**	**1.47**
150	3.91	3.06	2.67	2.43	2.27	2.16	2.07	2.00	1.94	1.89	1.85	1.82	1.76	1.71	1.64	1.59	1.54	1.47	1.44	1.37	1.34	1.29	1.25	1.22
	6.81	**4.75**	**3.91**	**3.44**	**3.14**	**2.92**	**2.76**	**2.62**	**2.53**	**2.44**	**2.37**	**2.30**	**2.20**	**2.12**	**2.00**	**1.91**	**1.83**	**1.72**	**1.66**	**1.56**	**1.51**	**1.43**	**1.37**	**1.33**
200	3.89	3.04	2.65	2.41	2.26	2.14	2.05	1.98	1.92	1.87	1.83	1.80	1.74	1.69	1.62	1.57	1.52	1.45	1.42	1.35	1.32	1.26	1.22	1.19
	6.76	**4.71**	**3.88**	**3.41**	**3.11**	**2.90**	**2.73**	**2.60**	**2.50**	**2.41**	**2.34**	**2.28**	**2.17**	**2.09**	**1.97**	**1.88**	**1.79**	**1.69**	**1.61**	**1.53**	**1.48**	**1.39**	**1.33**	**1.28**
400	3.86	3.02	2.62	2.39	2.23	2.12	2.03	1.96	1.90	1.85	1.81	1.78	1.72	1.67	1.60	1.54	1.49	1.42	1.38	1.32	1.28	1.22	1.16	1.13
	6.70	**4.66**	**3.83**	**3.36**	**3.06**	**2.85**	**2.69**	**2.55**	**2.46**	**2.37**	**2.29**	**2.23**	**2.12**	**2.04**	**1.92**	**1.84**	**1.74**	**1.64**	**1.57**	**1.47**	**1.42**	**1.32**	**1.24**	**1.19**
1,000	3.85	3.00	2.61	2.38	2.22	2.10	2.02	1.95	1.89	1.84	1.80	1.76	1.70	1.65	1.58	1.53	1.47	1.41	1.36	1.30	1.26	1.19	1.13	1.08
	6.66	**4.62**	**3.80**	**3.34**	**3.04**	**2.82**	**2.66**	**2.53**	**2.43**	**2.34**	**2.26**	**2.20**	**2.09**	**2.01**	**1.89**	**1.81**	**1.71**	**1.61**	**1.54**	**1.44**	**1.38**	**1.28**	**1.19**	**1.11**
∞	3.84	2.99	2.60	2.37	2.21	2.09	2.01	1.94	1.88	1.83	1.79	1.75	1.69	1.64	1.57	1.52	1.46	1.40	1.35	1.28	1.24	1.17	1.11	1.00
	6.64	**4.60**	**3.78**	**3.32**	**3.02**	**2.80**	**2.64**	**2.51**	**2.41**	**2.32**	**2.24**	**2.18**	**2.07**	**1.99**	**1.87**	**1.79**	**1.69**	**1.59**	**1.52**	**1.41**	**1.36**	**1.25**	**1.15**	**1.00**

Denominator df

Source: Reprinted by permission from STATISTICAL METHODS, Fifth Edition by William G. Snedecor and George W. Snedecor © 1956 by the Iowa State University Press, 2121 South State Avenue, Ames, Iowa 50010.

APPENDIX G
Chi Square Values

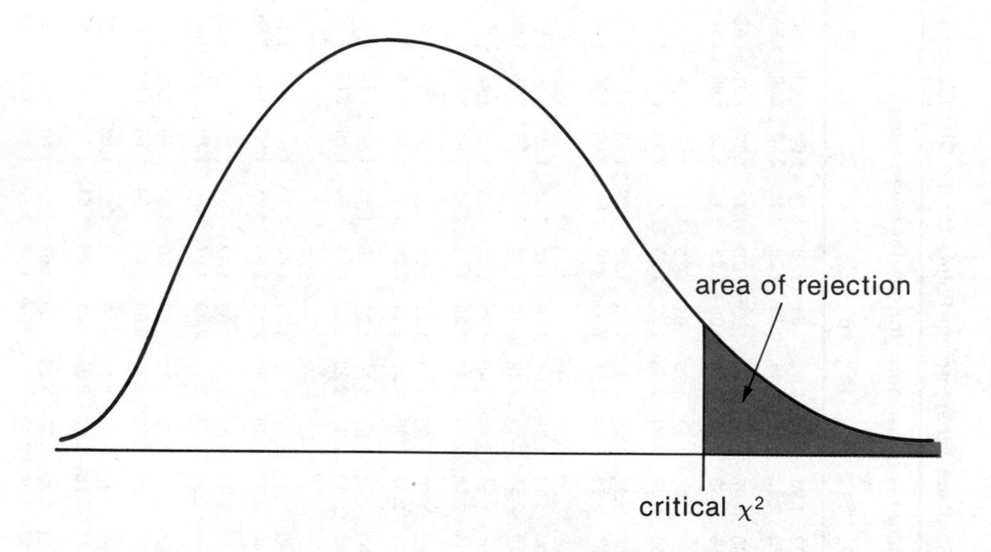

The degrees of freedom appear in the left-hand column. The column headings show the proportional area beyond the critical χ^2 (the area of rejection). Specific critical χ^2 values are given in the remainder of the table. These values correspond to the df and alpha level, respectively.

Critical Values of Chi Square (χ^2) at 0.05, 0.01, and 0.001 Levels of Significance

df	.05	.01	.001
1	3.841	6.635	10.827
2	5.991	9.210	13.815
3	7.315	11.341	16.268
4	9.488	13.277	18.465
5	11.070	15.086	20.517
6	12.592	16.812	22.457
7	14.067	18.475	24.322
8	15.507	20.090	26.125
9	16.919	21.666	27.877
10	18.307	23.209	29.588
11	19.675	24.725	31.264
12	21.026	26.217	32.909
13	22.362	27.688	34.528
14	23.685	29.141	36.123
15	24.996	30.578	37.697
16	26.296	32.000	39.252
17	27.587	33.409	40.790
18	28.869	34.805	42.312
19	30.144	36.191	43.820
20	31.410	37.566	45.315
21	32.671	38.932	46.797
22	33.924	40.289	48.268
23	35.172	41.638	49.728
24	36.415	42.980	51.179
25	37.652	44.314	52.620
26	38.885	45.642	54.052
27	40.113	46.963	55.476
28	41.337	48.278	56.893
29	42.557	49.588	58.302
30	43.773	50.892	59.703

Note: For larger values of df, the expression $\sqrt{2\chi^2} - \sqrt{2df - 1}$ may be used as a normal deviate with unit variance, remembering that the probability for χ^2 corresponds with that of a single tail of the normal curve.

Source: Appendix G is taken from Table IV of Fisher & Yates': *Statistical Tables for Biological, Agricultural and Medical Research* published by Longman Group UK Ltd. London (previously published by Oliver and Boyd Ltd, Edinburgh) and by permission of the authors and publishers'.

APPENDIX H

Fisher's *z* to *r* Transformations

Conversion of a Pearson *r* into a Corresponding Fisher's z_r Coefficient

r	*z*	*r*	*z*	*r*	*z*	*r*	*z*	*r*	*z*	*r*	*z*
.25	.26	.40	.42	.55	.62	.70	.87	.85	1.26	.950	1.83
.26	.27	.41	.44	.56	.63	.71	.89	.86	1.29	.955	1.89
.27	.28	.42	.45	.57	.65	.72	.91	.87	1.33	.960	1.95
.28	.29	.43	.46	.58	.66	.73	.93	.88	1.38	.965	2.01
.29	.30	.44	.47	.59	.68	.74	.95	.89	1.42	.970	2.09
.30	.31	.45	.48	.60	.69	.75	.97	.90	1.47	.975	2.18
.31	.32	.46	.50	.61	.71	.76	1.00	.905	1.50	.980	2.30
.32	.33	.47	.51	.62	.73	.77	1.02	.910	1.53	.985	2.44
.33	.34	.48	.52	.63	.74	.78	1.05	.915	1.56	.990	2.65
.34	.35	.49	.54	.64	.76	.79	1.07	.920	1.59	.995	2.99
.35	.37	.50	.55	.65	.78	.80	1.10	.925	1.62		
.36	.38	.51	.56	.66	.79	.81	1.13	.930	1.66		
.37	.39	.52	.58	.67	.81	.82	1.16	.935	1.70		
.38	.40	.53	.59	.68	.83	.83	1.19	.940	1.74		
.39	.41	.54	.60	.69	.85	.84	1.22	.945	1.78		

Source: Appendix H is taken from Table VII of Fisher & Yates': *Statistical Tables for Biological, Agricultural and Medical Research* published by Longman Group UK Ltd. London (previously published by Oliver and Boyd Ltd, Edinburgh) and by permission of the authors and publishers'.

Note: For all values of *r* below .25, $r = z$ to two decimal places.

APPENDIX I

ACCUSTAT Software Instructions

CONTENTS

GENERAL OVERVIEW

ACCUSTAT is a computerized statistical analysis system. It is designed to provide correct answers to selected chapter application exercises contained in PROBABILITY AND STATISTICS by Johnson.

The ACCUSTAT package consists of two major components—a User's Manual and several program diskettes. The User's Manual (Appendix I) contains step-by-step instructions for processing chapter application exercises and includes the following sections:

1. Operating Instructions describes how to operate your microcomputer in conjunction with the ACCUSTAT package.
2. Program Instructions describes how to operate each program available on the ACCUSTAT program diskettes.
3. Solvable Exercises identifies which program (and program option) to use to process each solvable chapter application exercise.
4. Error Messages lists all error messages that could appear on the screen and then suggests a corrective course of action.

The diskettes contain microcomputer programs that process and store the exercise data and solutions. All programs are menu-driven. Other features include input prompts that ask for specific data and suggest proper responses, edit routines that verify data accuracy, and error messages that suggest possible corrections.

ACCUSTAT has been developed for the Apple® IIe, Apple® IIc, and Apple® IIGS[1]; the IBM® PC[2] and IBM Personal System/2™[3]; and the Tandy® 1000[4] microcomputers. The minimum hardware

---

[1]Apple® II Plus, Apple® IIe, Apple® IIc, Apple® IIGS™, Applesoft®, and the Apple® logo are registered or pending trademarks of Apple Computer, Inc. Any reference to Apple II plus, Apple IIe, Apple IIc, Apple IIGS or Applesoft, or any depiction of the Apple logo, refers to this footnote.

[2]IBM® is a registered trademark of International Business Machines Corporation. Any reference to the IBM Personal Computer refers to this footnote.

[3]Personal System/2™ is a trademark of the International Business Machines Corporation. Any reference to the Personal System/2 refers to this footnote.

[4]Tandy® 1000 is a registered trademark of the Radio Shack Division of Tandy Corporation. Any reference to the Tandy 1000 or Radio Shack Microcomputer refers to this footnote.

requirements for each microcomputer system are listed below:

1. Apple IIe, Apple IIc, and Apple IIGS: 128K memory minimum, ProDOS® 8[5], 80-column card, monochrome monitor, one single-sided disk drive. Access to an 80-column printer is optional.
2. IBM PC: 128K memory minimum, DOS 2.0 or above, monochrome or color monitor, and one double-sided disk drive. Access to an 80-column printer is optional.
3. IBM Personal System/2: 640K memory minimum, DOS 3.3, and one double-sided disk drive ($3\frac{1}{2}''$ or $5\frac{1}{4}''$ format). Access to an 80-column printer is optional.
4. Tandy 1000: 128K memory minimum, MS-DOS 2.11 or above, and one double-sided disk drive ($3\frac{1}{2}''$ or $5\frac{1}{4}''$ format). Access to an 80-column printer is optional.

OPERATING INSTRUCTIONS

This section describes how to operate your microcomputer in conjunction with the ACCUSTAT package. It provides detailed descriptions of initializing/formatting data disks, start-up procedures, and specific screen formats.

Initializing/Formatting Data Disks

Apple (ProDOS):

One (1)-Drive System

1. Insert a ProDOS User's Disk into Drive 1.
2. Turn on the microcomputer. The Main Menu of the ProDOS User's Disk utility program will appear.
3. Select Option F (ProDOS Filer) by pressing the letter F.
4. Select Option V (Volume Commands) from the Filer Menu.
5. Select Option F (Format a Volume) from the Volume Commands Menu.
6. Remove the ProDOS User's Disk from the drive; insert a blank diskette into Drive 1.

---

[5]ProDOS® is a registered trademark of Apple Computer, Inc. Any reference to ProDOS refers to this footnote.

Apple Computer, Inc. makes no warranties, either express or implied, regarding the enclosed computer software package, its merchantability, or its fitness for any particular purpose. The exclusion of implied warranties is not permitted by some states. The above exclusion may not apply to you. This warranty provides you with specific legal rights. There may be other rights that you may have which vary from state to state.

7. Three prompts will appear on the screen. Respond to each prompt with the response indicated below.

   --FORMAT--
   THE VOLUME IS SLOT: **(6)**
   DRIVE: **(1)**
   NEW VOLUME NAME: **(/DATA**)

   Another prompt message is displayed after the new volume name is entered. Key-enter a (**Y**)es only if you are sure you want to format the disk. Formatting will erase all of the existing data on the disk.

   DESTROY "*volume name*"? (Y/N) **Y**

   If (**Y**)es is key-entered, the disk drive light will come on while the disk is being formatted. The message "Format Complete" is displayed after the process is complete.
8. Remove the diskette from the drive and turn off the computer.

Two (2)-Drive System

1. Insert a ProDOS User's Disk into Drive 1.
2. Turn on the microcomputer. The Main Menu of the ProDOS User's Disk utility program will appear.
3. Select Option F (ProDOS Filer) by pressing the letter F.
4. Select Option V (Volume Commands) from the Filer Menu.
5. Select Option F (Format a Volume) from the Volume Commands Menu.
6. Insert a blank diskette into Drive 2.
7. Three prompts will appear on the screen. Respond to each prompt by key-entering the information shown in bold.

   --FORMAT--
   THE VOLUME IS SLOT: **(6)**
   DRIVE: **(2)**
   NEW VOLUME NAME: **(/DATA**)

   Another prompt message is displayed after the new volume name is entered. Key-enter a (**Y**)es only if you are sure you want to format the disk. Formatting will erase all of the existing data on the disk.

   DESTROY "*volume name*"? (Y/N) **Y**

   If (**Y**)es is key-entered, the disk drive light will come on while the disk is being formatted. The message "Format Complete" is displayed after the process is complete.
8. Remove the diskettes from the drives and turn off the computer.

IBM PC, Personal System/2, and Tandy 1000

One (1)-Drive System

1. Insert a MS-DOS System diskette into Drive A.
2. Turn on the microcomputer. (Wait until the flashing cursor appears, a "beep" is emitted, the red light on the disk drive goes off, and the following appears on the screen: Enter new date: __. .)
3. Key the current date; strike ENTER. (Or simply strike ENTER.) (The following prompt appears on the screen: Enter new time: __.)
4. Key the time; strike ENTER. (Or simply strike ENTER.) (A> __ will be displayed on the screen.)
5. Key-enter **FORMAT A:** then strike ENTER. Remove the MS-DOS System diskette.
6. Insert a blank diskette into Drive A.
7. Strike the space bar (or any key) to continue. ("Format Complete" appears on the screen, followed by a prompt to format another data diskette.)
8. If you want to format another data diskette, strike **Y**(es) and respond to the prompts on the screen. If you do not want to format another data diskette, strike **N**(o).
9. Remove the diskette from the disk drive.
10. Turn off the microcomputer.

Two (2)-Drive System

1. Insert a MS-DOS System diskette into Drive A.
2. Turn on the microcomputer. (Wait until the flashing cursor appears, a "beep" is emitted, the red light on the disk drive goes off, and the following appears on the screen: Enter new date: __. .)
3. Key the current date; strike ENTER. (Or simply strike ENTER.) (The following prompt appears on the screen: Enter new time: __.)
4. Key the time; strike ENTER. (Or simply strike ENTER.) (A copyright message with A> __ on the last line appears on the screen.)
5. Key-enter **FORMAT B:** then strike ENTER.
6. Insert a blank diskette into Drive B.
7. Strike the space bar (or any key) to continue. ("Format Complete" appears on the screen, followed by a prompt to format another data diskette.)

8. If you want to format another data diskette, strike **Y**(es) and respond to the prompts on the screen. If you do not want to format another data diskette, strike **N**(o).
9. Remove the diskettes from the disk drives.
10. Turn off the microcomputer.

Start-Up Procedures

Apple IIe, Apple IIc, and Apple IIGS

1. Turn on the television/monitor.
2. Open the door to the disk drive. If your microcomputer has more than one drive, be sure to open the door to Drive 1.
3. With your thumb on the diskette label, carefully insert the ACCUSTAT program diskette into Drive 1 with the label facing upwards. The diskette is completely inserted when the label is past the entrance to the drive.
4. Gently close the disk drive door.
5. If you are using a driver diskette system, insert a formatted data diskette into Drive 2 and gently close the door.
6. If the computer is off, reach to the back left side of the machine and turn on the power switch. If your computer is already on, you must boot the diskette. On the Apple IIe, IIc, and IIGS hold down the CONTROL and OPEN APPLE () keys and at the same time press RESET. The power light on the computer will illuminate. The red IN USE light on the disk drive will come on, and you will hear a soft whirring sound. After a few seconds, the title screen will be displayed.
7. Press the RETURN key to continue until you have passed the three opening screens. After the third opening screen the ACCUSTAT Master Index Menu will appear. You are now ready to operate the software.

IBM PC, Personal System/2, and Tandy 1000

1. Carefully insert your copy of the DOS 2.0 or above diskette into Drive A with the label facing up. The diskette is completely inserted when the label is past the entrance to the drive.
2. If the computer is off, turn on the power switch. If the computer is already on, hold down the control (Ctrl), alternate (Alt), and the delete (Del) key at the same time. The cursor

will appear in the upper left corner of the screen and the red light on Drive A will come on. A prompt message will appear on the screen indicating that the current date should be key-entered.

3. Key-enter the date or just press the ENTER key. The prompt to key-enter the current time will then be displayed.
4. Key-enter a new time or just press the ENTER key. You should see the A> displayed on the screen.
5. Remove the DOS diskette from Drive A.
6. Insert the ACCUSTAT program diskette into Drive A.
7. Key-enter **CONTROL** and press the ENTER key.
8. Press the ENTER key to continue through the opening screens. The ACCUSTAT Master Index Menu will appear. You are now ready to operate the software.

Screen Format

Each screen display in ACCUSTAT conforms to a standard layout. Consider the screen display that appears in Figure I.1.

Figure I.1—Standardized Screen Format

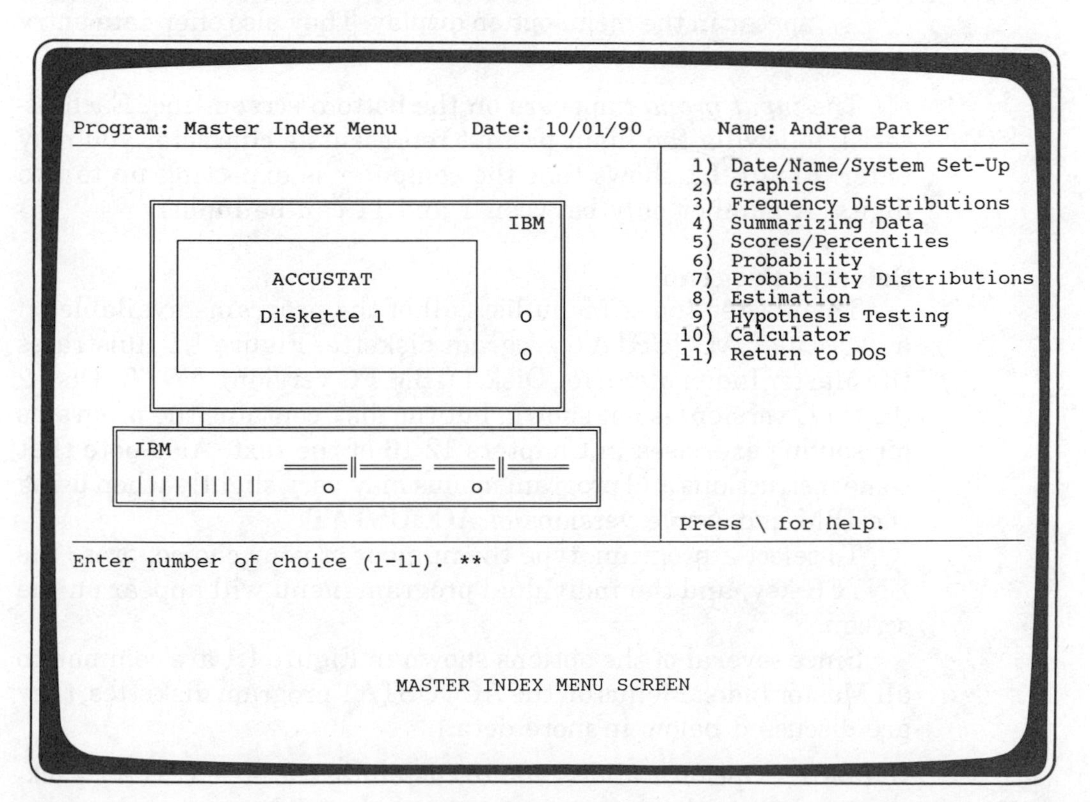

The *screen title* appears on the left side of the top screen line. Figure I.1 shows that you are executing the Master Index Menu program.

Next, the *current date* appears between the screen title and student name. Two digits must be entered for the month, two for the day, and two for the year. Separators (/ or slashes) are provided by the program. Figure I.1 shows October 1, 1990.

The *student name* appears on the right side of the top screen line. A maximum of 20 characters can be used.

The body of the screen (between the top and bottom solid rules) is divided into the *data-entry display* (left window) and the *menu-option display* (right window). This portion of the screen is used to display menu options and data-entry information.

You may press the backslash key (\) if the *Press \ for help* message is present on the screen.

1. A window will open in the data-entry display, and a limited narrative message will be visible. Enter **1** for more lines, **2** to print the entire narrative message, or **0** to exit the screen.
2. These narrative messages explain more fully the options that appear in the menu-option display. They also offer data-entry suggestions.

The *input prompt* appears on the bottom screen line. Each asterisk following the input prompt represents a character you may enter. Figure I.1 shows that the computer is expecting up to two digits. A number only between 1 and 11 can be input.

Select a Program

The Master Index Menu lists all of the programs available on a particular ACCUSTAT program diskette. Figure I.1 illustrates the Master Index Menu for Disk 1 (IBM PC version). ***NOTE:*** Disk 2 (IBM PC version) is not shown, but the disk contains the programs for solving exercises in Chapters 12-16 of the text. Also note that some instructions and program menus may vary slightly when using the IBM and Apple versions of ACCUSTAT.

To select a program, type the number of your choice, press the ENTER key, and the individual program menu will appear on the screen.

Since several of the options shown in Figure I.1 are common to all Master Index Menus on the ACCUSTAT program diskettes, they are discussed below in more detail.

Option 1. Option 1 allows you to customize the ACCUSTAT package to work with your computer hardware. When you select this option, the five prompts that follow appear, one at a time, on the input prompt line:

1. *Enter today's date (MMDDYY).* Enter the current date. The first two digits are for the month, the next two digits are for the day, and the last two digits are for the year; no slashes (/) or hyphens (-) are allowed. The key-entry for the date in Figure I.1 would be **100190**.
2. *Enter student name (20 characters).* Enter your name with a maximum of 20 characters.
3. *Enter student ID code (three uppercase letters/numbers).* Indicate a three-character ID code for yourself. This must be used consistently throughout the software.
4. *Enter data disk location (A, B, or C).* Press ENTER to accept the Drive B location, and the computer will store all exercise data on disk Drive B; enter **A** to store on disk Drive A, or enter **C** to store on disk Drive C.
5. *Enter 0 for no advance, 1 for automatic advance.* Press ENTER to accept the default value of 0, and the computer will print answers directly after each other (with no paging); or enter **1** and the computer will print separate problem summaries on single pages.
6. *Enter 0 for monochrome display, 1 for color display.* Press ENTER to accept the default value of 0 and the computer will display in black and white or monochrome. Enter **1** and the computer will display in color. (If a color monitor is not properly connected, the computer will default to monochrome, regardless of what you select.)

Option 10. Option 10 allows you to access an on-screen calculator. This calculator will add, subtract, multiply, and divide. It will also convert to negative numbers, represent in percentages, and calculate square roots.

Option 11. Option 11 allows you to leave the ACCUSTAT environment and return to DOS (the disk operating system). It is the most convenient way to end a work session.

Select a Program Option

The program menu lists all of the options available through a particular program. By selecting Option 3 from the Master Index Menu, a display similar to Figure I.2 will show on the screen.

To select a program option, enter the number of your choice, and press the ENTER key.

Figure I.2—Frequency Distributions Program Display

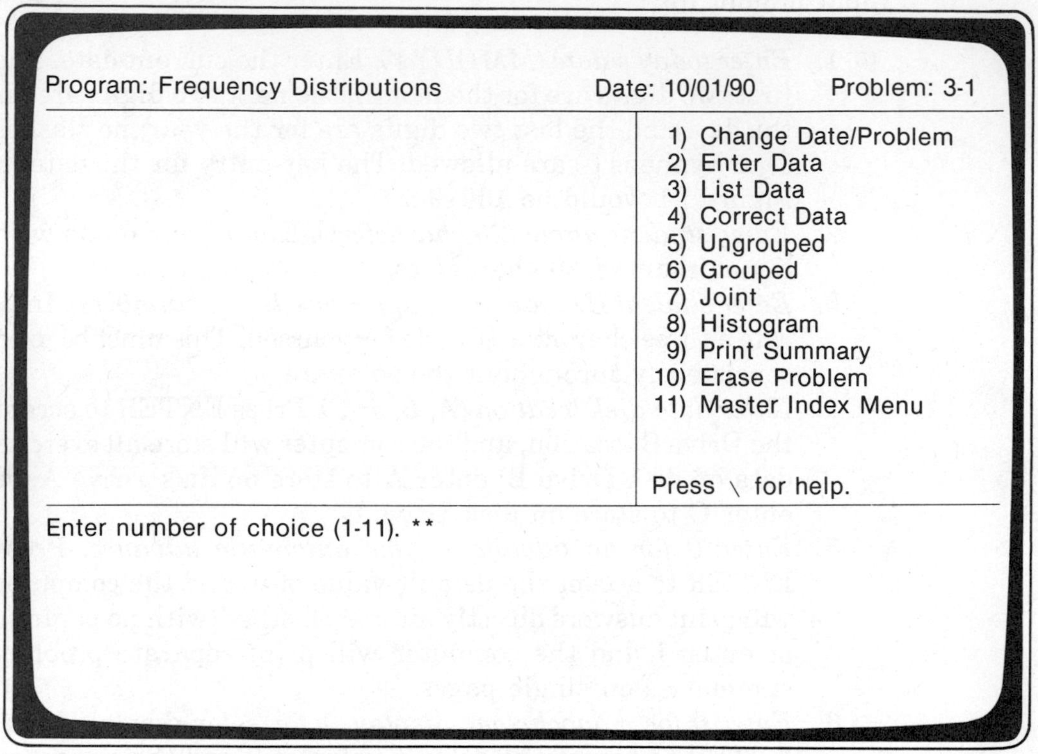

```
Program: Frequency Distributions          Date: 10/01/90        Problem: 3-1

                                           1) Change Date/Problem
                                           2) Enter Data
                                           3) List Data
                                           4) Correct Data
                                           5) Ungrouped
                                           6) Grouped
                                           7) Joint
                                           8) Histogram
                                           9) Print Summary
                                          10) Erase Problem
                                          11) Master Index Menu

                                          Press \ for help.

Enter number of choice (1-11). **
```

Once again several of the options shown in Figure I.2 are common to all program menus within the ACCUSTAT package, so let's examine them in more detail. ***NOTE:*** The name in the upper right-hand corner has been replaced by Problem number information.

Option 1. Option 1 allows you to enter today's date and an exercise number. When you select this option, the two prompts that follow appear, one at a time, on the input prompt line:

1. *Enter today's date (MMDDYY).* If you've completed this option on the Master Index Menu, the date shown should be correct, so simply press the ENTER key. Refer to the previous example for specific information on inputting the date.
2. *Enter problem number (5 characters).* Enter the application exercise number. The first two characters are for the chapter; the next character is a hyphen (-); and the last two characters are for the application exercise. Figure I.2 shows Chapter 3, Application Exercise 3.1.

Option 9. Option 9 allows you to display the application exercise data and solution on the printer. ***NOTE:*** Before attempting to use the line printer, be sure it is turned on and is on-line and ready to print; be sure there is an adequate supply of paper; be sure the paper is correctly fed into the printer; and be sure the printer is correctly connected to the computer.

Option 10. Option 10 allows you to remove the application exercise data and solution from the data disk. When you select this option, the following prompt appears on the input prompt line:

> *Caution. Enter 1 for blank page, 0 for exit.* Enter **1** and the computer will remove the data and solution; or enter **0** to return to the program menu.

Option 11. Option 11 allows you to return to the Master Index Menu.

You may press the backslash key (\) if the *Press \ for help* message is present on the screen. The help messages explain more fully the options that appear in the menu-option display.

PROGRAM INSTRUCTIONS

This section describes how to operate each program available on the ACCUSTAT program diskettes.

Graphics Program

The Graphics Program allows you to display on the monitor one or two groups of data in the form of a bar chart. When you select this program, the menu displayed in Figure I.3 appears.

Option 1. Option 1 allows you to enter today's date and an application exercise number.

The Change Date/Problem information for Application Exercise 2.1 would be entered as follows:

Enter today's date (MMDDYY). 100190 ⟨Enter⟩

As previously indicated, the date must contain two digits for the month, date, and year. If the date is correct, simply press ENTER to go on to the exercise number entry. Refer to the last section where the date entry is discussed in more detail.

Enter problem number (5 characters). 2-1 ⟨Enter⟩

Be careful to enter the exercise number correctly. The hyphen (-) must be entered between the chapter number and the application

Figure I.3—Graphics Menu-Option and Data-Entry Displays

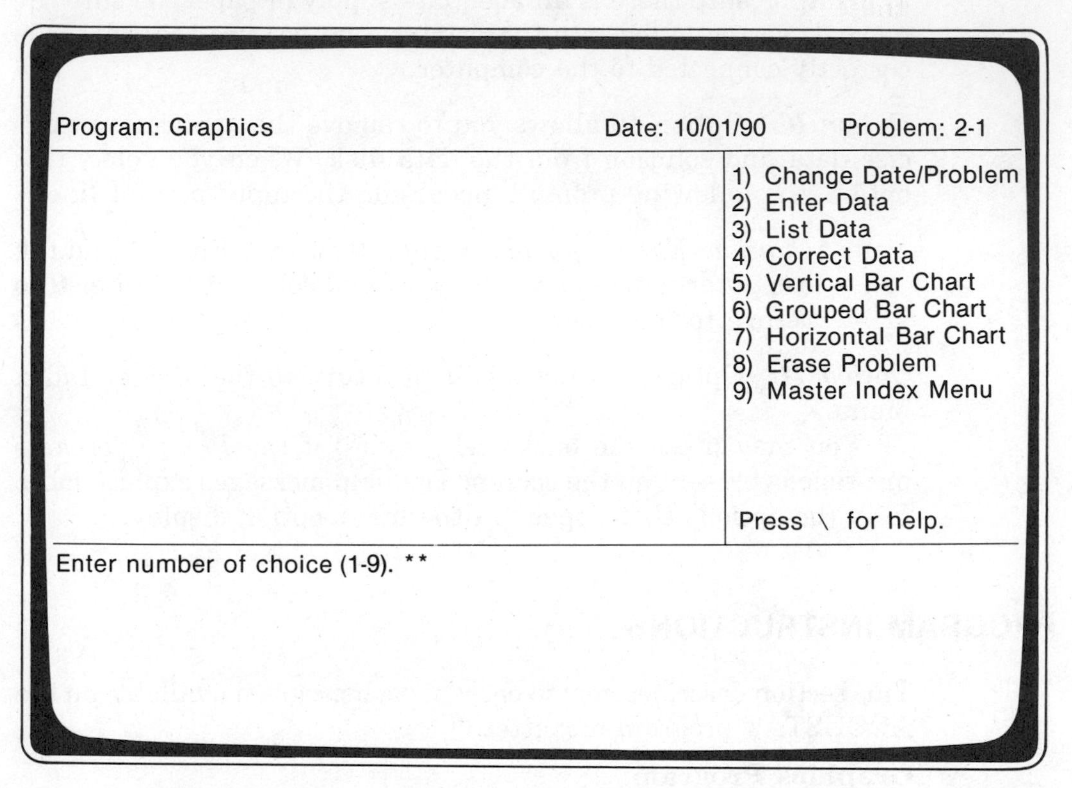

exercise number. The exercise data and solution are stored according to this number. As you move on to new exercises, be sure to change exercise numbers at the same time.

Option 2. Option 2 allows you to build a data set with up to five categories each in a maximum of two groups. When you select this option, the data-entry display will appear as shown in the *partial* screen that follows.

Program: Graphics Date: 10/01/90 Problem

Ln	Label	Freq 1	Freq 2	
				1) Change Date/Pro
				2) Enter Data
				3) List Data
1	[ ]	[.00]	[.00]	4) Correct Data
				5) Vertical Bar Cha
				6) Grouped Bar Ch
				7) Horizontal Bar C
				8) Erase Problem
				9) Master Index Me

Partial Screen

Brackets are used in data entry in order
those items to be input. (See the partial screen.) The data
cation Exercise 2.1 would be entered as follows:

Enter label (10 characters).	Very	⟨Enter⟩
Enter frequency one (1-99999.99).	39	⟨Enter⟩
Enter frequency two (1-99999.99).		⟨Enter⟩
Enter title-group 1 (10 characters).	Adults	⟨Enter⟩
Enter 1 for next item, 2 for correction, 0 for exit.	1	⟨Enter⟩

1. If you have successfully entered the item and want to enter the next item, enter the number **1**.
2. If you have made an error on an item, enter the number **2**. Each value will reappear. To accept the existing value, simply press ENTER; otherwise type the proper value and press ENTER.
3. If you have entered the last item, enter the number **0**.

The computer will ask for the title to Group 1 with the first item and for the title to Group 2 with the second item only.

Enter label (10 characters).	Fairly	⟨Enter⟩
Enter frequency one (1-99999.99).	41	⟨Enter⟩
Enter frequency two (1-99999.99).		⟨Enter⟩
Enter title-group 2 (10 characters).		⟨Enter⟩
Enter 1 for next item, 2 for correction, 0 for exit.	1	⟨Enter⟩
Enter label (10 characters).	Not Too	⟨Enter⟩
Enter frequency one (1-99999.99).	14	⟨Enter⟩
Enter frequency two (1-99999.99).		⟨Enter⟩
Enter 1 for next item, 2 for correction, 0 for exit.	1	⟨Enter⟩
Enter label (10 characters).	Not at all	⟨Enter⟩
Enter frequency one (1-99999.99).	3	⟨Enter⟩
Enter frequency two (1-99999.99).		⟨Enter⟩
Enter 1 for next item, 2 for correction, 0 for exit.	1	⟨Enter⟩
Enter label (10 characters).	No Opinion	⟨Enter⟩
Enter frequency one (1-99999.99).	3	⟨Enter⟩
Enter frequency two (1-99999.99).		⟨Enter⟩
Enter 1 for next item, 2 for correction, 0 for exit.	0	⟨Enter⟩

Option 3. Option 3 allows you to display the data set on the monitor. (Build the data set through Option 2 first.)

Option 4. Option 4 allows you to edit the data set. (First build the data set through Option 2.) When you select this option, the headings of the data-entry display shown in Figure I.3 reappear.

1. Enter the number of the line you want to edit. That line will appear on the monitor. To accept the existing value, simply press ENTER; otherwise enter the proper value and press ENTER.
2. If you want to change the title of Group one, edit line 1; and if you want to change the title of Group two, edit line 2.
3. When you are finished editing the data, enter **0** (zero) for line number, and press the ENTER key.

Option 5. Option 5 allows you to display a vertical bar chart on the monitor. (First build the data set through Option 2.)

Option 6. Option 6 allows you to display a grouped bar chart on the monitor. (Build the data set through Option 2 first.)

Option 7. Option 7 allows you to display a horizontal bar chart on the monitor. (First build the data set through Option 2.)

Option 8. Option 8 allows you to remove the exercise data and solution from the data disk. When you are finished with a particular exercise, erase it from the data disk. Only a limited number of problems will fit on a diskette (depending on the microcomputer), and you will want space on your data disk for future work.

Option 9. Option 9 allows you to return to the Master Index Menu.

NOTE: Each subsequent program description contains options that repeat for each program. These options are the Change Date/Problem (always Option 1), Print Summary, Erase Problem and Master Index Menu (usually the last three options). The options work the same on all programs. They will be omitted gradually in the subsequent program descriptions that follow.

Frequency Distributions Program

The Frequency Distributions Program allows you to display a single data set in the form of a frequency distribution. The menu-option display for this program is shown in Figure I.4.

Option 2. Option 2 allows you to build a data set with up to 80 values. Refer to Option 2 in the previous example for information on data entry.

Option 3. Option 3 allows you to display the data set on the monitor. (First build the data set through Option 2.)

Figure I.4—Frequency Distributions Menu-Option and Data-Entry Display

```
Program: Frequency Distributions          Date: 10/01/90     Problem: 3-2
--------------------------------------------------------------------------
                                          |  1) Change Date/Problem
                                          |  2) Enter Data
                                          |  3) List Data
                                          |  4) Correct Data
                                          |  5) Ungrouped
                                          |  6) Grouped
                                          |  7) Joint
                                          |  8) Histogram
                                          |  9) Print Summary
                                          | 10) Erase Problem
                                          | 11) Master Index Menu
                                          |
                                          | Press \ for help.
--------------------------------------------------------------------------
Enter number of choice (1-11). **
```

Option 4. Option 4 allows you to edit the data set. (Use Option 2 to first build the data set.)

Option 5. Option 5 allows you to display an ungrouped frequency distribution on the monitor.

Option 6. Option 6 allows you to build a grouped frequency distribution with between 5 and 11 class intervals. When you select this option, the data-entry display will appear as shown in the *partial* screen that follows.

```
Program: Frequency Distributions                  Date: 10/01/90
-----------------------------------------------------------------
                                                |  1)  Chan
Number of class intervals              [    ]   |  2)  Enter
Width of each class interval     [        .00]  |  3)  List
Lower limit of first interval    [        .00]  |  4)  Corr
                                                |  5)  Ungr
                                                |  6)  Grou
```

Partial Screen

The (raw) data for Application Exercise 3.2 would be entered through Option 2, while the (grouped frequency) data would be entered as follows:

Prompt	Entry	Key
Enter number intervals (5-11).	11	⟨Enter⟩
Enter interval width (0.01-99999.99).	25	⟨Enter⟩
Enter lowest limit (0.01-99999.99).	450	⟨Enter⟩
Enter 1 for computation, 0 for exit.	0	⟨Enter⟩

1. If you want to build another grouped frequency distribution (different number of intervals, different class width, or different beginning point), enter the number **1**.
2. If you want to return to the program menu, enter the number **0**.

Option 7. Option 7 allows you to build a joint frequency distribution with up to 10 intervals and 2 categories.

Option 8. Option 8 allows you to display a histogram on the monitor (assumes that Options 2 and 6 have been completed).

Option 9. Option 9 allows you to display the exercise data and solution on the line printer.

Option 10. Option 10 allows you to remove the exercise data and solution from the data disk.

Summarizing Data Program

The Summarizing Data Program allows you to compute the measures of central tendency and dispersion of a single data set. When you select this program, the menu-option display that follows will appear on the screen.

1) Change Date/Problem
2) Enter Data
3) List Data
4) Correct Data
5) Grouped
6) List Summary
7) Print Summary
8) Erase Problem
9) Master Index Menu

Option 2. Option 2 allows you to build a data set with up to 80 values.

Option 3 and Option 4. Options 3 and 4 are used similarly to those same options in previous examples.

Option 5. Option 5 allows you to build a grouped frequency distribution with between 5 and 11 class intervals. When you select this option, the *partial* data-entry display that follows will appear.

```
Program: Summarizing Data                    Date: 10/01/90
_____________________________________________________________
                                           | 1) Chan
Ln  Class interval                   Freq  | 2) Enter
                                           | 3) List
1   [            ] – [         .00]  [ 0 ] | 4) Corr
                                           | 5) Grou
                                           | 6) List
                                           | 7) Print
                                           | 8) Eras
                                           | 9) Mast
```

Partial Screen

The data for Application Exercise 5.11 would be entered as follows:

Enter lower limit (0-99999.99).	2	⟨Enter⟩
Enter upper limit (0.01-99999.99).	2	⟨Enter⟩
Enter frequency (0-99999).	1	⟨Enter⟩
Enter lower limit (0-99999.99).	3	⟨Enter⟩
Enter upper limit (0.01-99999.99).	3	⟨Enter⟩
Enter frequency (0-99999).	2	⟨Enter⟩
Enter lower limit (0-99999.99).	4	⟨Enter⟩
Enter upper limit (0.01-99999.99).	4	⟨Enter⟩
Enter frequency (0-99999).	5	⟨Enter⟩
Enter lower limit (0-99999.99).	5	⟨Enter⟩
Enter upper limit (0.01-99999.99).	5	⟨Enter⟩
Enter frequency (0-99999).	3	⟨Enter⟩
Enter lower limit (0-99999.99).	6	⟨Enter⟩
Enter upper limit (0.01-99999.99).	6	⟨Enter⟩
Enter frequency (0-99999).	2	⟨Enter⟩
Enter lower limit (0-99999.99).		⟨Enter⟩
Enter upper limit (0.01-99999.99.		⟨Enter⟩
Enter frequency (0-99999).		⟨Enter⟩
Enter 1 for computation, 0 for exit.	0	⟨Enter⟩

1. If you want to build another grouped frequency distribution (or correct the one shown on the monitor), enter the number **1**.
2. If you want to return to the program menu, enter the number **0**.

Option 6. Option 6 allows you to display the summary of the data on the monitor (assumes that Option 2 and then Option 5 have been completed).

Option 7. Option 7 allows you to display the exercise data and solution on the line printer (assumes that Option 2 and then Option 5 have been completed).

Scores/Percentiles Program

The Scores/Percentiles Program allows you to convert between raw scores, z-scores, T-scores, and percentile scores. When you select this program, the menu-option display that follows appears.

1) Change Date/Problem
2) Raw Score
3) z-Score
4) T-Score
5) Percentile
6) Print Summary
7) Erase Problem
8) Master Index Menu

Option 2. Option 2 allows you to convert a raw score into a z-score, a T-score, and a percentile score.

Option 3. Option 3 allows you to convert a z-score into a T-score and a percentile score.

Option 4. Option 4 allows you to convert a T-score into a z-score and a percentile score.

Option 5. Option 5 allows you to convert a percentile score into a z-score and a T-score.

Option 6. Option 6 allows you to display the exercise data and solution on a printer.

Probability Program

The Probability Program allows you to compute combinations and permutations and simple, joint, and conditional probability factors. When this program is selected, the menu that follows will be displayed.

1) Change Date/Problem
2) Simple
3) Joint
4) Conditional
5) Permutations
6) Print Summary
7) Erase Problem
8) Master Index Menu

Option 2. Option 2 allows you to compute the probability that one event will or will not occur.

Option 3. Option 3 allows you to compute the probability that either of two events will occur or that both events will occur.

Option 4. Option 4 allows you to compute the probability that one event will occur based on one or more other events.

Option 5. Option 5 allows you to compute the combinations and permutations of r things taken N at a time.

Probability Distributions Program

The Probability Distributions Program allows you to change one or more random variables with the binomial distribution, the Poisson distribution, and the normal curve. When you select this program, the menu that follows will appear.

1) Change Date/Problem
2) Binomial
3) Poisson
4) Normal
5) Print Summary
6) Erase Problem
7) Master Index Menu

Option 2. Option 2 allows you to compute binomial probabilities and to calculate the mean and standard deviation of a binomial distribution.

Option 3. Option 3 allows you to compute Poisson probabilities and to calculate the standard deviation of a Poisson distribution.

Option 4. Option 4 allows you to compute probabilities of random variables with the normal curve.

Estimation Program

The Estimation Program allows you to compute an estimate of the population mean and population standard deviation. When you select this program, the menu-option display that follows will appear.

1) Change Date/Problem
2) Probability
3) Standard Error
4) Arithmetic Mean
5) Standard Deviation
6) Print Summary
7) Erase Problem
8) Master Index Menu

Option 2. Option 2 allows you to compute the chance of a random variable.

Option 3. Option 3 allows you to compute the standard error of the mean (standard deviation of the sampling distribution).

Option 4. Option 4 allows you to compute the point or interval estimate of the population arithmetic mean.

Option 5. Option 5 allows you to compute the point estimate of the population standard deviation.

Hypothesis Testing Program

The Hypothesis Testing Program allows you to conduct a one- or two-tailed hypothesis test involving one population mean. When you select this program, the menu-option display that follows will appear.

1) Change Date/Problem
2) Probability
3) Critical Value
4) Decision
5) Arithmetic Mean
6) Print Summary
7) Erase Problem
8) Master Index Menu

Option 2. Option 2 allows you to compute the chance of a random variable.

Option 3. Option 3 allows you to compute the critical z-value.

Option 4. Option 4 allows you to decide whether or not to reject a null hypothesis.

Option 5. Option 5 allows you to conduct a one population arithmetic mean test.

Correlation Program

The Correlation Program allows you to conduct a two-tailed hypothesis test involving the population product-moment correlation coefficient. When you select this program, the menu-option display that follows appears.

1) Change Date/Problem
2) Enter Data
3) List Data
4) Correct Data
5) Scattergram
6) Coeff of Correlation
7) Decision
8) Confidence Intervals
9) Print Summary
10) Erase Problem
11) Master Index Menu

Option 2. Option 2 allows you to build a bivariate data set with up to 20 values.

Option 3. Option 3 allows you to display the data set on the monitor.

Option 4. Option 4 allows you to edit the data set.

Option 5. Option 5 allows you to display a bivariate plot (scattergram) on the monitor.

Option 6. Option 6 allows you to compute the Pearson product-moment correlation coefficient.

Option 7. Option 7 allows you to decide whether or not to reject a null hypothesis.

Option 8. Option 8 allows you to compute the interval estimate of the population product-moment correlation coefficient.

Regression Program

The Regression Program allows you to construct a (simple) least-squares regression line. When you select this program, the menu-option display that follows appears.

1) Change Date/Problem
2) Enter Data
3) List Data
4) Correct Data
5) Scattergram
6) Equation
7) Decision
8) Predicted Value
9) Print Summary
10) Erase Problem
11) Master Index Menu

Option 2. Option 2 allows you to build a bivariate data set with up to 20 items.

Option 5. Option 5 allows you to display a bivariate plot (scattergram) on the monitor.

Option 6. Option 6 allows you to construct a least-squares regression line from a bivariate distribution.

Option 7. Option 7 allows you to decide whether or not to reject a null hypothesis.

Option 8. Option 8 allows you to compute the predicted *Y*-value with a least-squares regression line.

t-Test Program

The *t*-Test Program allows you to conduct a one- or two-tailed hypothesis test involving two populations when samples are relatively small. When you select this program, the menu-option display that follows will appear.

1) Change Date/Problem
2) Enter Data
3) List Data
4) Correct Data
5) Correlated
6) Independent
7) Print Summary
8) Erase Problem
9) Master Index Menu

Option 2. Option 2 allows you to build a data set with up to 20 items each in a maximum of two groups.

Option 5. Option 5 allows you to conduct a two-population correlated mean test.

Option 6. Option 6 allows you to conduct a two-population independent mean test.

Analysis of Variance Program

The Analysis of Variance Program allows you to conduct a two-tailed hypothesis test involving three or four population means. When you select this program, the menu-option display that follows appears.

1) Change Date/Problem
2) Enter Data
3) List Data
4) Correct Data
5) ANOVA Table
6) Scheffé Model
7) Print Summary
8) Erase Problem
9) Master Index Menu

Option 2. Option 2 builds a data set with up to 20 items each in a maximum of four groups.

Option 5. Option 5 allows you to construct an Analysis of Variance table and decide whether or not to reject a null hypothesis (constructed from the data set or entered separately).

Option 6. Option 6 allows you to conduct a pairwise comparison of groups with the Scheffé method (constructed from the data set or entered separately).

Chi-Square Program

The Chi-Square Program allows you to conduct a one-tailed hypothesis test involving goodness-of-fit and independence. When you select this program, the menu-option display that follows will appear.

1) Change Date/Problem
2) Critical Value
3) Decision
4) One Variable
5) Two Variables
6) Print Summary
7) Erase Problem
8) Master Index Menu

Option 2. Option 2 allows you to compute the critical chi-square value.

Option 3. Option 3 allows you to decide whether or not to reject a null hypothesis.

Option 4. Option 4 allows you to conduct a goodness-of-fit test.

Option 5. Option 5 allows you to conduct a test for independence.

SOLVABLE EXERCISES

This section identifies the chapter number and then which program (and program option) to use to process each solvable application exercise. It also offers helpful hints for data entry.

One word of caution: Some answers generated by ACCUSTAT will differ slightly from hand calculations. These differences are due to rounding and table values.

Chapter 1 (Introduction to Statistics). All of the application exercises in Chapter 1 are essay and not solvable with ACCUSTAT.

Chapter 2 (Displaying Data). The following application exercises in Chapter 2 are solvable with the Graphics Program:

1. Option 2, then Option 5.
2. Option 2, then Option 6. Hint: combine the "F" and "Don't Know" grades into one category.
6. Option 2, then Option 7.
7. Option 2, then Option 6. Hint: male is Group 1 and female is Group 2.

Chapter 3 (Frequency Distributions). All but one of the following application exercises in Chapter 3 are solvable with the Frequency Distributions Program:

1. Option 2, then Option 5.
2. Option 2, then Option 6.
3. For 3a, Option 6, then Option 8. No solution for 3b and 3c.
4. Option 7. Hint: men are Group 1 and women are Group 2.
5. Option 7.
6. Use the Correlation Program to solve this exercise. Option 2, then Option 5. Hint: height is the *X*-variable; weight is the *Y*-variable.

Chapter 4 (Summarizing Data: Measures of Central Tendency). The following application exercises in Chapter 4 are solvable with the Summarizing Data Program:

3. Option 2, then Option 6.
4. Option 2, then Option 6.
5. Option 2, then Option 6.
6. Option 2, then Option 6.
11. For 11a, Option 2, then Option 6.

Chapter 5 (Summarizing Data: Measures of Dispersion). The following application exercises in Chapter 5 are solvable with the Summarizing Data Program:

6. For 6e, 6f, 6g, and 6h, Option 2, then Option 6. Hint: enter **0.001** for the 0 (zero) value.
7. Option 2, then Option 6.
11. Option 5, then Option 6. Hint: lower and upper limits equal.
12. Option 5, then Option 6. Hint: lower and upper limits equal.
13. Option 2, then Option 6. Hint: process the *U*-variable as Exercise 5-13 and the *V*-variable as Exercise 5-90.

Chapter 6 (Describing Individual Performances). The following application exercises in Chapter 6 are solvable with the Scores/Percentiles Program:

1. Option 3.
2. Option 3.
3. Option 3.
4. Option 3.
5. Option 4.
6. For 6a, 6b, 6c, and 6d, use Option 3. For 6e, 6f, and 6g, use Option 4. Hint: 6d and 6g must be solved in two parts.
7. Option 2. Hint: multiply probabilities by the total number of scores for individual counts.
8. Option 5.
9. Option 5.
10. Option 3. Hint: must be solved in two parts.
11. Option 5.
12. Option 2.
13. Option 2.
14. Option 2. Hint: assume a mean of 100 points and a standard deviation of 15.

15. Option 2. Hint: solve for Eugene and Sally separately, then compare results.
16. Option 2. Hint: first calculate the mean (= 8) and standard deviation (= 3) of the X-variable.
17. Option 2.
18. Option 2. Hint: use the 100 points found in 18a to calculate probability in 18b, then multiply probability by 30 students.

Chapter 7 (Elementary Probability). The following application exercises in Chapter 7 are solvable with the Probability Program:

1. Option 2.
2. Option 2.
3. Option 2.
4. Option 2.
5. Option 2.
6. Option 2. Hint: will calculate probability but not odds.
7. Option 3.
8. Option 3. Hint: $p(\text{A}) = 0.3333$, $p(\text{B}) = 0.4666$, $p(\text{C}) = 0.2000$.
9. Option 3. Hint: $p(\text{A}) = 0.2500$, $p(\text{B}) = 0.5000$.
10. Option 3. Hint: $p(\text{A}) = 0.2105$, $p(\text{B}) = 0.3158$, $p(\text{X}) = 0.2632$, $p(\text{Y}) = 0.2105$.
11. Option 3. Hint: will calculate probability but not odds.
12. Option 3.
13. Option 3. Hint: $p(\text{A}) = 0.2000$, $p(\text{X}) = 0.5250$.
16. Option 4. Hint: $p(\text{A}) = 0.2000$, $p(\text{A and X}) = 0.0600$.
20. Option 3. Hint: $p(\text{A}) = 0.5000$, $p(\text{X}) = 0.2500$.
21. Option 5.
22. Option 5.
23. Option 5.
24. Option 5.
28. Option 4. Hint: $p(\text{A}) = 0.4000$, $p(\text{A and X}) = 0.1500$.

Chapter 8 (Probability Distributions). The following application exercises in Chapter 8 are solvable with the Probability Distributions Program:

3. Option 2. Hint: trials = 10.
4. For 4a and 4c, Option 2. Hint: trials = 7. No solution for 4b.
5. Option 2. Hint: trials = 6.

6. Option 2. Hint: trials = 8.
7. Option 4.
8. Option 4.
9. Option 4.
10. Option 4.
14. Option 3. Hint: success = 0.
17. Option 3. Hint: success = 0.

Chapter 9 (Applied Sampling). All of the exercises in Chapter 9 are essay and not solvable with ACCUSTAT.

Chapter 10 (Estimation). The following application exercises in Chapter 10 are solvable with the Estimation Program:

1. Option 3.
3. For 3a, Option 4. For 3b, Option 5. No solution for 3c.
4. For 4b and 4c, Option 4.
6. Option 2. Hint: $N = 1$ for 6a, $N = 100$ for 6b.
7. For 7a, 7b, Option 2. For 7c, Option 4.
10. For 10a, Option 4. For 10b and 10c, Option 2. Hint: $N = 1$.

Chapter 11 (Hypothesis Testing). The following exercises in Chapter 11 are solvable with the Hypothesis Testing Program:

1. For 1a, 1b, and 1d, Option 3. No solution for 1c.
2. For 2a, 2b, 2d, and 2e, Option 4. No solution for 2c and 2f.
5. For 5a, Option 2. For 5b, Option 5.
6. Option 5.
7. Option 5.
8. For 8a, Option 5. No solution for 8b.

Chapter 12 (Correlation). The following application exercises in Chapter 12 are solvable with the Correlation Program:

3. For 3a, Option 2, then Option 6. For 3b, Option 5.
4. Option 7.
5. Option 7.
7. Option 6.
8. Option 8.
9. Option 8.
11. For 11a, Option 2, then Option 6. For 11b and 11c, Option 7. For 11d, Option 8. No solution for 11d(1).

Chapter 13 (Regression Analysis). The following application exercises in Chapter 13 are solvable with the Regression Program:

4. Option 8.
5. For 5a, 5b, 5c, 5d, and 5f, Option 8. No solution for 5e.
6. Option 6.
7. For 7a, 7b, 7c, and 7f, Option 2, then Option 6. For 7d, Option 8. For 7e, Option 7. No solution for 7g.
8. For 8a and 8b, Option 2, then Option 6. No solution for 8d and 8e. For 8c, change to exercise number 8-90, then Option 2 and Option 6.

Chapter 14 (t-Test). The following application exercises in Chapter 14 are solvable with the *t*-Test Program:

3. Option 6.
4. Option 6.
6. Option 2, then Option 5.
7. Option 2, then Option 6.
8. Option 2, then Option 6.
9. Option 2, then Option 5.
10. Option 6.

Chapter 15 (Analysis of Variance). The following application exercises in Chapter 15 are solvable with the Analysis of Variance Program:

3. Option 2, then Options 5 and 6.
4. Option 2, then Options 5 and 6.
5. Option 6.

Chapter 16 (Chi Square). The following application exercises in Chapter 16 are solvable with the Chi-Square Program:

1. Option 3. Hint: row = 1, column = 10.
2. Option 3. Hint: row = 1, column = 6.
3. For 3a and 3b, Option 2. No solution for 3c, 3d, 3e.
4. Option 4.
5. Option 4.
6. Option 4.
8. Option 4. Hint: expected frequencies 60, 90, 30.
9. Option 4. Hint: expected frequencies 30, 10, 5.
11. Option 3. Hint: row = 2, column = 4.

12. Option 3. Hint: row = 4, column = 6.
13. Option 5.
14. Option 5.
16. Option 5.
17. Option 5.
18. Option 5.

ERROR MESSAGES

This section lists all error messages that could appear on the screen and suggests a corrective course of action.

01 Disk drive door is open. The computer is not able to read from the program or data disk because the disk drive door is open.

Operator action—Close the door on each disk drive.

02 Diskette is write-protected. The computer is not able to write to the program or data disk because it contains a write-protect label.

Operator action—Remove the write-protect label from the disk.

06 Same as background color. The color you chose for either the foreground, menu display, message prompt, or cursor is identical to the color you chose for the background. If permitted, portions of the screen would be invisible (Date/Name/System Set-Up Program).

Operator action—Request a color that is dissimilar to the color chosen for the background.

07 Not a color monitor. The computer is not able to display in color because your monitor is not a color monitor (Date/Name/System Set-Up Program).

Operator action—If the monitor is a color monitor, be sure it is properly connected to the computer. Otherwise request black/white display option.

11 Line printer is not ready. The computer is not able to print because the line printer is not ready.

Operator action—Be sure the printer is properly connected to the computer; be sure the printer is on line; be sure the printer has a good supply of paper.

20 ***(File name) (problem number) not found.*** The computer is not able to load the specified file because it was not found on the data disk.

Operator action—Be sure the program and data disks are in the correct drives; be sure the exercise number is correct. If the message appears again, reenter the data through the appropriate program option.

21 ***(File name) (problem number) is full.*** The computer is not able to add more data to the specified file because it is full.

Operator action—Remove all unnecessary data items.

23 ***May not exceed line n.*** The line number you chose to correct may not exceed the number of data items in the file.

Operator action—Request a line number between 1 and *n* only.

30 ***May not exceed number of events.*** The number of items taken each time may not exceed the number of possible events (Probability Program).

Operator action—Reenter both numbers.

31 ***Must be greater than point one.*** The value of point two must be greater than the value of point one (Probability Program, Probability Distributions Program, and Estimation Program).

Operator action—Reenter both numbers.

32 ***Response not within limits.*** The value is below the lower limit or is above the upper limit.

Operator action—Reenter response within stated limits.

33 ***Must have standard deviation.*** Either the population or sample standard deviation must be supplied (Hypothesis Testing Program).

Operator action—Reenter one or both of the numbers.

GLOSSARY

A priori Probability. The probability of the occurrence of an event determined on the basis of the physical properties of the trial.

Absolute Value. The magnitude of a numerical value without regard to the algebraic sign; for example, $|-7| = 7$.

Addition Rule of Probability. If two events A and B are mutually exclusive, the addition rule for finding the probability of A or B is $p(\text{A}) + p(\text{B})$.

Alpha Level α. A predetermined level of probability for rejecting a null hypothesis; the probability of falsely rejecting a null hypothesis (committing a type I error).

Alternate Hypothesis. The hypothesis that is supported when the null hypothesis is rejected.

Analysis of Variance (ANOVA). A statistical test for detecting differences among two or more population means.

A posteriori (post hoc) Test. A follow-up procedure used when an ANOVA has indicated significant differences among more than two groups. The purpose is to identify which pairs of means are significantly different.

Apparent Limits. The real, obtainable numbers used as the upper and lower score values for a class interval.

Area(s) of Rejection. The area or areas of a probability distribution that equal alpha proportion of the entire area of the curve; the area(s) in the "tail(s)" of a probability distribution beyond the critical (tabled) baseline value(s).

Arithmetic Mean. The sum of the values divided by N, the number of values; generally called the mean for brevity.

Average. A general term referring to any of the several measures of central tendency.

Axis. A line in space perpendicular to a second line (axis) that forms a frame of reference or origin for coordinates or for graphically expressed quantities.

Bar Graph. A graphic portrayal of frequencies (or relative frequencies) represented by a set of parallel bars of equal widths and lengths proportional to the frequencies.

Bell-Shaped Curve. A frequency or relative frequency curve with large frequencies in the central portion and few in the tails; *see* Normal Curve.

Beta β. The Greek letter used to denote the probability of a type II statistical error; also used to symbolize a standardized regression coefficient.

Between-Groups Variance. Variability between the means of separate groups and the overall grand mean that is attributable to treatment or nonrandom differences between groups.

Biased Sample. A subset of a population that is not representative of the population with respect to relevant variables.

Binomial Distribution. The distribution of the probability of a specified number of successes in a designated number of binomial trials.

Binomial Theorem. A general formula or set of rules for writing any power of a binomial without the necessity of completing the multiplication.

Bivariate. Refers to two variables, two measures per case.

Box-and-Whisker Plot. A graphic representation showing the median, the interquartile range, and the outliers in a distribution.

Cell. The intersection, or combination, of a row and a column.

Central Limit Theorem. States that if a population N has a variance (σ^2) and a mean (μ), then the sampling distribution of means for randomly drawn samples of size n approaches a normal distribution with a mean of μ and a standard deviation of $\frac{\sigma}{\sqrt{N}}$.

Central Tendency. The middle portion of a numerical distribution or a typical value for the distribution.

Chi Square. A statistical test to determine if an observed frequency distribution differs significantly from the expected frequency distribution.

Circle Graph. A circular-shaped graph, sometimes called a pie graph, in which the sizes of the central angles are proportional to the amount of the quantity being represented.

Class Interval. Range of values bounded by upper and lower limits that measurements can assume.

Coefficient of Determination r^2. The amount of variability in one variable that can be statistically explained by the variability in another variable.

Combinations. The number of ways of combining several objects when there is no need to order or sequence the objects; *see* Permutations.

Conditional Probability. The probability that an event A will occur, given that another event B has occurred; written $p(\mathrm{A}|\mathrm{B})$.

Confidence Interval. A pair of values enclosing the middle $100(1 - \alpha)\%$ of a sampling distribution; the range used to estimate the amount of error present when estimating the population parameter.

Contingency Table. A two-way table showing the joint frequency distribution of two discrete variables.

Continuous Variable. A variable that may assume any value in the range but is restricted in a practical sense by the precision of measurement.

Correlation. A set of statistical procedures for determining the strength and direction of the relationship between two variables.

Critical Value(s). The point or points on the baseline of a probability distribution that determine the area(s) of rejection; the point(s) on the baseline of a probability distribution beyond which the probability of occurrence of the random variable is less than alpha (α) or $\frac{1}{2}$ alpha.

Cumulative Frequency. A table showing the number of cases with scores equal to or below a particular value.

Curve. A smooth-line representation of a frequency polygon, relative frequency polygon, or relative cumulative frequency polygon.

Curvilinear. A nonlinear relationship between two continuous variables.

Data. Numerical values assigned to performances, traits, or characteristics of individuals or objects.

Datum. A single numerical value.
Degrees of Freedom. A phrase meaning "freedom to vary"; an integer necessary to fully describe a probability distribution.
Dependent Variable. The variable in a research design whose values presumably depend on the value of another (independent) variable; in regression, the variable to be predicted by the regression model.
Descriptive Statistics. The branch of statistics dealing with organizing and summarizing numerical data.
Deviation Score. x; the difference between a score and the arithmetic mean of the distribution; $x = X - \overline{X}$.
Discrete Variable. A discontinuous variable made up of distinct values.
Dispersion. The degree or amount of spread (scatter) of a distribution of numerical data; the homogeneity of data.

Estimation. The process of inferring the value of an unknown parameter from a measured or computed value of a sample statistic.
Expected Frequencies. A hypothetical frequency distribution that would be most likely to occur under assumptions and conditions specified by the problem or researcher.

F-test. A statistic method for testing the null hypothesis of no difference between the variances of two sets of data.
Frequency. The number of observations that fall in a cell of a table or in a class interval.
Frequency Distribution. A tabular presentation of the number of occurrences of the various values in a set of numbers.
Frequency Polygon. A broken-line graph that visually shows a frequency (or relative frequency) distribution.

Geometric Mean. A measure of central tendency; the nth root of the product of n positive factors.
Geometric Probability. A technique for studying the probability of outcomes that can be represented by points in a plane or on a line through the use of plane and analytic geometry.
Goodness of Fit. A chi square design using only a single variable; *see* Chi Square.
Grouped Bar Graph. A type of bar graph showing various quantities of subcategories within each of the major classifications of the nominal variable.
Grouped Frequency Distribution. A frequency distribution for which measurements have been grouped or condensed into categories or intervals larger than one unit of measurement.

Harmonic Mean. A measure of central tendency for computing averages of rates such as speeds traveled for certain distances; the reciprocal of the arithmetic mean of the reciprocals of a limited series of numbers.
Histogram. A bar graph displaying a frequency (or relative frequency) distribution.
Homogeneity of Variance. An expression indicating that there is no significant difference between variances of two or more groups on a continuous variable.
Homoschedasticity. Refers to uniform scatter of bivariate points about a regression line; equal spread.
Hypothesis. A conjectural statement about the relationship between two variables; *see* Null Hypothesis.

Independent Events. Occurs when one event does not affect the probability of the occurrence of the other event.

Independent Variable. The variable in a research design to be manipulated and whose effects are being studied; in regression, the variable referred to as the predictor.
Inference. Generalizing statistical results from a sample (statistics) to a population (parameters).
Intercept. The point on the axis of the dependent variable where the linear regression line intersects the axis; the value of the dependent variable when the value of the independent variable is zero.
Intersection of Two Sets. The intersection of two sets, A and B, is a set containing the elements in both A and B, denoted $A \cap B$.
Interval Estimation. Same as a confidence interval.
Interval Scaling. Units of measurement that are assumed to represent equal intervals along a baseline or equal amounts of the variable with no absolute zero point.

Joint Events. The co-occurrence of two or more outcomes of a trial experiment.
Joint Frequency Distribution. A bivariate frequency distribution of categorical variables showing the intersection of joint events in the cells of a contingency table.
Joint Probability. The probability of the occurrence of a joint event.

Leaf. The next digit following the stem in a stem-and-leaf diagram.
Level of Confidence. The probability that a confidence interval captures the population parameter.
Linear. The tendency to follow a straight line.

Marginals. The sums of the cells in the rows and columns of a joint distribution.
Mean. A measure of central tendency; usually refers to the arithmetic mean $\overline{X}$ or μ, but includes geometric and harmonic means.
Mean Deviation. A measure of dispersion; the mean (average) deviation between numerical values in a distribution and the distribution mean.
Mean Square. Variance; within an ANOVA context, the total mean square is partitioned into between-groups and within-group variance.
Measurement. The process of assigning numbers to objects, events, performances, or people according to specified rules.
Median. The value that divides a numerical distribution into two equal parts such that half the scores are above and half below the point; also called the 50th percentile.
Mode. The measure of central tendency determined by the value occurring most frequently.
Multiplication Rule of Probability. The method for determining the probability of joint events; $p(\text{A and B}) = p(\text{A}) \times p(\text{B})$.
Mutually Exclusive. Two events that cannot co-occur (happen) on a single trial of an experiment; for example, drawing a jack and a queen from a deck of cards on a single draw.

Negative Correlation. The systematic tendency for one variable to decrease in size as another variable increases in size.
Negative Skew. A frequency distribution with a gradual slope to the left.
Nominal Scaling. Numbers used only to identify or name individuals or sets of individuals (objects, events, people, etc.).
Nonparametric. Data that do not permit meaningful interpretation of deviation from the distribution mean and sum of

squares of deviation scores; scaled on less than an interval level.

Normal Curve. A hypothetical bell-shaped frequency curve used to model many measured and random variables in nature; it is symmetric about the mean, median, and mode.

Null Hypothesis. A conjectural statement of what is expected to occur in an experiment based entirely on random chance; generally a statement that is subject to statistical testing of no difference between groups or of no relationship between variables.

Observed Frequencies. A frequency distribution formatted according to a chi square design; the frequencies that are actually counted by the researcher.

Odds. The probability of success divided by the quantity one minus the probability of success.

Ojive (Ogive). A curved-line graph in the shape of a "lazy s" that shows a cumulative frequency or a relative cumulative frequency distribution.

One-tailed Test. A statistical test of the null hypothesis in which alpha proportion of the probability curve is in one tail.

Ordinal Scaling. Numerical values that are capable of being ranked in ascending or descending order of size.

Outlier. An extremely large or small score in a distribution.

Parameter. A characteristic, usually an unknown, of a population.

Percentile. The point on a baseline of scores below which falls that percentage of the scores (students, subjects, objects, etc.).

Permutations. The number of ways several objects can be combined when order is considered; *see* Combinations.

Pictogram. A method of visually portraying values or frequencies with the use of pictures or icons.

Point Estimate. The value of a sample statistic used as an estimate of a population parameter.

Points of Inflection. The points on the normal curve where the curve changes from concave upward to concave downward.

Population. A well-defined collection of elements (objects, people, events, etc.); the universal set that is sampled and to which inferences are made.

Positive Skew. A frequency distribution with a gradual slope to the right.

Post hoc Test. An analysis to determine where differences occur when an ANOVA shows overall significance among three or more groups.

Probability. The science of dealing with the uncertainty and possibility of certain events occurring; or the chances, expressed in decimal fraction or percentage, that an event will occur.

Probability Distribution. Used in identifying critical values for areas of rejection for testing null hypotheses (t, z, F, binomial, Poisson, etc.).

Pseudo-random Numbers. Digits generated by a deterministic formula that have varying degrees of the characteristics of randomness.

Random Sampling. A procedure for selecting elements from a population in which each element (person, thing, etc.) has an equal probability of selection.

Random Variable. A variable whose value is the chance result of a given experiment.

Range. A measure of dispersion indicating the difference between the minimum and maximum values in a distribution.

Ratio Scaling. A set of interval-scaled numbers with an absolute zero point.

Raw Score. A measured quantity before it has been transformed into a standard score, percentile, grade-equivalent, or other scaled score such as the number correct on an exam; the percentage correct.

Real Limits. The true or hypothetical upper and lower limits of a class interval of a continuous variable.

Regression Coefficient. The slope of the regression equation for a least-squares analysis.

Regression Equation. The equation of the least-squares line of regression (best fit); $Y' = a + bX$.

Relative Cumulative Frequency. The proportion of cases falling at or below a particular value or class of values; $\text{rcf} = \frac{\text{cf}}{N}$

Relative Frequency. The proportion of cases in a distribution having a particular value or falling within a class interval; $\text{rf} = \frac{\text{f}}{N}$

Sample. A subset of a defined population.

Sample Space. The set of all possible outcomes of a probability experiment.

Sampling. A set of procedures for selecting a subset of a population to form a sample.

Sampling Distribution. A theoretical distribution consisting of measures on an infinite number of samples from a population.

Scattergram. A bivariate scatterplot graph of two continuous variables; each point represents two measures, that is, the coordinates of an ordered pair.

Significance Level. Alpha; the probability of making a type I error when the null hypothesis is rejected; the prespecified probability of rejecting a null hypothesis.

Skewed. The lack of symmetry in a distribution; the side of the graph of a frequency distribution with the more gradual slope (the skew).

Slope. The vertical rise (or decline) per horizontal unit of a linear regression line expressed as a regression coefficient.

Stacked Bar Graph. A type of bar graph showing the cumulative quantities of subcategories within each of the major classifications of the nominal variable.

Standard Deviation (s or σ). A measure of dispersion or spread; the square root of the variance.

Standard Error. The standard deviation of a hypothetical sampling distribution.

Standard Error of Estimate. The standard deviation of the errors of prediction in a regression analysis.

Standard Scores. Scores that have been transformed so that the distribution has a particular mean and standard deviation.

Statistic. A quantity calculated on sample data used to estimate a parameter.

Statistical Significance. A phenomenon or event that departs so radically from what would be expected to occur by chance alone that its probability of occurrence is small; when the probability of occurrence of an event by chance is less than some small, predetermined value of alpha; the rejection of a null hypothesis at a particular alpha level.

Statistics. The science dealing with organizing, summarizing, analyzing, and interpreting numerical data; a set of measures taken on a sample.

Stem-and-Leaf Diagram. A display of numbers with certain digits (for example, tens place) serving as the "stem" and the remaining digit(s) (for example, ones place) forming the "leaves."

Sum of Squares. Refers to the sum of the squares of the deviations from the mean, that is, Σx^2.
Symmetric. A characteristic of a curve meaning that the shape of one half is the same as the shape of the other half; that is, one half of the curve is a mirror image of the other half.
Symmetric Frequency Curve. A smooth-line representation of a frequency polygon that is shaped the same on both sides of the middle.

Tree Diagram. A diagram showing the various alternative probability paths; a device used to enumerate all possible outcomes of a sequence of probability trials.
t-Test. A statistical procedure for testing the difference between two means for statistical significance.
Two-tailed Test. A statistical test of the null hypothesis in which alpha proportion of the probability curve is divided into two tails.
Type I Error. The rejection of a null hypothesis that should not be rejected.
Type II Error. The failure to reject a null hypothesis that should be rejected.

Unbiased Estimate. A point estimate of a population parameter that has no tendency to be too large or too small.
Unbiased Sample. A subset of a population that is assumed to be representative of the population on all pertinent variables.
Ungrouped Frequency Distribution. A tabular presentation of raw score values with class intervals of one unit.
Union of Two Sets. The union of two sets, A and B, is a set containing the elements in A or in B or in both and is symbolized A ∪ B.

Variable. A property or quality on which individuals differ among themselves.
Variance (s^2 or σ^2). A measure of dispersion or spread in a distribution; the average of the sum of squares.
Vertical Bar Graph. One of several types of graphs known as bar graphs in which the quantities are scaled along the vertical axis of the graph.

Within-group Variance. Error or unexplained variance resulting from individual (as opposed to group) differences.

Yates' Correction. A modification of the chi square formula to correct for discontinuity.

z-Score. A standard score providing a description of how far the score is from the distribution mean and whether it is above or below the mean.
z-Score Distribution. The standard normal distribution having a mean of zero and a standard deviation of one.
z-Transformation. Used for the purpose of "normalizing" a distribution of correlation coefficients; commonly called Fisher's r to z transformation.

INDEX

D

Q

R